G PROTEIN-COUPLED RECEPTORS

G PROTEIN-COUPLED RECEPTORS

Edited by

Tatsuya Haga, Ph.D.

Professor of Neurochemistry
Department of Neurochemistry
Faculty of Medicine
University of Tokyo
Hongo, Tokyo

Gabriel Berstein, M.D., Ph.D.

Scientist
Millennium Pharmaceuticals, Inc.
Cambridge, Massachusetts

CRC Press
Boca Raton London New York Washington, D.C.

Library of Congress Cataloging-in-Publication Data

G protein-coupled receptors / Tatsuya Haga and Gabriel Berstein, editors.
     p.   cm. -- (Signal transduction series)
     Includes bibliographical references and index.
     ISBN 0-8493-3384-9 (alk. paper)
     1. G proteins--Receptors Laboratory manuals. I. Haga, Tatsuya. II. Berstein, Gabriel. III. Series.
QP552.G16G17   1999
572'.6--DC21

                                      99-26359
                                         CIP

No claim to original U.S. Government works
International Standard Book Number 0-8493-3384-9
Library of Congress Card Number 99-26359
Printed in the United States of America 1 2 3 4 5 6 7 8 9 0
Printed on acid-free paper

CRC METHODS IN SIGNAL TRANSDUCTION SERIES

Joseph Eichberg, Jr., Series Editor

SERIES PREFACE

In the past few years, the field of signal transduction has expanded at an enormous rate and has become a major force in the biological and biomedical sciences. Indeed, the importance of this area can hardly be exaggerated and still continues to grow. The knowledge gained has vastly increased and in some instances revolutionized our perceptions of the vast panoply of signaling pathways and molecular mechanisms through which healthy cells respond to both extracellular and intracellular cues, and how these responses can malfunction in a variety of disease states.

The increase in our understanding of signal transduction mechanisms has relied upon the development and refinement of new and existing methods. Successful investigations in this field now often require an integrated approach which utilizes techniques drawn from molecular biology, cell biology, biochemistry, genetics, and immunology. The overall aim of this series is to bring together the wealth of methodology now available for research in many aspects of signal transduction. Since this is such a fast moving field, emphasis will be placed wherever possible on state-of-the-art-techniques. Each volume is assembled by one or more editors who are expert in their particular topic and leaders in research relevant to it. Their guiding principle is to recruit authors who will present detailed procedures and protocols in a crucial yet reader friendly format which will be of practical value to a broad audience, including students, seasoned investigators, and researchers who are new to the field.

The study of G protein-coupled receptors (GPCRs) has long been and continues to be an important aspect of signal transduction research. The chapters in this volume provide theoretical treatments and practical descriptions of techniques related to three themes in this field. First, in Chapters 1 through 4, techniques for the characterization and identification of ligands are presented. This is an area which is crucial to expanding our understanding of receptor function, as well as to furthering advances in drug development. Second, research into GPCR coupling to G proteins and mechanisms of GPCR desensitization now utilizes a broad variety of approaches, some of which are detailed in Chapters

5 through 9. Finally, Chapters 10 through 14 present methods for generation of large quantities of receptors and for determination of the three-dimensional structures of these membrane proteins. Overall, it is hoped that this volume will not only provide useful information for laboratory investigations, but will also serve to illustrate trends and indicate future directions for increasing our knowledge of receptor structure and function.

Joseph Eichberg, Ph.D.
Advisory Editor for the Series

Preface

G proteins (heterotrimeric GTP-binding proteins) have been said to be "involved in everything from sex in yeast to cognition in humans."* G protein-coupled receptors (GPCRs), which activate G proteins when bound to specific ligands, constitute one of the largest superfamilies of proteins. In *Caenorhabditis elegans*, an estimated 1049 genes which code for GPCRs are present among 19,099 genes (5.5%).** In the human genome, 252 GPCR genes were identified at the time when 15% of the genome was sequenced, suggesting that approximately 1700 GPCR genes may be present.*** Ligands of GPCRs are extremely diverse and include hormones, neurotransmitters, autacoids, cytokines, chemotactic factors, odorants, pheromones, and taste substances. Rhodopsin and cone pigments in retina, which respond to light, are also GPCRs, as are targets of many drugs including those for pain, hypertension, duodenal ulcer, schizophrenia, and asthma. An enormous variety of chemical species comprise GPCR ligands, among which are amines, amino acids, peptides, proteins, nucleosides, nucleotides, Ca^{2+} ions, and lipids, such as prostaglandins and leukotrienes. In addition, 1000 distinct GPCR odorant receptors are thought to discriminate over 100,000 kinds of odorants, which have diverse chemical structures.

This book covers current techniques used to study GPCRs, particularly methods related to studies that become possible by the molecular cloning of GPCRs. Most chapters include step by step protocols which are intended to be useful for new researchers as well as specialists in the field. The book consists of three sections:

Section I
  Novel Ligands: Ligand Screening Systems and Orphan Receptors
Section II
  Function and Regulation: Receptor-G Protein Coupling and Desensitization
Section III
  3D Structure: Large-Scale Expression and Electron- and X-Ray Crystallography

Section I should be particularly useful for researchers who are interested in drug discovery. Basic theory of receptor-ligand interactions and different kinds of ligand screening systems are introduced. Historically, receptors and receptor subtypes were functionally identified by the action of specific ligands, exogenous agonists and antagonists. For example, muscarinic acetylcholine receptors were discriminated from nicotinic acetylcholine receptors by the actions of nicotine and muscarine and were further subdivided into M_1, M_2, and M_3 subtypes through the use of specific

---

* Gilman, A.G., *Science* 266, 368, 1994.
** Bargman, C., *Science* 282, 2028, 1998.
*** Henikoff, S. et al., *Science* 278, 609, 1997.

antagonists. Molecular cloning revealed the existence of two additional subtypes, M_4 and M_5. A current challenge is to find specific ligands for the M_4 and M_5 receptor subtypes. A similar requirement exists for almost all GPCRs. This kind of "reverse pharmacology" is more important for novel GPCRs, for which no specific ligands have been identified, and for "orphan receptors," for which endogenous ligands have not been identified. Identification of subtype-specific ligands or an endogenous ligand for a given GPCR will be a key step in elucidating receptor functions and developing novel drugs. Chapter 1 provides fundamental knowledge and techniques concerning the interaction between GPCRs and iso- or allosteric ligands. Chapters 2 and 3 outline ligand screening systems that utilize yeast which express mammalian GPCRs and frog melanophores, respectively. Methods used to screen orphan receptors for endogenous ligands are described in Chapter 4.

Section II is concerned with the interactions of GPCRs with G proteins and with GPCR desensitization. Interactions between GPCRs and G proteins can be analyzed using *in vitro* reconstitution of purified proteins in lipid membranes as presented in Chapter 5, or by means of expression of mutated proteins in cultured cells, as described in Chapter 6. These approaches have provided mechanistic insights into the initial biochemical events that are triggered by receptor activation, and they continue to be useful strategies in the characterization of novel receptors. Attenuation of receptor-mediated responses following prolonged exposure to agonists has long been known as desensitization. This is an active area of research because it is critical to understand how molecular mechanisms of desensitization control the duration and amplitude of receptor-mediated signals. Chapter 7 introduces general methods used to study desensitization of GPCRs and related phenomena of GPCR phosphorylation. Chapter 8 discusses a specific form of desensitization, namely internalization of GPCRs. Chapter 9 describes the application of electrophysiology to studies of GPCR desensitization.

Section III addresses the three-dimensional structure of GPCRs. The determination of these membrane protein structures will unquestionably have a strong impact on our understanding of the interaction between GPCRs and G proteins and may also lead to theoretical prediction of effective ligands. Currently no three-dimensional structures of GPCRs are available, except a low resolution structure of rhodopsin. The first difficulty in structural determination is that except rhodopsin, sufficient amounts of GPCRs cannot be readily attained. Large-scale expression and purification of GPCRs is therefore a prerequisite for structural studies. Chapter 10 presents a method for expression of functional GPCRs in *E. coli*. Chapter 11 extensively surveys and compares different expression systems for GPCRs, placing particular emphasis on expression in yeast. Examples for determination of the atomic structure of membrane proteins are described in Chapters 12 through 14. The structure of bacteriorhodopsin, which has seven transmembrane segments and is often assumed to be a model for GPCRs, has been elucidated using both electron crystallography (Chapter 12) and X-ray crystallography (Chapter 14). Principles and applications of two-dimensional crystallization and cubic phase crystallization are introduced in Chapters 12 and 14, respectively. Another example of the structural determination of a membrane-bound receptor both with and without its ligand-induced conformational change is provided in Chapter 13.

Thus we believe that this book will be useful for those who wish to develop new drugs targeted to GPCRs, find endogenous ligands for orphan receptors, elucidate the molecular mechanisms underlying the function and regulation of GPCRs, or determine the three-dimensional structure of GPCRs.

Finally, we would like to sincerely thank each of the authors for taking time out of their busy schedules to write these excellent chapters.

Tatsuya Haga
Gabriel Berstein
June 18, 1999

Contributors

Robert S. Ames
SmithKline Beecham Pharmaceuticals
King of Prussia, PA

Derk J. Bergsma
SmithKline Beecham Pharmaceuticals
King of Prussia, PA

Gloria H. Biddlecome
Howard Hughes Medical Institute
University of Iowa College of Medicine
Iowa City, IA

Nigel J.M. Birdsall
Division of Physical Biochemistry
National Institute for Medical Research
Mill Hill, London, UK

Mark R. Boyett
Department of Physiology
University of Leeds
Leeds, UK

Jon K. Chambers
SmithKline Beecham Pharmaceuticals
New Frontiers Science Park
Harlow, Essex, UK

Mark L. Chiu
Department of Chemistry
Seton Hall University
South Orange, NJ

Jeffrey S. Culp
SmithKline Beecham Pharmaceuticals
King of Prussia, PA

James J. Foley
SmithKline Beecham Pharmaceuticals
King of Prussia, PA

Yoshinori Fujiyoshi
Department of Biophysics
Faculty of Science
Kyoto University
Kyoto, Japan

Reinhard Grisshammer
MRC Laboratory of Molecular Biology
Cambridge, UK

John R. Hadcock
Pfizer Central Research
Division of Cardiovascular/Metabolic
  Diseases
Groton, CT

Mac E. Hadley
Departments of Cell Biology and
  Anatomy
College of Medicine
University of Arizona
Tucson, AZ

Brian D. Hellmig
SmithKline Beecham Pharmaceuticals
King of Prussia, PA

Jan Jakubik
Laboratory of Bioorganic Chemistry
NIDDK
National Institutes of Health
Bethesda, MD

Iftikhar A. Khan
Department of Physiology
University of Leeds
Leeds, UK

Evi Kostenis
Laboratory of Bioorganic Chemistry
NIDDK
National Institutes of Health
Bethesda, MD

Hitoshi Kurose
Laboratory of Pharmacology and
  Toxicology
Graduate School of Pharmaceutical
  Sciences
University of Tokyo
Tokyo, Japan

Jelveh Lameh
Molecular Research Institute
Palo Alto, CA
Departments of Biopharmaceutical
  Sciences and Pharmaceutical Chemistry
University of California
San Francisco, CA

Ehud M. Landau
Biozentrum
University of Basel
Basel, Switzerland

Sebastian Lazareno
MCR Collaborative Centre
1-3 Burtonhole Lane
Mill Hill, London, UK

Kaspar P. Locher
Department of Microbiology
Biozentrum
University of Basel
Basel, Switzerland
present address:
Division of Chemistry
California Institute of Technology
Pasadena, CA

Michele Loewen
Department of Biology
Massachusetts Institute of Technology
Cambridge, MA

Dean E. McNulty
SmithKline Beecham Pharmaceuticals
King of Prussia, PA

Dino Moras
Laboratorie de Biologies Structurale
  Institut de Génétique et de Biologie
Moléculaire at Cellulaire
CNRS/INSERM/ULP
Illkirch Cedex, France

Alison I. Muir
SmithKline Beecham Pharmaceuticals
New Frontiers Science Park
Harlow, Essex, UK

Peter Nollert
University of California
San Francisco, CA

Janet E. Park
SmithKline Beecham Pharmaceuticals
New Frontiers Science Park
Harlow, Essex, UK

Mark H. Pausch
Wyeth-Ayerst Research
Molecular Genetic Screen Design Group
Princeton, NJ

J. Mark Quillan
Departments of Biopharmaceutical
  Sciences and Pharmaceutical Chemistry
University of California
San Francisco, CA

Bernard Rees
Laboratorie de Biologies Structurale
  Institut de Génétique et de Biologie
Moléculaire at Cellulaire
CNRS/INSERM/ULP
Illkirch Cedex, France

Helmut Reiländer
Max Planck Institut für Biophysik
Abt. Mol. Membranbiologie
Frankfurt/M, Germany

Christoph Reinhart
Max Planck Institut für Biophysik
Abt. Mol. Membranbiologie
Frankfurt/M, Germany

Jurg P. Rosenbusch
Department of Microbiology
Biozentrum
University of Basel
Basel, Switzerland

Henry M. Sarau
SmithKline Beecham Pharmaceuticals
King of Prussia, PA

Torsten Schöneberg
Institut fur Pharmakologie
Fachbereich Humanmedizin
Freie Universität Berlin
Berlin, Germany

Zhigang Shui
Department of Physiology
University of Leeds
Leeds, UK

J. Randall Slemmon
SmithKline Beecham Pharmaceuticals
King of Prussia, PA

Jeffrey M. Stadel
SmithKline Beecham Pharmaceuticals
King of Prussia, PA

Agnes Szmolenszky
Max Planck Institut für Biophysik
Abt. Mol. Membranbiologie
Frankfurt/M, Germany

Julie Tucker
MRC Laboratory of Molecular Biology
Cambridge, UK
present address:
Laboratory of Molecular Biophysics
Oxford, UK

Jürgen Wess
Laboratory of Bioorganic Chemistry
NIDDK National Institute of Health
National Institutes of Health
Bethesda, MD

Christine Widmer
Biozentrum
University of Basel
Basel, Switzerland

Shelagh Wilson
SmithKline Beecham Pharmaceuticals
New Frontiers Science Park
Harlow, Essex, UK

Fu-Yue Zeng
Laboratory of Bioorganic Chemistry
NIDDK
National Institutes of Health
Bethesda, MD

Xiangyang Zhu
Laboratory of Bioorganic Chemistry
NIDDK
National Institutes of Health
Bethesda, MD

List of Protocols

Abbreviations

ABT	Aminobenztropine
ADH1	Alcohol dehydrogenase gene
AlF_4	Aluminum tetrafluoride
AOX1	Alcohol oxidase I gene
AT	Aminotriazole
ATCC	American Type Culture Collection
βARK	β-adrenergic receptor kinase
BHA	Benzhydrylamine
Boc	*Tert*-butyloxycarbonyl
bR	Bacteriorhodopsin
BSA	Bovine serum albumin
cAMP	Adenosine 3′: 5′-cyclic monophosphate
cDNA	Complementary deoxyribonucleic acid
$C_{12}E_{10}$	Polyoxyethylene 10 lauryl ether
CCD	Charge-coupled device
CCK	Cholecystokinin
CHAPS	3-[(3-cholamidopropyl)dimethylammonio]-1-propane sufonate
CHO cells	Chinese hamster ovary cells
CHS	Cholesteryl hemisuccinate
CMC	Critical micellar concentration
CNS	Central nervous system
CRLR	Calcitonin receptor-like receptor
DMEM	Dulbecco's modified Eagle medium
DMSO	Dimethyl sulfoxide
DSP	Dithiobis(succinimidylpropionate)
DTT	Dithiothreitol
EDTA	Ethylenediaminetetraacetic acid
EGTA	Ethylenebis(oxyethylenenitrilo)tetraacetic acid
ELISA	Enzyme-linked immunosorbent assay
EM	Electron microscopy
ET	Endothelin
FBS	Fetal bovine serum
FLIPR	Fluorescence imaging plate reader
Fmoc	9-fluorenylmethoxycarbonyl
GAP	Glyceraldehyde-3-phosphate dehydrogenase (Chapter 10)
GAP	GTPase-activating protein (Chapter 5)
GDP	Guanosine diphosphate
GFP	Green fluorescent protein

GHRH	Growth hormone releasing hormone
GMP	Guanosine monophosphate
Gpa1	*S. cerevesiae* G protein α subunit
GPCR	G protein-coupled receptor
GRK	G protein-coupled receptor kinase
GRP	Gastrin-releasing peptide
GTP	Guanosine triphosphate
GTPγS	Guanosine 5′-3-*O*-(thio)triphosphate
HA	Hemagglutinin
HBSS	Hanks balance salts solution
HEK 293 cells	Human embryonic kidney 293 cells
HEPES	4-(2-hydroxyethyl)-1-piperazineethanesulfonic acid
HIS4	Histidine dehydrogenase gene
5-HT	5-hydroxytryptamine, or serotonin
HTS	High-throughput screening
$i_{K,ACh}$	Acetylcholine-activated K^+ current
IBMX	3-isobutyl-1-methylxanthine
IMAC	Immobilized metal affinity chromatography
IP_1	Inositol monophosphate
IP_3	Inositol 1, 4, 5-trisphosphate
IPMS	Infinite periodic minimal surfaces
IPTG	Isopropyl β-thiogalactopyranoside
kDa	Kilodaltons
KPi	Potassium phosphate
KRH	Krebs Ringer Henseleit
LAPS	Light addressable potentiometric sensor
LHC-II	Light harvesting complex II
LM	n-dodecyl-β-D-maltoside
LPA	Lysophosphatidic acid
M_1 receptor	M_1 muscarinic acetylcholine receptor
MAP kinase	Mitogen-activated protein kinase
MBHA	p-methyl-benzhydrylamine
MBP	*E. coli* maltose-binding protein
MD	Minimal dextrose medium
MEK	MAP kinase kinase
MEM	Minimum essential medium
MI	Melanophore index
MSH	Melanocyte stimulating hormone
MUPL	Multi-use peptide library
mut^S	Methanol utilization sensitive mutant
NECA	5′-*N*-Ethylcarboxamidoadenosine
Ni^{2t}-NAT	Nickel-nitrilotriacetic acid
NK	Neurokinin
nmt	No message in thiamine
NPY	Neuropeptide Y
NT	Neurotensin

NTR	High-affinity neurotensin receptor
PACAP	Pituitary adenylate cylase activating polypeptide
PAK	p20-activated kinase
PAO	Phenylarsine oxide
PBS	Phosphate buffer saline
PCA	Perchloric acid
PE	Phosphatidylethanolamine
PEG	Polyethylene glycol
PEI	Polyethyleneimine
Pi	Inorganic phosphate
PI	Phosphatidylinositol
PI	Protease inhibitors
PIA	$R(-)N_6$-(2-Phenylisopropyl) adenosine
PIP_2	Phosphatidylinositol 4, 5-bisphosphate
PKA	Protein kinase A
PKC	Protein kinase C
PLC	Phospholipase C
PMSF	Phenylmethylsulfonylfluoride
PRB1	Vacuolar endopeptidase B gene
PS	Phosphatidylserine
QNB	1-quinuclidinyl benzylate methyl chloride
RGS	Regulators of G protein signaling
SDS-PAGE	Sodium dodecyl sulfate-polyacrylamide gel electrophoresis
SMPS	Simultaneous multiple peptide synthesis
S/N	Signal-to-noise
T43NTR	Truncated NTR starting at Thr-43
TCA	Trichloroacetic acid
TFA	Trifluoroacetic acid
Tris	Tris(hydroxymethyl)aminomethane
TrxA	*E. coli* thioredoxin
VIP	Vasoactive intestinal polypeptide
YNB	Yeast nitrogen base
YPD	Yeast extract peptone dextrose medium
YPDS	Yeast extract peptone dextrose sorbitol medium

Table of Contents

Section I

---

*Novel Ligands: Ligand Screening
Systems and Orphan Receptors*

1 Measurement of Competitive and Allosteric Interactions in Radioligand Binding Studies

Sebastian Lazareno and Nigel J.M. Birdsall

CONTENTS

0-8493-3384-9/00/$0.00+$.50

1.1   INTRODUCTION

Radioligand binding studies are relatively easy to perform, often very accurate, and
can usually be interpreted according to a mechanistic model to yield fundamental
chemical quantities, such as the affinity of a ligand for a receptor or the number of
receptors in a tissue. For these reasons, binding studies are used in a very wide range

of applications, from the simple detection of active ligands in a screening assay to detailed mechanistic studies of ligand binding, receptor activation, and receptor regulation. There are problems and pitfalls with any technique, and the power of binding studies to provide quantitative estimates of molecular parameters implies a special need for rigor in the interpretation of binding data. The basic protocols for binding studies are straightforward in general, but the specific details, and the quantities that can be estimated, depend crucially on the system under investigation.

This chapter is concerned with the binding of ligands to G protein-coupled receptors, and has a "top down" structure. It begins with basic theory, outlining the fundamental concepts of ligand–receptor interaction, and progressing to a consideration of the binding of agonists and allosterically acting agents. We then consider some practical aspects and pitfalls of binding studies, such as buffer composition, incubation time, and radioligand depletion. Next we describe five different types of binding assay: radioligand dissociation, radioligand association, radioligand saturation, inhibition of radioligand binding by an unlabeled competitor, and interactions of an allosterically acting ligand with both a radioligand and an unlabeled competitor. Finally, we provide detailed protocols for culturing CHO cells expressing muscarinic receptors, preparation of membranes and their characterization with respect to protein and receptor concentration, and a screening assay for allosteric effects on unlabeled acetylcholine (ACh) binding. In order to aid clarity the initial, or most salient, mention of important terms is indicated in bold. There are a number of excellent reviews of binding methodology which provide more detailed coverage of aspects that can be covered only briefly in this chapter[1–10] and the GraphPad site on the World Wide Web (www.graphpad.com) contains many useful documents.

1.2 THEORETICAL CONSIDERATIONS

The binding of a radioligand is comprised of two components, a "specific" component, binding to the receptor(s) of interest, and a "**nonspecific**" component to other sites. The nonspecific component of radioligand binding is measured in the presence of a concentration of unlabeled defining ligand sufficient to prevent binding of the radioligand to the binding sites under study, and only to those sites. The defining agent binds to the receptors of interest and, ideally, it should be chemically different from the radioligand, potent, and polar. Nonspecific binding is usually, but not always, a linear function of radioligand concentration, reflecting binding to nonsaturable components (e.g., lipid) or to high capacity, low affinity binding sites, as well as some binding to filters or occluded radioligand in centrifugation assays. Nonspecific binding is discussed further in Reference 3. **Specific** binding is measured as the difference between **total** binding observed in the absence of the defining agent and nonspecific binding. In the following discussion only specific radioligand binding is considered.

1.2.1 SIMPLE LAW OF MASS ACTION BINDING

The simplest model is a reversible bimolecular reaction between a single type of ligand and receptor.

$$L + R \; \underset{k_{off}}{\overset{k_{on}}{\rightleftharpoons}} \; LR \qquad\qquad \textbf{Scheme 1}$$

The ligand, L, binds to a receptor, R, with an **association rate constant k_{on}**, to form a receptor–ligand complex, LR, which is stable for a time but eventually dissociates to free, unchanged receptor and ligand, with a **dissociation rate constant k_{off}**. The units of k_{off} are time^{-1}, and k_{on} has units of M^{-1}·time^{-1}.

Since no catalytic processes are involved, the total concentrations of ligand and receptor are constant.

$$R + LR = R_t \qquad\qquad (1.1a)$$

$$L + LR = L_t \qquad\qquad (1.1b)$$

where R and L are the free concentrations of receptor and ligand, respectively, R_t and L_t are the total concentrations of receptor and ligand, respectively, and LR is the concentration of ligand–receptor complexes, or bound ligand. The following derivations assume that $L_t \gg R_t$, so that free and total ligand concentrations are essentially the same, i.e. there is no ligand depletion. See Section 1.3.10 below for a consideration of ligand depletion.

1.2.1.1   Non-Equilibrium Binding

The rate of change in the concentration of the ligand–receptor complex is given by the difference between the rates of formation and breakdown of ligand–receptor complexes:

$$\frac{\mathrm{d}(LR)}{\mathrm{d}t} = k_{on} \cdot L \cdot R - k_{off} \cdot LR \qquad\qquad (1.2)$$

Solution of this equation[11,12] allows the rates of association and dissociation to be defined.

After initiating a binding reaction, the concentration of ligand–receptor complexes is given by

$$LR = LR_{eq} \cdot \left(1 - e^{-k_{obs} \cdot t}\right) \qquad\qquad (1.3a)$$

$$k_{obs} = L \cdot k_{on} + k_{off} \qquad\qquad (1.3b)$$

where LR_{eq} is the concentration of LR at equilibrium, k_{obs} is the **observed association rate constant** and t is the time after the initiation of the reaction. The time to reach 50% of equilibrium binding is $\ln(2)/k_{obs}$. Note that k_{obs} is a function of L.

The rate of dissociation of ligand from existing ligand–receptor complexes can be studied by suddenly preventing the rebinding of the ligand (see Section 1.4.1.1). The concentration of bound ligand at time t thereafter is given by

$$LR = LR_0 \cdot e^{-k_{off} \cdot t} \tag{1.4}$$

where LR_0 is the concentration of bound ligand at the start of dissociation. Fifty percent dissociation occurs at a time of $\ln(2)/k_{off}$.

1.2.1.2   Equilibrium binding

Binding reaches equilibrium when the rate of formation of bound receptor is equal to the rate of dissociation of bound receptor, i.e

$$\frac{d(LR)}{dt} = 0 \tag{1.5}$$

and by rearrangement of Equation 1.2

$$k_{on} \cdot L \cdot R = k_{off} \cdot LR \tag{1.6}$$

Equation 1.6 can be rearranged to give two equilibrium binding constants:

$$\frac{LR}{L \cdot R} = \frac{k_{on}}{k_{off}} = K_A \tag{1.7a}$$

$$\frac{L \cdot R}{LR} = \frac{k_{off}}{k_{on}} = K_D \tag{1.7b}$$

K_A is the **equilibrium association constant**, or **affinity constant**, and has units of M^{-1}.

K_D, the reciprocal of K_A, is the **equilibrium dissociation constant** (also referred to as the **dissociation constant** or **Kd**) with units of M.

As well as being a ratio of rate constants, the K_D is also the ligand concentration giving 50% receptor occupancy, since when $R = LR$ the left side of Equation 1.7b becomes L. The K_D (and K_A) can therefore be measured in two quite different, independent ways, with kinetic assays and with equilibrium assays.

Equation 1.1a can be combined with Equation 1.7a or 1.7b to define the concentration of **receptor-bound ligand at equilibrium**:

$$LR = \frac{Rt \cdot L \cdot K_A}{1 + L \cdot K_A} \tag{1.8a}$$

$$LR = \frac{Rt \cdot L}{L + K_D} = \frac{Rt}{1 + \dfrac{K_D}{L}} \tag{1.8b}$$

These equations describe the binding of any ligand to a receptor. If the ligand is radiolabeled, then bound ligand can be measured directly, and the K_D of the radioligand can be measured by constructing a **saturation curve** (see Section 1.4.3) with a number of radioligand concentrations. If the ligand is unlabeled, then its binding must be measured indirectly, by competition with a radioligand.

The binding of a radioligand in the presence of an unlabeled ligand is given by

$$LR = \frac{Rt \cdot L \cdot K_A}{1 + L \cdot K_A + B \cdot K_B} \tag{1.9a}$$

$$LR = \frac{Rt}{1 + \dfrac{K_D}{L} \cdot \left(1 + \dfrac{B}{K_i}\right)} \tag{1.9b}$$

where L is the concentration of a radioligand with an affinity of K_A or a dissociation constant of K_D, and B is the concentration of an unlabeled ligand with an affinity of K_B or a dissociation constant of K_i.

The progressive inhibition of radioligand binding with increasing concentrations of unlabeled ligand is known as an **inhibition curve** (see Section 1.4.4), and the concentration of unlabeled ligand which inhibits initial radioligand binding by 50% is referred to as the IC_{50}. **Fractional inhibition** of radioligand binding by antagonist has the same form as Equation 1.8b:

$$\text{fractional inhibition} = \frac{B}{B + IC_{50}} \tag{1.10}$$

Just as the inhibitor competes with the radioligand, so the radioligand competes with the inhibitor, and the IC_{50} will usually be larger than the dissociation constant of the inhibitor. This value can be easily related to the IC_{50} when it is considered that binding in the presence of the IC_{50} concentration of unlabeled ligand (Equations 1.9a and b) is half the binding of radioligand alone (Equations 1.8a and b). Thus

$$K_B = \frac{1 + L \cdot K_A}{IC_{50}} \tag{1.11a}$$

$$K_i = \frac{IC_{50}}{\dfrac{L}{K_D} + 1} \tag{1.11b}$$

Equation 1.11b is often referred to as the **Cheng-Prusoff equation.**[13]

Equations 1.8, 1.9, and 1.10 are the starting point for the analysis of equilibrium binding data. They are forms of the Langmuir isotherm and define a specific, hyperbolic shape for the binding curve. At a ligand concentration of $K_D/10$ the fractional receptor occupancy will be 0.09, and at a ligand concentration 10 times higher than K_D the occupancy will be 0.91, so a simple **law of mass action** occupancy curve goes between about 10% and 90% over a hundredfold range of ligand concentrations. Similarly, a simple law of mass action inhibition curve spans about 10% and 90% inhibition between antagonist concentrations of $IC_{50}/10$ and $10 \cdot IC_{50}$. Equation 1.10 is a particular case of a more general **logistic** function:

$$\text{fractional inhibition} = \frac{B^n}{B^n + IC_{50}{}^n} \tag{1.12}$$

where n is the **Hill slope** or **Hill slope factor**. Simple law of mass action curves have Hill slopes of 1. Curves that cover 10 to 90% occupancy over a smaller range of ligand concentrations are said to have **steep** slopes, >1, while curves that extend over more than a hundredfold range of ligand concentrations are said to be **flat** or **shallow**, with slopes <1. Steep or shallow slopes indicate that the mechanism of binding is more complex than the simple law of mass action.

1.2.2 AGONIST BINDING

Agonist binding is often more complex than antagonist binding. This can be useful for distinguishing agonists from antagonists and for providing clues about the molecular nature of agonist efficacy. If agonist binding is studied by competition with an antagonist radioligand then the complete concentration dependency of agonist binding can be observed. If agonist binding is measured directly with an agonist radioligand, then only the highest affinity states of the binding may be observed, either because of very rapid dissociation of radioligand from low affinity sites in a filtration assay, or because of the unfavorable specific:nonspecific binding ratio of higher radioligand concentrations in a centrifugation assay. All affinity states may be detected by some peptide agonist radioligands (see, for example, Reference 14). Complexities of agonist binding at G protein-coupled receptors include that:

1. Agonists may bind at much lower concentrations than are required to cause functional effects.
2. Agonist inhibition of antagonist radioligand binding may extend over a wide concentration range, leading to "flat" curves and Hill slopes <1.
3. Agonists may show greater potency in inhibiting agonist radioligands than antagonist radioligands.
4. Agonist binding may be modulated by submillimolar concentrations of guanine nucleotides such as guanosine triphosphate (GTP), guanosine diphosphate (GDP), GppNHp (guanylylimidodiphosphate, GMP-PNP), and GTPγS. Direct agonist radioligand binding may be reduced, often reflecting a change in maximal binding capacity. Agonist inhibition curves

of antagonist radioligand binding may show a rightward shift, sometimes with steepening of a flat Hill slope to a value approaching 1. The increase in agonist IC_{50} in the presence of guanine nucleotide is referred to as the **GTP shift**.

1.2.2.1   The Ternary Complex Model

In order to account for such observations, the simple Law of Mass Action model has been extended by including an interaction between the receptor and a G protein, the **ternary complex model**[15] in which the receptor can bind simultaneously an agonist and a second, membrane-bound ligand, a G protein, to form a ternary complex:

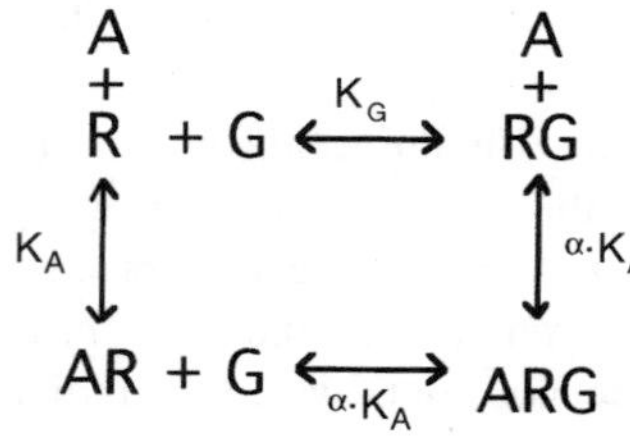

$$
\begin{array}{ccc}
A & & A \\
+ & & + \\
R + G & \xrightleftharpoons{K_G} & RG \\
\Big\updownarrow K_A & & \Big\updownarrow \alpha \cdot K_A \\
AR + G & \xrightleftharpoons{\alpha \cdot K_A} & ARG
\end{array}
$$

Scheme 2

where A represents an agonist ligand with an affinity of K_A, G represents a G protein with an affinity of K_G for the receptor, and α is a cooperativity factor reflecting the change in affinity of each receptor ligand (A or G) when the receptor is occupied by the other ligand.

According to this model, the G protein is activated when bound to the receptor, and binding of an agonist to the receptor causes G protein activation by increasing the affinity of the receptor for the G protein. A corollary of the positive cooperativity between agonist and G protein is that the agonist has a higher affinity for the G protein-coupled receptor than for the free receptor. The **efficacy** of the agonist is determined by the degree to which it increases the affinity of the receptor for the G protein.[16] The G protein contains a binding site for guanine nucleotides, and it is assumed that binding of a guanine nucleotide destabilizes the ternary complex, so that in the presence of guanine nucleotide the high affinity ternary complex is transient, and only binding of agonist to the uncoupled, low affinity state of the receptor is detected. The shape of agonist inhibition curves derived from the model in the absence of guanine nucleotide depends on the stoichiometry of receptor to G protein.[17] If there are many more receptors than G proteins, then only a small proportion of the receptors can participate in the ternary complex and agonists will bind mainly to the free receptor. If there are many more G proteins than receptors, then all receptors can form the high affinity ternary complex, and the agonist will bind with high affinity and a slope of 1. Flat inhibition curves are only predicted by the model if receptors and G proteins occur in roughly similar numbers, so the free concentration of G proteins decreases as the ternary complex accumulates.

The ternary complex model of agonist–receptor–G protein coupling can account for most of the observed binding and functional phenomena, at least qualitatively.

It accounts for the high affinity component of agonist binding in the absence of guanine nucleotides, the reduction in agonist potency and steepening of the agonist inhibition in the presence of guanine nucleotides, and the often observed correlation between the efficacy of agonists (usually estimated by the size of the maximal effect of the agonist in a functional assay relative to other agonists) and the size of the GTP shift,[18] because both phenomena are related to the increase in affinity of the agonist-occupied receptor for the G protein.

In recent years it has been possible to create preparations that exhibit **constitutive activity**, i.e. an agonist-like receptor-mediated effect in the absence of an agonist, by overexpressing wild-type receptors, G proteins, or expressing constitutively active mutant receptors.[19] The ternary complex model accounts for constitutive activity by the increased concentration of reactants (wild-type receptors or G proteins) or by a presumed increased affinity of a mutant receptor for the G protein. The availability of constitutively active preparations has revealed that some antagonists, so-called **inverse agonists**, do not simply prevent access of an agonist to the receptor: they exert their own effects, which are opposite to the effects of agonists. This phenomenon was predicted by binding studies which found that under some conditions antagonist radioligand binding could be enhanced by guanine nucleotides even after stringent efforts to eliminate any contaminating agonists[20] (the possible presence of agonist contaminants remains a problem in interpreting apparent inverse agonist data[21]). Inverse agonism is readily accounted for by the ternary complex model by assuming that binding of an inverse agonist reduces the affinity of the receptor for the G protein.[22]

1.2.2.2 The Extended Ternary Complex Model

There is, however, one phenomenon that cannot be explained by the ternary complex model. It is found that agonists bind with higher affinity to constitutively active receptors than to the wild-type, even in the absence of G proteins, and that the size of this difference depends on the efficacy of the agonist. In order to account for these observations, Samama et al.[23] developed the **extended ternary complex model** (Leeb-Lundberg and Mathis[14] proposed a related model to account for kinetic data with ^{3}H-brady kinin).

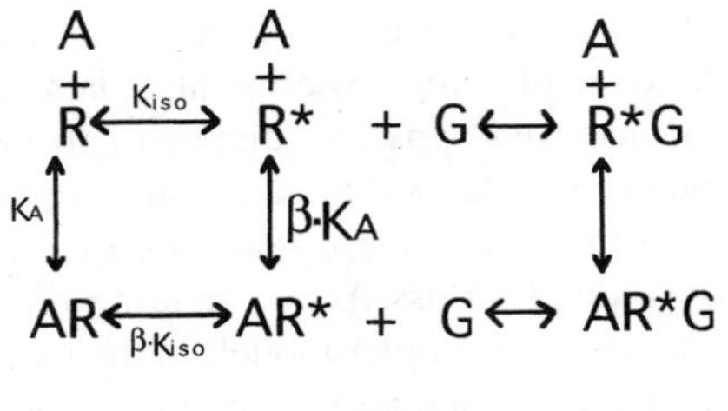

$$
\begin{array}{ccccc}
A & & A & & A \\
+ & & + & & + \\
R \xrightleftharpoons{K_{iso}} & R^* & + G \longleftrightarrow & R^*G \\
\Big\uparrow K_A & & \Big\updownarrow \beta \cdot K_A & & \Big\updownarrow \\
AR \xrightleftharpoons[\beta \cdot K_{iso}]{} & AR^* & + \ G \longleftrightarrow & AR^*G
\end{array}
$$

Scheme 3

According to this model, the receptor shuttles between a **ground** state (R) and **active** state (R^*), the proportions of which are determined by an **isomerization constant** (k_{iso}). Only the active state can form a high affinity and effective ternary complex with a G protein. An agonist has higher affinity for the active than for the ground state, so it "selects" the active state and thereby shifts the receptor population to the active state. β is a cooperativity factor which determines the difference in the affinity of the agonist for the ground and active state. Another, perhaps more intuitive, way of thinking about the model is that β determines the increase in the isomerization constant resulting from occupancy of the receptor by an agonist. In the unliganded receptor k_{iso} will usually be <<1, so that only a small fraction of receptors are active, while in the agonist-occupied receptor $\beta \cdot k_{iso}$ will be >>1 so most of them will be active. The greater the increase in the isomerization constant, or affinity for the active state, the greater the efficacy of the ligand. Inverse agonists have higher affinity for the ground state and reduce the isomerization constant ($\beta < 1$).

An important implication of this model is that the apparent affinity of a ligand for the receptor, in the absence of G proteins, depends on the both the true affinity and the efficacy of the ligand, and also on the value of the isomerization constant, and the true affinity constant cannot be measured directly.[23-25] Samama et al.[23] could account for the efficacy-related increase in agonist binding affinity for a constitutively active mutant β_2 adrenoceptor, as well as the functional changes, by assuming that the isomerization constant determining the fraction of receptors in the ground and active states is larger in the mutant receptor. This model, unlike the ternary complex model, also accommodates functional inverse effects which do not involve G proteins.[26]

Although the model is presented in terms of only one active state, it is possible that a receptor can adopt a number of different active states, which might differ in their affinity for different classes of G proteins, and that different agonists could stabilize different subsets of active states, giving rise to "agonist-dependent trafficking."[19,27] In support of this possibility, a constitutively active mutant receptor may show different changes in affinity with different chemical classes of agonist,[28] and there is evidence that the rate of dissociation of G from the $AR*G$ ternary complex is different with different agonists, which is not consistent with a single active receptor state.[29]

1.2.2.3 Ternary Complex Model vs. Extended Ternary Complex Model

The extended ternary complex model and more complete versions of it have received much attention recently,[19,30,31] but in many cases it has no advantages over the simpler ternary complex model for explaining experimental data. Both models associate agonist efficacy with an increased apparent affinity of the agonist-bound receptor for the G protein. Although the extended model assumes that the free receptor can exist in a low and a high affinity state, these states are freely interconverting, so the extended model generates simple Mass Action binding curves (in the absence of G protein interactions), as does the simpler model. In the extended model the apparent binding affinity of an agonist is actually a function of the receptor isomerization constant and the agonist efficacy, as well as the true affinity, but these are all receptor-specific parameters and so should be constant for a particular ligand–receptor pair, giving the same apparent affinity value as the simpler model. The models do not differ in their capacity to accommodate the binding of many classes of G protein to a single receptor, or the possibility that different agonists may stimulate different patterns of G protein activation. The extended model would seem to be of use mainly when the receptor is subject to experimental modification, such as mutation or chemical modification.[32]

1.2.2.4   Identifying Agonists

It is sometimes possible, and useful, to estimate the efficacy of a ligand using binding studies. There are three approaches.

1. If both an agonist and an antagonist radioligand are available, then unlabeled agonists may show higher affinity for the agonist than for the antagonist radioligand.[33,34]
2. If the affinity of agonists for inhibiting antagonist radioligand binding is reduced by guanine nucleotides, then the size of the GTP shift is an index of efficacy. We have found that the size of the GTP shift with a particular agonist may vary from day to day, but the relative shift, compared to the GTP shift of a standard agonist measured in the same assay, is very reproducible (unpublished observations using muscarinic M_2 receptors in rat heart and oxotremorine as the reference agonist). GDP and GTP have similar effects on agonist binding. These nucleotides are used in the 10^{-6} to 10^{-3} M range, but the lower concentrations may become less effective at longer incubation times because of hydrolysis of the nucleotides by endogenous ATPases and GTPases. Guanyl nucleotides that are less susceptible to enzymatic degradation, for example GTPγS, GppNHp, and GDPβS can substitute for GTP and GDP in binding assays. It should be noted that the ability of an agonist to exhibit a GTP shift may be related to its chemical structure as well as to its efficacy,[35,36] and that not all G protein-coupled receptor systems show GTP shifts (for example, see References 37 and 38).
3. Flat agonist inhibition curves may be analyzed by fitting the data to the ternary complex model, to estimate the agonist affinities at the free and coupled receptor (actually, the high affinity component from such a fit is a composite value[39]), but more commonly such curves are fitted to a two-site model, yielding estimates of two affinity states and the fraction of total binding comprising each state.[17] The values derived from a two-site fit may be similar to the values obtained by fitting the same data to the ternary complex model.[17,39] However, there may be many causes of complex inhibition curves, and unless there is good reason to interpret flat curves according to the ternary complex model the results of such fits may be misleading.[17]

1.2.3   TERNARY COMPLEX MODEL–ALLOSTERIC LIGANDS

In some receptor systems there are ligands that bind to the receptor but do not take the place of the endogenous ligand at the **primary** or "orthosteric" site; rather, they bind to a different, **allosteric**, site simultaneously with the binding of a ligand at the primary site, and the effect is to alter the affinity of the primary ligand. The best known allosteric ligands are perhaps the benzodiazepine tranquilizers such as diazepam, which act allosterically at some subtypes of the $GABA_A$ receptor to enhance the affinity of gamma-aminobutyric acid (GABA).[40,41] Some G protein-coupled receptors also have allosteric sites, and the most extensively studied system is the muscarinic receptor for ACh, for which there are allosteric ligands that enhance or inhibit the affinity of primary ligands.[42–44] Allosteric agents can also alter the kinetics of directly acting ligands.

The simplest model to account for allosteric action at $GABA_A$ and muscarinic receptors is the ternary complex allosteric model:[45–47]

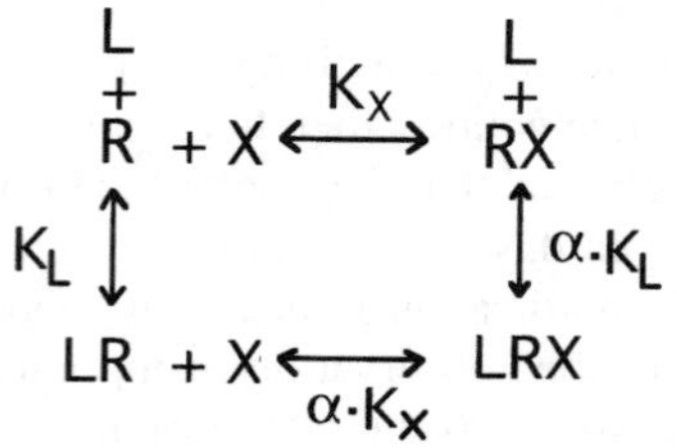

$$
\begin{array}{ccc}
\begin{array}{c} L \\ + \\ R \end{array} + X & \underset{}{\overset{K_X}{\longleftrightarrow}} & \begin{array}{c} L \\ + \\ RX \end{array} \\[2ex]
K_L \updownarrow & & \updownarrow \alpha \cdot K_L \\[2ex]
LR + X & \underset{\alpha \cdot K_X}{\overset{}{\longleftrightarrow}} & LRX
\end{array}
$$

Scheme 4

where L is a primary ligand, X is an allosteric ligand, K_L and K_X are affinity constants for L and X, respectively, and α is the **cooperativity factor**: if $\alpha < 1$ then there is **negative cooperativity**, and if $\alpha > 1$ then there is **positive cooperativity** between the allosteric and primary ligands.

Note that the interaction between the primary and allosteric ligands is reciprocal, i.e., the degree to which the affinity of the primary ligand is modified by the allosteric ligand is identical to the degree to which the affinity of the allosteric ligand is modified by the primary ligand. Note also that for both types of ligand the cooperativity factor, α, is the ratio of affinities at the liganded and free receptor.

This model is very similar to the ternary complex model of receptor–G protein interaction (see Scheme 2 above), but with the important difference that the free concentration of allosteric agent is known, can be manipulated, and can be assumed to be the same as the total concentration under all conditions, whereas the free concentration of G proteins is not known experimentally but becomes progressively lower as more receptor–G protein complexes are formed. One consequence is that the ternary complex G protein model can generate flat or biphasic binding curves, whereas the ternary complex allosteric model does not.

An important feature of this model is that the direction and magnitude of the allosteric interaction depends on the primary as well as the allosteric ligand, so an allosteric agent may show positive cooperativity with one primary ligand and negative cooperativity with another primary ligand. This implies that if allosteric agents are to be developed as therapeutic agents, then they should be studied using the endogenous ligand which, in most cases, will be unlabeled.

Another feature of the model is that it allows **neutral cooperativity**, where the agent binds to the receptor but does not modify the affinity of the primary ligand.

It is often difficult to find primary ligands with strong selectivity for one receptor subtype among a family of subtypes. The existence of neutral cooperativity suggests that it may be possible to find allosteric agents that show **absolute subtype selectivity**, with positive or negative cooperativity with a particular primary ligand at one receptor subtype and neutral cooperativity at the other subtypes.

It is difficult to distinguish between strong negative cooperativity and a competitive interaction. The difference is that a cooperative effect has a limit, defined by the cooperativity factor, but this will only become apparent when high concentrations of ligand are used.

Allosteric agents alter the affinity of primary ligands, and the affinity is the ratio of association and dissociation rate constants, so allosterically acting agents can often be detected by their effects on radioligand kinetics, especially dissociation. According to the model,[46] the EC_{50} or IC_{50} of an agent for modifying the dissociation rate of a radioligand in many cases corresponds to the K_D of the agent for the liganded receptor. Since cooperativity is the ratio of affinities for the liganded and free receptor, it follows that an agent that is negatively cooperative with a radioligand will be more potent in inhibiting equilibrium binding than in modifying the dissociation rate.

If an allosteric agent slows radioligand dissociation, as is usually found with muscarinic receptors, then the same concentration range of the agent will slow the association kinetics of the radioligand in an equilibrium assay. For strongly negatively cooperative agents this kinetic slowing will occur at much higher concentrations than inhibition of radioligand binding and will therefore have no consequences. For agents showing positive, neutral, or small negative cooperativity, however, the kinetic slowing and the cooperative effect will occur over a similar concentration range, and high concentrations of allosteric agent may slow radioligand kinetics so much that binding equilibrium is not attained. This will result in an inhibition of radioligand binding which may be mistaken for negative cooperativity or a competitive interaction, but will actually be a **kinetic artefact**.

Since the concentration of the allosteric agent is known and can be varied, use of the ternary complex allosteric model allows the affinity of the agent and the cooperativity with both labeled and unlabeled primary ligands to be measured experimentally. The model also allows the kinetic effects of allosteric ligands to be related quantitatively to the equilibrium effects, thus providing added stringency to the assessment of whether the behavior of a test agent is consistent with the model. Equations derived from the model are presented below (Section 1.4.5).

1.3 PRACTICAL CONSIDERATIONS

In the literature there is a bewildering variety of conditions used to perform binding assays. It is not always obvious why specific conditions are selected. On one hand, the choice of conditions may have been based on a prior detailed study to optimize an assay in order to obtain specific quantitative information; at the other extreme, it may just be based on a historical precedent. In the following sections, we discuss some of the variables that are important in setting up the conditions for binding assays and how these choices may affect the information obtained. Within the constraints of length of this chapter, it is not possible to provide an encyclopedic coverage, but we attempt to cover many of the important principles and techniques. For more detailed information the reader is referred to specialist books (for example Reference 2).

1.3.1 BUFFER

Cationic, anionic, and zwitterionic buffers are all used: typically these are Tris, phosphate, and HEPES, respectively. These are used at concentrations between 5 and 50 mM but generally in the 10 to 20 mM range. There has been relatively little detailed study on the perturbant effects of different buffers on binding, but there are reports in the literature that Tris, a relatively small organic cation, can have effects on binding. However, these may well be specific for certain receptors.[48,49]

The pH used is normally in the range 6.5 to 8.5 (mainly 7.0 to 7.7) but optimization has only been carried out in a few initial characterization studies of a new radioligand. One should be aware that the radioligand, competing ligand, receptor, and other interacting proteins may all be protonated/deprotonated with change in pH and this can change the affinity and nature of the interactions being monitored.[50] In some instances a proton can behave as a competitive antagonist and in others the structure binding relationships and the subtype selectivity of drugs can be pH sensitive.

1.3.2 IONS

Ions not only contribute to the ionic strength of the incubation medium, but may also have specific effects on the receptor system. In many systems it has been found that low ionic strength can favor radioligand binding, giving a higher ratio of specific to nonspecific binding. However these are "nonphysiological conditions" and detailed quantitative comparisons of binding affinities with those derived from functional studies may be compromised. It has been found that different incubation conditions can even change the rank order of affinities of antagonists.[51]

Monovalent cations generally have some inhibitory effects on antagonist binding and more pronounced inhibitory effects on agonist binding. In the literature this latter phenomenon is sometimes ascribed to a "sodium effect" on the uncoupling of receptors from G proteins (thereby reducing agonist but not antagonist binding), but relatively few studies have separated the "sodium effect" from a less specific ionic strength effect by comparing the effects of equivalent concentrations of Na^+ and K^+ on agonist binding.[52,53]

Divalent cations may inhibit antagonist binding, but magnesium ions have a specific action on G protein-coupled receptors by increasing their coupling to G proteins and hence increasing agonist binding. The optimal concentration is between 1 and 10 mM; higher concentrations result in decreased agonist binding.[54] Other divalent cations may mimic the effect of magnesium,[54] for example manganese (II), but caution should be used in the use of other, more "exotic" divalent cations, as they may interact with amino acid residues on the receptor and possibly chemically modify the receptor, for example either directly by interacting with cysteine residues or indirectly by generating free radicals by lipid peroxidation.

In general, we use a 20 mM HEPES buffer (pH 7.4) containing 100 mM NaCl and 3 or 10 mM MgCl$_2$ as this is of moderately high ionic strength, facilitates receptor–G protein coupling, and allows direct comparison with functional GTPγ^{35}S assays carried out under identical conditions.[55]

1.3.3 PROTEASE INHIBITORS

Peptide ligands may be susceptible to degradation by a number of endogenous proteases and inclusion of appropriate peptidase inhibitors may slow down degradation, but it should be checked that the inhibitors do not affect the receptor directly. In studies using radiolabeled peptides it should always be checked that the peptide is chemically intact at the end of the incubation.

It should also be noted that receptors and G proteins are susceptible to proteolysis and interactions may be modified or abolished. Prolonged incubations and high incubation temperatures favor proteolysis, which is a problem particularly in studies using soluble receptor preparations.

In general the two most effective ways to inhibit proteolysis is to use membranes washed with 10 mM ethylene diamine tetraacetate (EDTA), with 1 mM EDTA in the assay (to inhibit metalloproteases and calcium-activated proteases) and phenylmethyl sulfonyl fluoride (final concentration 0.5 mM, to inhibit serine proteases). Other peptidase inhibitors which may be of use are bacitracin (100 µg/ml), and chymostatin, leupeptin, antipain, aprotinin, and pepstatin (all at 0.5 to 5 µg/ml). Sulfhydryl reagents, e.g., N-ethylmaleimide, should be avoided because they chemically modify the SH groups on many receptors and change the binding properties.

1.3.4 REDUCING AGENTS

A number of ligands, for example catecholamines, are susceptible to oxidation and precautions need to be taken to stabilize the ligand and avoid potential artifacts resulting from the generation of new (often highly reactive) species. A number of antioxidants, e.g., sodium metabisulfite, ascorbate, and dithiothreitol, have been used.

1.3.5 RECEPTOR PREPARATION

The main choices are to carry out the binding studies on membrane preparations, whole cells, tissue sections, soluble and purified receptors. For detailed descriptions of the preparations see Reference 56. Most studies are carried out on membrane preparations and one should be aware that such studies are carried out under "nonphysiological" conditions, but equilibrium studies are possible. Binding studies on whole cells, especially those involving agonists, are complicated by post binding events involving the receptor, for example phosphorylation, segregation, internalization, degradation, and interaction with other proteins as well as synthesis of new receptors.[57]

Autoradiographic binding studies on tissue sections give information on the anatomical localization of the receptor, and studies on the soluble and particularly the purified receptor tend to be biased towards more biophysical investigations of receptor structure and function.

1.3.6 TEMPERATURE

The choice of temperature is often a compromise between the stability of the receptor and ligands, the kinetics of binding of the ligand, optimization of the ratio of specific to nonspecific binding, and the wish to relate the binding results to those of functional studies. In general, binding assays are carried out in the 20 to 37°C range. Studies at low temperatures may not be at equilibrium unless incubation times are prolonged (see next section). Studies on agonist binding to whole cells are sometimes carried out at low temperatures to minimize the post binding events discussed in the previous section.[58] It should be noted that affinity constants for agonists and antagonists can be quite sensitive to temperature and that, for example, antagonist affinities can increase or decrease as the temperature is lowered.[51]

1.3.7 INCUBATION TIME

An equilibrium binding study should, of course, reach equilibrium. In order to ensure this, binding is often measured at a single radioligand concentration over a range of times in order to select a suitable incubation time, but this procedure is not adequate to ensure that all samples attain equilibrium.

By combining Equations 1.3b and 1.7b it can be seen that the observed rate of radioligand binding depends on the radioligand concentration relative to the K_D and on the dissociation rate constant:

$$k_{obs} = k_{off} \cdot \left(1 + L/K_D\right) \qquad (1.13)$$

Equation 1.13 is quite useful. It allows an estimate of the observed association rate constant, k_{obs}, from a knowledge of the radioligand concentration, L, its dissociation rate constant, k_{off}, and its K_D. The binding half life, or time for half the free receptors (relative to equilibrium binding) to become bound, is $\ln(2)/k_{obs}$, and 5 half lives gives about 97% of true equilibrium binding, which is usually considered to be adequate. The following generalizations can be made from Equation 1.13:

1. If [L] << K_D then $k_{obs} \approx k_{off}$
2. If [L] = K_D then $k_{obs} = k_{off} \cdot 2$
3. If [L] >> K_D then $k_{obs} \approx k_{off} \cdot [L]/K_D$

If the radioligand concentration that was used to determine the appropriate incubation time was equal to or greater than the K_D, then lower concentrations of radioligand used in saturation studies (see Section 1.4.3) will probably not have reached binding equilibrium.

The presence of a competing ligand will also affect the time to equilibrium.[59] If the competing ligand has a slower k_{off} than the radioligand, then radioligand binding will initially "overshoot" and then decrease to the equilibrium level as the antagonist binding reaches its own equilibrium. If the competing ligand has much faster kinetics than the radioligand, as is often the case, then the rate of radioligand binding will decrease with increasing antagonist concentrations. This can be understood intuitively when it is considered that the effect of a competitive antagonist is to increase the apparent K_D of the radioligand, and from Equation 1.13 it can be seen that the observed rate of radioligand binding, k_{obs}, will decrease as apparent K_D increases, with k_{off} as the limiting value. If the radioligand concentration used in the assay, and also used to determine the appropriate incubation time, was equal to or greater than the K_D, then binding in the presence of high concentrations of antagonist will probably not be in equilibrium.

In order for saturation and inhibition assays to be completely in equilibrium, the incubation time should be at least 5 *dissociation* half lives. It is necessary and useful to measure the rate of association; it is the dissociation rate, however, that should guide the choice of incubation time.

1.3.8 COLLECTION

Most radioligand binding assays, but not all (see Section 1.3.9), require methods of separating the radioligand bound to the receptor preparation from free ligand. In the case of membranes or cells this is normally accomplished by rapid filtration, although centrifugation may be used in certain cases. Examples of the latter are when the radioligand binds strongly to filters but not to microcentrifuge tubes[60] or when high speed centrifugation (>100,000 g hours) is used to ensure that all the ligand-bound receptor is collected. Assays of soluble receptors use gel filtration, adsorption of unbound radioligand on charcoal, filtration on polyethyleneimine (PEI)-treated filters, or precipitation with, e.g., ammonium sulfate. Detailed descriptions of the different separation methodologies are provided in a number of chapters of Reference 2.

The most popular technique uses a filtration manifold or cell harvester which allows 24 to 96 samples to be filtered simultaneously and for the filters to be washed rapidly and efficiently to remove any weakly bound radioligand that is present on the filter or on the membranes. Most filters used are made of glass fiber (pore size 0.7 to 1.2 μm). All assay protocols involve a series of compromises and need to be optimized individually, depending on the radioligand, the receptor, and the information to be obtained.

Care has to be taken to ensure that the washing procedure removes the weakly bound radioactivity without allowing significant dissociation of the radioligand from the receptor. This may often be accomplished by two or three rapid washes with ice-cold buffer but the procedure should be optimized. There is also a limitation on the amount of membrane protein that can be filtered. This is about 1 mg for a 1.5 cm diameter filter area and proportionately less for smaller areas; higher levels of protein result in clogging of the filters and unacceptably slow filtration and washing rates.

It should be noted that the filtration procedure does not trap all the membrane fragments; up to 50% of the receptors can pass through glass fiber filters. This problem can be overcome partially by the use of filters with a smaller pore size, but this results in a slower filtration rate. Depending on the particular aim of the experiment and the experimental conditions this factor should be borne in mind, especially where a substantial fraction of the radioligand is binding to the receptor (see Section 1.3.10) and the free concentration of the radioligand cannot therefore be calculated from the difference between total radioactivity added and bound radioactivity.

In the case of centrifugation assays, the supernatant is removed by suction or by inverting the tube and the pellet rapidly and superficially washed to remove unbound radioactivity. Any radioactivity bound or entrapped within the pellet will tend not to be removed.

1.3.9   COUNTING

β-radioactivity on filters is generally counted using liquid scintillation spectrometry. The filter areas are punched out into vials, an appropriate aqueous scintillation cocktail added, and the vials left to stand, with occasional agitation, until all the radioactivity is extracted (2 to 15 hours). It is not necessary to dry the filters thoroughly but it is our experience that radioactivity on damp filters is extracted more slowly than that on dry filters. Drying may be accelerated by placing the filters on a thermostatically regulated hotplate, set to about 90°C.

A solid-phase scintillant (e.g., Meltilex, Wallac) can also be used. This is melted into the glass fiber filter strip and allows the samples on a whole strip to be counted in a suitable counter, which may be formatted to allow parallel acquisition of data (up to 12 samples at a time).

Radioactivity in pellets in 1.5 ml microcentrifuge tubes can be counted after first dissolving the pellet in 1 M NaOH (10 µl) at 65°C for about 40 min and then neutralizing with an equal volume of 1 M HCl before adding 1.2 ml aqueous scintillant.

It is now possible to measure radioactivity bound to membranes without separation of unbound radioligand using the scintillation proximity assay technique (SPA). In this assay the receptor preparation, membrane bound or soluble, is adsorbed to the surface of microspheres containing scintillant. Any decay of a radioligand bound to the receptor preparation is sufficiently close to the scintillant to trigger light emission, whereas this is not true for the bulk unbound radioligand. This technique can be applied to both kinetic and equilibrium studies. Unfortunately these beads are not generally available to nonindustrial organizations other than in specific assay kits.

Gamma-radioactivity can be counted directly on a gamma counter.

1.3.10   RADIOLIGAND DEPLETION

The equations normally used to interpret radioligand binding studies are expressed in terms of the free concentrations of the ligands, but usually only the total ligand concentration is known with accuracy. If a substantial proportion of the added ligand is bound to the receptor preparation, then saturation and inhibition curves will become steeper, ligand potencies will be underestimated, and other parameter estimates will be inaccurate. For example, the IC_{50} of a competitive inhibitor is increased by a factor $D/(1-D)$, where D is the fractional depletion in the absence of inhibitor.[3]

Ligand depletion of up to 10 to 15% is tolerable. Higher levels of depletion will occur with low ligand concentrations if the receptor concentration is more than 20% of the K_D and/or there is a high level of nonspecific binding. Ligand depletion is inversely related to the concentration of ligand and is reduced if the incubation volume is increased or the tissue concentration is decreased.

Depletion of unlabeled ligands will usually be detected indirectly, by measuring the effects of altering incubation volume or tissue concentration on the shape and potency of an inhibition curve. Depletion may be measured directly by taking a sample of the supernatant from one assay and testing its inhibitory potency in a second assay. The problem may be avoided by increasing the incubation volume, decreasing the tissue concentration, or raising the concentration of the radioligand well above its K_D. Since the nonspecific binding of the unlabeled ligand is not usually known, it is not possible to estimate the free concentration of the unlabeled ligand computationally. An exception is the case of homologous inhibition, where the inhibitor is an unlabeled form of the radioligand: this problem has been solved analytically by Swillens,[61] and it is handled computationally by the program LIGAND[62] (available from Biosoft, Cambridge, U.K.).

It is easy to detect depletion of the radioligand, since the concentrations of both added and bound material are measured routinely. The possibility of radioligand depletion should always be considered. The consequences of radioligand depletion, and computational methods for coping with them, have been considered in detail.[3,10,63] As discussed below, computational adjustments are unlikely to be completely accurate. If possible, it is preferable to avoid the problem by increasing the radioligand concentration or the incubation volume, or decreasing the tissue concentration, but practical limitations may sometimes make it impossible to prevent depletion, especially with very potent tritiated radioligands.

There are two types of computational approach to radioligand depletion: using equations that are formulated in terms of total, rather than free, radioligand, or the calculation of free radioligand in each tube. The former approach can be analyzed with any general purpose curve-fitting program and is statistically correct, but it is limited to the few models for which analytical solutions have been derived. The latter approach requires a curve fitting program such as SigmaPlot (SPSS, Erkrath, Germany) which can handle more than one independent variable (except with saturation curves, which have only one independent variable) and errors in the dependent variable (bound) propagate to one independent variable (free radioligand), but the approach can be used with any binding model.

1.3.10.1 Depletion in Saturation Curves

Equations describing radioligand binding as a function of added radioligand have been presented.[3,10,61] These can be used with any curve-fitting program that allows user-defined functions.

The calculation approach is described in Section 1.4.3. Briefly, free radioligand in each tube is calculated as the difference between added and bound radioactivity (dpm). The relationship between free radioligand and nonspecific binding is determined, and this is used to estimate nonspecific binding in the "total binding" tubes.

The trouble with this method is that errors in the measurement of bound radioligand cause a corresponding inverse error in the estimation of free radioligand. An advantage of the method is that it is simple to understand, and it can be used with Scatchard/Hofstee analysis as well as with nonlinear regression.

A problem of both these approaches is that they assume that nonspecific binding is a linear function of free radioligand, which is usually true but may not be. A second, more serious, problem is that both methods assume that the measured binding is the same as the binding that occurred during the incubation, but, for filtration assays, this does not allow for loss of binding during the washing procedure (see Section 1.3.11.3); binding will be lost because of dissociation of radioligand from receptors, loss of membranes through the filter, and from the loss of loosely bound, "washable" nonspecific binding, which may be substantial at high membrane concentrations.[64] These losses of binding will cause the degree of radioligand depletion to be underestimated, leading to an increase in apparent K_D in some conditions.[64]

Both problems can be avoided by the empirical approach of preparing two identical sets of tubes, filtering one set to measure bound radioligand, and centrifuging the other set to measure the free radioligand in the supernatant.

1.3.10.2   Depletion in Inhibition Curves

It is more difficult to adjust for the effects of radioligand depletion in inhibition curves than in saturation curves, and the best solution is to avoid the problem empirically. Nevertheless, equations describing radioligand binding to one or two sites in the presence of an inhibitor as a function of added radioligand have been presented.[3,10]

Alternatively, free radioligand in each tube is calculated by subtracting bound from added, and the nonlinear regression analysis uses the concentrations of both the unlabeled inhibitor and the free radioligand as independent variables: this analysis requires curve-fitting programs such as SigmaPlot which allow more than one independent variable.

Both methods will underestimate depletion in filtration assays because of inevitable losses of binding during the washing process.

1.3.11   PROBLEMS THAT MAY ARISE DURING THE PERFORMANCE
OF SOME BINDING ASSAYS

Previous sections have dealt with general approaches to the performance of binding assays. In this section we deal with some specific problems associated with subsets of receptor systems, ligands, or assays.

1.3.11.1   Adsorption of Ligands to Pipette Tips,
Plastic and Glass

This is a problem that is associated with certain peptides, proteins, and some hydrophobic and amphipathic compounds. The adsorption of the radioligand is readily detectable but that of competing ligands can be insidious.

If ^{125}I is the radioisotope, adsorption can be detected directly by γ-counting of the containers or pipette tips. For ^3H and ^{35}S, adsorption is detected by measurement of deviations between observed and expected levels of radioactivity in samples. Slow adsorption to containers can be measured as a time-dependent decrease in the radioactivity present in aliquots; for example, we have observed that when one radioligand was taken up into a repetitive pipette, the amount of radioactivity dispensed decreased with each aliquot such that the level in the fifth aliquot was only 50% of that in the first. Fast (and slow) adsorption of the radioligand can be detected by losses in measured radioactivity associated with repetitive transfers of the same sample from container to container.

In many cases only a percentage of the radioactivity is adsorbed. The adsorption is proportional to surface area and it shows some characteristics of a saturation phenomenon. For example we have observed that the second use of a pipette tip can result in up to 50% more radioactivity being dispensed than on the first pipetting.

The adsorption of a competing ligand to surfaces is sometimes manifest by competition curves having Hill coefficient >1, by the potency being sensitive to the dilution protocol for the ligand, or by deviations from an apparently competitive behavior. Experiments involving the use of repetitive transfers and exposures to surfaces can detect adsorption as a change in the inhibitory potency of solutions.

There are a number of protocols that have been found to be useful in minimizing adsorption losses, but it is not possible to predict how individual ligands will behave.

1. Minimize the number of transfers of solutions.
2. Minimize the surface area exposed to the solutions.
3. Keep the concentration of ligand as high as possible during handling processes prior to the assay.
4. Test different surfaces for adsorption.
5. Silanize the surfaces.
6. Test whether adsorption is reduced in the presence of different buffers, high(er) ionic strength solutions, other solvents for dilutions (e.g., 50% ethanol, 10 to 100% dimethylsulfoxide, DMSO, for hydrophobic and/or highly insoluble ligands), or the presence of protein (e.g., 0.001 to 0.1% bovine serum albumin, BSA). Note that the adsorption of some radioligands is also reduced by the presence of membranes!

1.3.11.2 Adsorption of Ligands to Filters

For most purposes, the filter may be treated as another surface and some of the protocols suggested above may be applicable. The exception is silanization which dramatically reduces the permeability of the filters. Adsorption to the filter can be measured directly and not indirectly, as in general in Section 1.3.11.1.

The most common method for reducing nonspecific binding to filters is to soak the filters for a short period of time in PEI (polyethyleneimine 0.05 to 0.5% in water). This is particularly effective for positively charged ligands but also for some neutral polar ligands. It also has the added benefit in some instances of increasing the effectiveness and reproducibility of the filters to entrap membranes. It should not (for obvious reasons) be used for binding studies involving negatively charged ligands—the adsorption of GTPγ^{35}S to PEI-treated filters is almost quantitative. In this context one should be aware that PEI residues can contaminate harvesters and cause high and variable binding of GTPγ^{35}S to filters. It can be quite difficult to remove this contamination.

In some instances binding of a radioligand to filters can be inhibited by ligands independently of their effect on the binding of the radioligand to the receptor. This can give rise to artifactual results in inhibition studies (e.g., inhibition of total binding decreasing below the estimate of "nonspecific binding"). Determinations of nonspecific binding at each concentration of nonradioactive ligand may be required. This phenomenon may be exploited to reduce filter binding.

1.3.11.3   Loss of Specific Binding During Washing

Aside from filters not trapping all the membranes (Section 1.3.8), there may be losses of bound radioactivity from the receptor caused by the washing procedure. The avoidance of this artifact is a compromise between the time taken to reduce nonspecific binding to the lowest level and the dissociation rate constant of the radioligand from the receptor.

Washing should ideally be carried out with the incubation buffer at the temperature of the assay in order not to perturb the system. In many cases, however, the washing is carried out with cold solutions (0 to 4°C, in order to reduce the off rate of the radioligand) and using, for convenience, water or a saline solution. However, precautions should be taken to ensure that the washing solution (if the pH or ionic strength is different) does not perturb the receptor such that the dissociation rate is increased. Note that in order to ensure that the washing takes place at a low temperature, the filters should be rinsed with the cold buffer before initiation of filtration of the membranes.

Two to three washes (1 to 5 ml) are generally sufficient to reduce nonspecific binding to a plateau level and this can take 5 to 20 s. A radioligand that has a dissociation rate of about 5×10^{-3} s^{-1} (*under the washing conditions*) will dissociate by 5 to 10% during the washing and this represents a limit value for acceptability for the application of the rapid filtration method for measuring radioligand binding. Allosteric ligands may increase the dissociation rate of a radioligand[65] and caution should be exercised that the dissociation rate is not enhanced to such an extent that accurate measurement of bound radioactivity is compromised.

As a general comment, if the nonspecific binding to the filters is relatively high, the washing conditions are critical, or there is a variability between filters in their retention properties, then it is advisable to perform a single set of measurements on one filter and if necessary to perform a replicate set of measurements on a second filter. By these means interfilter variability can be accounted for in the data analysis.

1.3.11.4   Endogenous Ligands

If the endogenous ligand for the receptor is present in high concentration in the tissues or cells to be examined or is generated and/or released during incubation, it can interfere with the radioligand binding assay. Examples of such ligands are

adenosine, dopamine, 5HT and GABA. Extensive washing during membrane preparation may be of some use (but not in the case of adenosine, which is released from
membranes during incubation). The ligands may be removed substantially by treating
with an appropriate enzyme (e.g., adenosine deaminase for adenosine).

A subtle manifestation of contamination by endogenous ligands is their ability,
under some conditions, to become "locked" into the receptor as extremely slowly
dissociating agonist–receptor–G protein complexes. This phenomenon was first
described for beta-adrenoceptors but is found for a number of other G protein-
coupled receptors, including the adenosine A_1 receptor.[66,67] This locking process can
take place both during the membrane preparation and during the incubation. It
generates slow kinetics and can also result in an underestimation of *Bmax* under
some conditions. The locked complexes can be dissociated by guanine nucleotides.
Caution should therefore be taken, for example, to exclude a locking mechanism
before concluding that a guanine nucleotide increase in antagonist binding is evidence of receptor-precoupling/constitutive activity/inverse agonism.

1.3.11.5 Precautions in the Use of Radioiodinated Ligands

[125] Iodine is a convenient isotope for labeling peptides, proteins and some nonpeptide
ligands because there are easy methods of chemically labeling the tyrosine (or aromatic) residues. Because of its relatively short half life, ^{125}I has a higher specific activity
than tritium for the monosubstituted molecule and can therefore detect lower concentrations of receptors. However, it is not always appreciated that the iodinated molecules
have different chemical and possibly pharmacological properties from the unmodified
ligand. It is therefore important to separate the radioiodinated species from the precursor and to separate the individual radiolabeled species when there is the possibility
of more than one site of iodination. In the case of a protein or a large peptide, separation
of the different species may not be possible and caution should be taken in the
interpretation of experiments, especially with regard to estimates of *Bmax*.

Estimation of the K_D and *Bmax* of the radioligand requires either the use of large
amounts of radioactivity or the preparation of pure iodinated nonradioactive ligand
for use in self inhibition studies.

1.4 BASIC ASSAYS

There are four general designs of assay for the characterisation of labeled and
unlabeled ligands: dissociation, association, saturation, and inhibition. The assays
are interpreted initially by fitting the data to a mechanistic model, the simple Law
of Mass Action (Scheme 1), or in the case of inhibition curves to an empirical model,
a logistic function. Before computers were generally available, it was necessary to
transform the data with a function that yields a straight line when applied to simple
Law of Mass Action data, and then use linear regression analysis with the transformed data. The most commonly used transformations will be presented for historical reasons, and because the linear plots are useful intuitively and for detecting
deviations from the model. Wherever possible, however, parameter estimates should
be obtained by fitting the untransformed data directly to the model using nonlinear

regression analysis. The appropriate equations will be presented, although most curve-fitting programs have the equations already set up.

The analysis of allosteric interactions with labeled and unlabeled primary ligands (Scheme 4) is more complicated than the standard assays because it involves interactions between three ligands, and because allosteric interactions with the unlabeled primary ligand may be difficult to visualize if the allosteric agent has a strong effect on the radioligand. In addition, the allosteric agents may slow radioligand binding and prevent binding equilibrium from being attained. Two assays will be presented: a screening assay in which the data are transformed to reveal allosteric effects, and an assay suitable for quantitation using nonlinear regression.

1.4.1   ASSAY 1: DISSOCIATION RATE MEASUREMENTS AND ANALYSIS

The objective is to measure the pattern and extent of radioligand dissociation as a function of time and hence determine the dissociation rate constant(s).

1.4.1.1   Experimental Procedure

Receptors are prelabeled with the radioligand, usually (but not necessarily) until equilibrium is reached. Further binding of the radioligand is then prevented, by adding a high concentration of unlabeled competing ligand and/or by dilution of the sample (at least 100-fold). Both processes prevent the radioligand from rebinding to the receptor after it has dissociated.

At various times thereafter binding is measured. In addition, samples that have not dissociated are required (thus providing a measure of the binding at time zero) as well as determinations of nonspecific binding at time zero (and at subsequent times if there is any dilution of the prelabeled sample).

The assay is straightforward if a filtration manifold or a number of microcentrifuges are used and the dissociation rate is slow ($t > 5$ min): the above procedure may be carried out in a single container and aliquots removed for analysis at various times. However, if the dissociation rate is fast or a cell harvester is used, all the samples have to be filtered at the same time. This necessitates a "reverse time" strategy, with the assay being started at different times for the different time points.

Ideally, especially with slowly dissociating ligands, the assay should be planned so that all the samples are exposed to radioligand for the same time before dissociation is initiated. However if "equilibrium" binding remains stable over the total time of the assay, as may be the case with more rapidly dissociating radioligands, then prelabeling can be started with all the tubes at the same time and only the initiation of dissociation needs to be timed individually.

1.4.1.2   Linearizing Transformation of the Data

After subtraction of nonspecific binding from all the data points, $ln(B_t/B_0)$ is plotted against time, where B_0 is the specific binding before initiation of dissociation and B_t is the binding remaining at time t (Figure 1.1). For dissociation from a uniform population of labeled receptors, the graph is a straight line, the negative slope of the straight line being the dissociation rate constant, k_{off}. The plot is curved in more complex kinetic situations.

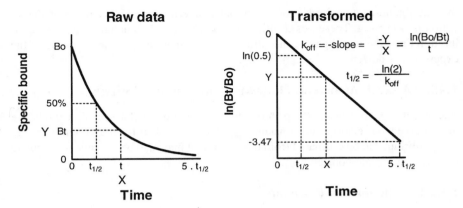

FIGURE 1.1   Dissociation assay—raw data and linearizing transformation.

1.4.1.3   Nonlinear Regression Analysis of the Data

After subtraction of nonspecific binding from all the data points, the specific binding
is fitted initially to the equation:

$$B_t = B_0 \cdot \exp\left(-k_{off} \cdot t\right) \tag{1.14}$$

where B_0 and B_t are the observed specific binding at times 0 and t, respectively,
and k_{off} is the dissociation rate constant.

If the data appear to asymptote above zero at long times, the data can be fitted
to the equation:

$$B_t = \left(B_0 - asymptote\right) \cdot exp\left(-k_{off} \cdot t\right) + asymptote \tag{1.15}$$

If there appear to be two kinetic components, the data can be fitted to:

$$B_t = B_0 \cdot \left[q \cdot exp\left(-k_{off1} \cdot t\right) + (1 - q) \cdot exp\left(-k_{off2} \cdot t\right)\right] \tag{1.16}$$

where k_{off1} and k_{off2} are the dissociation rate constants of components 1 and 2
and q is the fraction of initial binding attributable to component 1.

1.4.1.4   Interpretation of Analysis

If the simple model adequately fits the data, the results are consistent with a uniform
population of labeled receptors. The half time for the radioligand to dissociate, t, is
given by $\ln(2)/k_{off}$.

If a more complex model is required to fit the data, an empirical model should
be considered unless there is independent evidence to support a mechanistic interpretation. Possible explanations of complex curves may be receptor heterogeneity,

that the ligand-occupied receptor undergoes isomerization, or that receptor degradation is occurring. These possibilities may be explored by changing the pre-equilibration time, the radioligand concentration, the temperature, or other aspects of the experimental design.

1.4.2   ASSAY 2: ASSOCIATION RATE MEASUREMENTS AND ANALYSIS

The objective is to measure the pattern and rate of radioligand association and hence the observed association rate constant, k_{obs}. The time taken for the system to come to equilibrium can be estimated from the data and it may also be possible to calculate the association rate constant, k_{on}.

1.4.2.1   Experimental Procedure

A fixed concentration of radioligand is added to the receptor preparation and bound radioactivity measured at defined time points. In addition the assay contains measures of nonspecific binding at some or all of the time points.

In the design of the assay, the concentration of radioligand to be used has to be carefully considered. The choice depends on the information required. On one hand it is important to know the time course of binding of the radioligand concentration used in other studies such as inhibition assays (see below), a concentration which is often around the K_D level or below. On the other hand, for an accurate estimate of k_{on} it is necessary to use a much higher radioligand concentration (see Section 1.3.7 and below), provided the nonspecific binding is not too high. It is recommended that at least two radioligand concentrations be used.

If a cell harvester is used, then the binding interaction has to be initiated at different times if filtration for the different times of association is to be carried out simultaneously (a reverse time strategy). A decision must also be made whether to store the receptor preparation on ice or under the incubation conditions prior to the initiation of binding: receptors are likely to degrade less at lower temperatures, but liganded receptors may degrade less (or possibly more) than unliganded ones at the incubation temperature. If detailed kinetic studies are to be carried out, it is necessary to measure receptor degradation in order to avoid the artefact of increases in binding during association being obscured by decreases associated with degradation.

1.4.2.2   Linearizing Transformation of the Data

After subtraction of nonspecific binding from all data points, the "equilibrium" level of specific binding, B_{eq}, is estimate by eye and $\ln(B_{eq}/(B_{eq} - B_t))$ is plotted against time, where B_t is the specific binding at time t (see Figure 1.2). In the case of a simple association process and a good choice of B_{eq}, the slope of the resulting straight line is an estimated observed association rate constant k_{obs}.

1.4.2.3   Equations for Nonlinear Regression

After subtraction of nonspecific binding from all data points, the data are fitted to the equation

$$B_t = B_{eq} \cdot \left(1 - exp\left(-k_{obs} \cdot t\right)\right) \tag{1.17}$$

where the symbols have the meaning given above.

If there appear to be two binding components, 1 and 2, with observed association rate constants of k_{obs1} and k_{obs2}, then the data are fitted to:

$$B_t = B_{eq} \cdot \left[1 - q \cdot exp\left(-k_{obs1} \cdot t\right) - (1 - q) \cdot exp\left(-k_{obs2} \cdot t\right)\right] \tag{1.18}$$

where q is the fraction of equilibrium binding attributable to component 1.

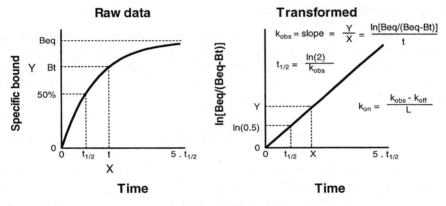

FIGURE 1.2   Association assay—raw data and linearizing transformation.

1.4.2.4   Interpretation

For a simple monoexponential binding process, the time for specific binding of the radioligand concentration used in the assay to reach 50% of its equilibrium level, $t_{1/2}$, is given by $ln(2)/k_{obs}$.

The association rate constant k_{on} is calculated from the values of the radioligand concentration L, the observed association rate constant at that concentration k_{obs}, and the dissociation rate constant k_{off}:

$$k_{on} = \left(k_{obs} - k_{off}\right)/L \tag{1.19}$$

For the data to fit the simple model (Scheme 1) well, similar values of k_{on} should be obtained with different radioligand concentrations. In addition, the ratio, k_{off}/k_{on}, which defines the equilibrium dissociation constant, K_D, should also agree with the value obtained with equilibrium saturation analysis (see below).

It should be noted that the value of k_{obs} approaches k_{off} at low $[L]$ (see Equation 1.13), and so, for an accurate estimation of k_{on}, values of $[L] > K_D$ should be used.

If a more complex model is required to fit the data, then it should be considered to be an empirical model unless and until there is independent evidence to support a mechanistic interpretation. As considered in the previous section on interpretation of dissociation rate measurements (Section 1.4.1.4), possible explanations of complex curves include ligand-occupied receptor isomerization, receptor heterogeneity, and receptor degradation. These possibilities may be addressed by suitable modification of the experimental design.

1.4.3   ASSAY 3: SATURATION ANALYSIS

The objective is to measure the amount of specific binding with a range of radioligand concentrations and to estimate the affinity constant, K_A (or dissociation constant, K_D) of the radioligand, and the concentration of binding sites (*Bmax*).

1.4.3.1   Experimental Procedure

Receptors are incubated with a range of radioligand concentrations in the absence and presence of unlabeled ligand used to define nonspecific binding. The concentration of unlabeled defining ligand should be at least 100 times its apparent dissociation constant, $Ki_{apparent}$, in the presence of the highest concentration of radioligand, L, where

$$Ki_{apparent} = K_i \cdot \left(1 + L/K_D\right) \tag{1.20}$$

The assay is incubated for sufficient time for binding equilibrium to be reached for *all* concentrations of radioligand (see Section 1.3.7).

If possible, concentrations of radioligand should span at least a 100-fold range, be equally spaced on a log scale, and should include concentrations in the range of the estimated K_D.

1.4.3.2   Linearizing Transformation of the Data

The calculation of specific binding depends on whether or not the free radioligand (added dpm minus bound dpm) differs by more than about 15% between samples measuring total and nonspecific binding. If the free radioligand concentration is effectively constant, specific binding at each radioligand concentration is calculated by subtracting nonspecific from total binding.

If the free radioligand differs between "total" and "nonspecific" samples, calculate free radioligand as the difference between added dpm and bound dpm. Perform a linear regression of *nonspecific bound* on free and, from the results, calculate the nonspecific binding for the values of free radioligand found in the "total" samples. The specific binding at each radioligand concentration may then be calculated by subtracting the *estimated* nonspecific from total binding.

The free radioligand concentration, the difference between added dpm and total bound dpm, is converted into nM.

The Scatchard plot portrays specific/free vs. specific (Figure 1.3) and the K_D and *Bmax* values can be estimated from linear regression analysis. It is easier and statistically more accurate to calculate them from the reverse plot of specific vs.

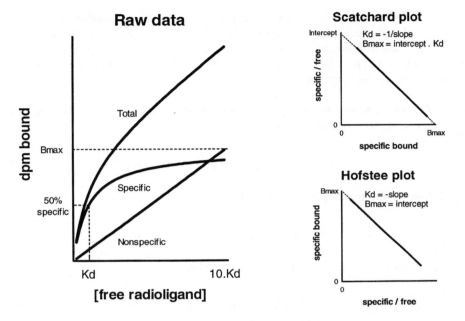

FIGURE 1.3   Saturation assay—raw data and linearizing transformations.

specific/free, the Hofstee plot. Linear regression of specific on specific/free gives the intercept as *Bmax* and the slope as $-K_D$.

1.4.3.3   Nonlinear Regression of the Data

There are three types of nonlinear regression program which may be used to analyze saturation data.

1. LIGAND[62] (available from Biosoft, Cambridge, U.K.) is designed to analyze these types of data and automatically handles nonspecific binding and radioligand depletion. Therefore, the user has only to follow the instructions for data input.
2. Programs that allow only a single independent variable in the analysis should be provided with values of free radioligand and specific bound concentrations, calculated as described above. The data are fitted to the equation

$$specific = Bmax/\left[1+\left(K_D/free\right)\right] \tag{1.21}$$

For a two-site fit, the following equation should be used:

$$specific = Bmax1/\left(1+K_D1/free\right)+Bmax2/\left(1+K_D2/free\right) \tag{1.22}$$

3. Programs that can handle more than one independent variable and complex models (e.g., SigmaPlot, SPSS, Erkrath, Germany) should be provided with the added dpm and the *measured* total and nonspecific bound dpm. The *free* dpm at each point is calculated from the difference between added and bound dpm.

The total and nonspecific binding data are fitted simultaneously. Total binding is fitted with the function

$$total(free) = specific(free) + nsb(free) \tag{1.23}$$

Nonspecific binding is fitted with the function

$$nsb(free) = k_{nsb} * free + background \tag{1.24}$$

and specific binding is defined by the function

$$specific(free) = Bmax/(1 + K_D/free) \tag{1.25}$$

As an alternative, the data may be fitted to the equations of Swillens,[61] which also allow for radioligand depletion.

For a two-site fit to the data, the following equation should be used:

$$specific(free) = Bmax1/(1 + K_D1/free) + Bmax2/(1 + K_D2/free) \tag{1.26}$$

1.4.3.4   Interpretation

It is always best to analyze experimental data using nonlinear regression, but the Scatchard or Hofstee plots provide a useful visual representation of the goodness of fit of the saturation data to the simple model. If the radioligand binds obey a simple binding equation (Scheme 1), then the Scatchard and Hofstee plots will be straight lines. Curvature of the plot indicates that the Hill slope is flat (i.e. < 1, Scatchard plot is concave) or steep (i.e. > 1, Scatchard plot is convex). A nonlinear Scatchard plot requires additional experiments in order to interpret the data.

It should be noted that, while all the methods described above can allow for radioligand depletion, to the extent that it can be seen, they cannot take into account any loss of binding during the filtration procedure (both specific and nonspecific), which may lead to overestimation of the K_D under some conditions (see Section 1.3.10).

1.4.4   ASSAY 4: INHIBITION ANALYSIS

The objective is to measure the degree and pattern of inhibition of the binding of a fixed concentration of radioligand in the presence of a range of concentrations of unlabeled ligand. The concentration of inhibitor which inhibits specific binding by

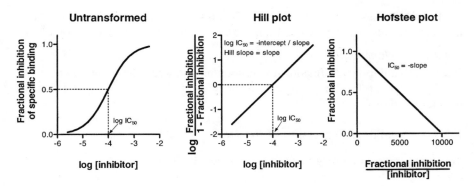

FIGURE 1.4 Inhibition assay—raw data and linearizing transformations.

50%, the IC_{50}, and the slope of the inhibition curve, b, will be estimated. It may also be possible, by calculation, to determine the dissociation constant of the inhibitor, Ki.

1.4.4.1   Experimental Procedure

Receptors are incubated with a single concentration of radioligand, alone and in the presence of increasing concentrations of the unlabeled ligand for sufficient time for binding equilibrium to be reached (see Section 1.3.7). Nonspecific binding is also measured, if necessary in the simultaneous presence of both the ligand used to define nonspecific binding and the higher concentrations of unlabeled ligand (in case the test ligand inhibits nonspecific binding). In general, the radioligand concentration should be as low as possible but with less than 15% of added radioligand bound (see Section 1.3.10), and concentrations of inhibitor should span at least a 100-fold range.

1.4.4.2   Linearizing Transformations

After subtraction of nonspecific binding from all data points, the fractional inhibition of specific binding at each inhibitor concentration is calculated using the equation:

$$fractional\ inhibition = 1 - \frac{B}{B_0} \tag{1.27}$$

where B is specific binding in the presence of a given concentration of inhibitor and B_0 is the specific binding measured in the absence of inhibitor.

1.4.4.2.1   Hill plot

For each inhibitor concentration, log (fractional inhibition /(1 − fractional inhibition)) is calculated and this value is plotted vs. log [inhibitor] (Figure 1.4).

Using only the fractional inhibition values in the range 0.1 to 0.9 a linear regression analysis of values of log (fractional inhibition /(1 − fractional inhibition)) on log [inhibitor] is performed. The gradient is the Hill slope and the log IC_{50} is −intercept/slope.

If the Hill slope is 1, then the equilibrium dissociation constant of the inhibitor, Ki, can be calculated from the IC_{50} using the Cheng-Prusoff equation (Equation 1.11).

1.4.4.2.2   Hofstee plot

Fractional inhibition vs. (fractional inhibition/inhibitor concentration) is plotted.

The slope of the linear regression analysis of data (in the same fractional inhibition range) is $-IC_{50}$.

1.4.4.3   Nonlinear Regression of the Data

Two types of program may be used to analyze inhibition data.

1. LIGAND[62] is designed specifically to analyze these data and automatically handles nonspecific binding and radioligand depletion. Therefore the user has only to follow the instructions for data input.
2. General nonlinear regression programs can be used to fit the data either to a logistic function or to a one- or two-site model. Nonspecific binding can either be subtracted before the analysis or included as a parameter in the model (treating observed nonspecific binding as that obtained in the presence of a very high antagonist concentration). For generality it will be assumed that nonspecific binding is included.

1.4.4.3.1   The logistic function, an empirical model

The observed data are fitted to the logistic equation

$$B = \frac{B_0 - nsb}{1 + 10^{(\log(I) - \log(IC_{50})) \cdot b}} + nsb \tag{1.28}$$

where

B = binding in the presence of inhibitor
B_0 = binding in the absence of inhibitor
nsb = nonspecific binding
I = the concentration of inhibitor
b = the slope factor, equivalent to the Hill slope
IC_{50} = the inhibitor concentration giving 50% inhibition

The fitting procedure is more robust if the log IC_{50} is estimated, rather than the IC_{50}.

1.4.4.3.2   Mechanistic models

For a one-site model, the observed data are fitted to the equation

$$B = \frac{B_0 - nsb}{1 + 10^{(\log(I) - \log(IC_{50}))}} + nsb \tag{1.29}$$

(which is the same as the logistic function with a slope factor of 1).

An estimate of the Ki value may be calculated from the IC_{50} using the Cheng-Prusoff equation (Equation 1.11).

Alternatively, the antagonist Ki can be estimated directly by fitting the data to the equation:

$$B = \frac{B_0 - nsb}{1 + 10^{\left(\log([I]) - \log(Ki)\right)} \cdot \dfrac{K_D}{K_D + L}} + nsb \tag{1.30}$$

where L and K_D are the known concentration and dissociation constant, respectively, of the radioligand.

For a two-site model, the observed data are fitted to the equation

$$B = (B_0 - nsb) \cdot \left(\frac{fraction1}{1 + 10^{\left(\log(I) - \log(IC_{50}1)\right)}} + \frac{1 - fraction1}{1 + 10^{\left(\log(I) - \log(IC_{50}2)\right)}} \right) + nsb \tag{1.31}$$

where $IC_{50}1$ and $IC_{50}2$ are the IC_{50} values of the two components and *fraction 1* is the fraction of control specific binding arising from the first component. As above, the IC_{50} values may be converted to estimates of Ki using the Cheng-Prusoff equation (Equation 1.11).

1.4.4.4 Interpretation of the Analysis

If the inhibition curves have Hill slopes of 1, the data are consistent with competitive action of the radioligand and unlabeled inhibitor at a single site. Accordingly the use of the Cheng-Prusoff correction to estimate a Ki value is valid.

Inhibition curves with slopes less than 1 are not consistent with simple competitive binding at a single site. One possible explanation for such curves is that the inhibitor recognizes two independent populations of radiolabeled sites with different affinities. However, it should be noted that a flat inhibition curve which can be fitted to a two-site model does not prove that this model is correct—many other models would also fit the data—and more experiments are required to establish the basis for the observed inhibition curves.

If the radioligand is known to bind to two sites with different known affinities, the Cheng-Prusoff correction can be used with the two K_D values and the two IC_{50} values to estimate the two Ki values, *provided* it is possible to link a given K_D value with the equivalent IC_{50} value.

If the radioligand itself appears to bind to a single site but a two-site fit has been obtained with an inhibitor, then either (1) the radioligand is not selective, in which case the Cheng-Prusoff correction is valid, or (2) the selectivity of the radioligand is too small to measure accurately, in which case there will be some error, probably small, in the estimated Ki values.

If a logistic fit to data from an experiment has a slope different from 1, then the Cheng-Prusoff correction may be used to account for the competitive effect of the radioligand concentration, but the Ki value derived from such a fit is only an estimate of the IC_{50} value which would probably be obtained with a very low radioligand concentration, rather than an estimate of the intrinsic dissociation constant of the inhibitor.

1.4.5   ASSAY 5: ALLOSTERISM ANALYSIS

The following assay is based on the allosteric model presented above (Scheme 4) and allows quantitative measurement of the interactions between an allosteric agent and both a radioligand and a second, unlabeled, primary ligand, under both equilibrium and nonequilibrium conditions. Analysis of the data with nonlinear regression provides estimates of the affinity of the allosteric agent, K_X; the cooperativity between the allosteric agent and the radioligand, α; and the cooperativity between the allosteric agent and the unlabeled primary ligand, β.

The data from a particular form of the assay may be transformed, and "affinity ratio plots" produced to provide a visual, semiquantitative representation of the allosteric effect of an agent on both primary ligands without the need for nonlinear regression analysis, providing an efficient screening assay (see Section 1.5, Protocol 1-5, and Figure 1.5).

1.4.5.1   Experimental Procedure

If the unlabeled ligand is an agonist, then GTP (0.2 mM) should be included in the assay. The design depends on whether the value of the K_D of the radioligand is assumed or measured in each assay (recommended). In the latter case the assay contains an additional high concentration of radioligand.

Prepare tubes to measure:

1. Total and nonspecific binding of a low (about the K_D) and high (about $K_D \cdot 5$, optional) concentration of radioligand alone;
2. Inhibition curves of the unlabeled ligand, alone and in the presence of three or more concentrations of test agent, or at least five concentrations of test agent alone and in the presence of at least one concentration of unlabeled ligand.

Incubate the tubes for sufficient time for binding equilibrium to be attained (at least 5 radioligand dissociation half-lives). If slowing of radioligand association is predicted, then either increase the incubation time or be prepared for kinetic effects when analyzing the data.

1.4.5.2   Equations for Nonlinear Regression

This analysis requires a program such as SigmaPlot which can handle more than one independent variable.

Subtract the appropriate value of nonspecific binding. The following equations refer to specific binding. For each data point, make available to the fitting procedure the specific dpm bound, the free concentration of radioligand, the concentration of unlabeled ligand, and the concentration of test agent. Fitting the data to the following equation provides estimates of the pharmacological parameters of all three ligands, unless binding of a high radioligand concentration was not measured, in which case the affinity of the radioligand (K_L or $1/K_D$) must be provided.

$$B_{eq} = \frac{Bmax \cdot L \cdot K_L \cdot \left(1 + X \cdot K_X \cdot \alpha\right)}{1 + X \cdot K_X + \left(A \cdot K_A\right)^n \cdot \left(1 + X \cdot K_X \cdot \beta\right) + L \cdot K_L \cdot \left(1 + X \cdot K_X \cdot \alpha\right)} \quad (1.32)$$

where B_{eq} is specific binding at equilibrium; L, A, and X are the concentrations of radioligand, unlabeled ligand, and test agent, respectively; K_L, K_A, and K_X are the affinities of radioligand, unlabeled ligand, and test agent, respectively, for the free receptor; α and β are the cooperativity factors for the interactions of the test agent with the radioligand and unlabeled ligand, respectively; and n is a slope factor to account for the shape of the inhibition curves of the unlabeled ligand. If only one concentration of unlabeled ligand was used, then n should be set to 1. If the concentration dependence of the effect of test agent appears to be steep or shallow, then a slope factor, b, can be introduced by substituting $(X \cdot K_X)^b$ for $X \cdot K_X$ in the above equation.

If the test agent inhibits the dissociation of the radioligand, then it is possible that binding in the presence of high concentrations of test agent was not at equilibrium. In this case, given the incubation time, t, and the dissociation rate constant of the radioligand, k_{off}, the general equation for nonequilibrium radioligand binding is given by

$$B_t = B_{eq} + \left(B_0 - B_{eq}\right) \cdot \exp\left[-t \cdot k_{obsAX}\right] \quad (1.33a)$$

and

$$k_{obsAX} = \frac{k_{off} \cdot \left(1 + \alpha \cdot X \cdot K_X \cdot \gamma\right)}{1 + X \cdot K_X \cdot \alpha} + \frac{k_{off} \cdot \left(1 + \alpha \cdot X \cdot K_X \cdot \gamma\right) \cdot L \cdot K_L}{1 + X \cdot K_X + \left(A \cdot K_A\right)^n \cdot \left(1 + X \cdot K_X \cdot \beta\right)} \quad (1.33b)$$

where
B_t = nonequilibrium binding;
B_{eq} = the predicted binding at equilibrium, defined above;
B_0 = the initial amount of radioligand binding, set to 0 in this case;
t = the incubation time;
k_{off} = the known dissociation rate constant of the radioligand;
γ = the maximal fold change in k_{off}.

As above, a slope factor, b, can be introduced to account for steep or shallow binding curves for the test agent if required.

1.4.5.3 Interpretation

A test agent may perturb the binding of labeled and unlabeled primary ligands through a number of mechanisms, so it is particularly important with this model to ensure by visual inspection that the fitted model provides a good fit to the data. If it is necessary to introduce slope factors (n for the unlabeled ligand A, and b for the test agent X), then the model changes from a mechanistic model to a mixed mechanistic/empirical model, and the mechanistic interpretation of the respective affinity estimates (K_A and K_X) is unclear.

If the test agent is known to inhibit radioligand dissociation completely or almost completely, then γ should be set to 0. If the fitted value of γ is non-zero, then it should agree with the value measured directly using radioligand dissociation assays.

1.4.5.4 Visualization

Allosteric effects on the binding of the radioligand are easy to see from the raw data, but effects on the binding of the unlabeled ligand may be unclear. In order to visualize the allosteric effects of the agent with both primary ligands on the same scale, calculate affinity ratios (see Protocol 1-5) from the parameters of the fit:

$$r_L = \frac{1 + \alpha \cdot X \cdot K_X}{1 + X \cdot K_X} \tag{1.34}$$

$$r_A = \frac{1 + \beta \cdot X \cdot K_X}{1 + X \cdot K_X} \tag{1.35}$$

where r_L and r_A are the affinity ratios of the labeled and unlabeled ligand respectively; α and β are the cooperativity factors for the interactions of the test agent with the labeled and unlabeled ligand, respectively; X is the concentration of test agent, and K_X is its affinity. Plot r_L and r_A vs. the log concentration of the test agent.

1.5   BASIC PROTOCOLS

In this section we describe in detail the procedures we use to prepare and characterize membranes from CHO cells expressing muscarinic receptors and their use in a screening assay for allosteric ligands.

Protocol 1-1: Tissue Culture

1. Seed a 175 cm² flask containing 80 ml culture medium (α-MEM with ribonucleosides and deoxyribonucleosides (Gibco) + 10% newborn calf serum + 50 units/μl penicillin + 50 μg/ml streptomycin + 2 mM glutamine) with cells ($10^6 - 10^7$).
2. Incubate for about 7 d at 37°C in an atmosphere of 5% CO_2 in air with 96% humidity until confluent.

3. Decant the medium, wash with 10 ml phosphate-buffered saline and add 3 ml trypsin (0.05%) + EDTA (0.02%) in Modified Puck's saline A. Incubate for 5 min.
4. Gently agitate the flask to free the cells from the surface.
5. Add at least 27 ml of culture medium to the flask.
6. Distribute the cells to four 175 cm^2 flasks each containing 80 ml growth medium.
7. Incubate approximately 4 d until the cells are confluent.
8. Repeat steps 3 to 5.
9. Transfer the contents of each flask to four 600 cm^2 square plates containing 125 ml medium.
10. Incubate for 3 to 6 d until confluent.

Protocol 1-2: Membrane Preparation

1. Decant the medium from each plate.
2. Wash each plate with 10 ml of ice cold homogenizing buffer (20 mM HEPES + 10 mM EDTA, pH 7.4).
3. Detach the cells from the plate with a Teflon scraper and pipette the suspension into a 35 ml centrifuge tube on ice.
4. Wash the plate with 4 ml buffer, again scraping the plate and collecting the suspension.
5. Homogenize the cell suspension with a Polytron homogenizer, using two 6-s bursts at setting 6, leaving the centrifuge tube to cool on ice for 30 s between each burst.
6. Add homogenizing buffer to balance the tubes in pairs.
7. Centrifuge the tubes at 40,000 × g for 10 min at 4°C.
8. Decant the supernatant and resuspend the pellet in 30 ml cold storage buffer (20 mM HEPES + 0.1 mM EDTA).
9. Repeat steps 7 and 8.
10. Centrifuge the tubes and resuspend each pellet in 5 ml storage buffer.
11. Pool the suspensions. The membrane preparation can be assayed for protein at this stage or stored at −80°C, with no loss of activity, in which case two aliquots (of 50 μl and 500 μl) are stored separately from the bulk of the homogenate, to be used for protein estimation and binding characteristics.

Protocol 1-3: Protein Estimation and Final Membrane Preparation

1. If necessary, thaw the small aliquot of membranes.
2. Prepare a stock bovine serum albumin (BSA, Sigma) solution containing 50 μg/ml in water.
3. Dilute 20 μl membrane suspension with 80 μl water. Add 20 μl 1M NaOH. The suspension should rapidly become clear as the NaOH dissolves the protein. Leave for 10 min. Then add 880 μl water to give a 1/50 membrane dilution.

4. For the standard curve, prepare 3.5 ml polystyrene tubes, in duplicate, with varying volumes of water and 50 μg/ml BSA stock solution to give a total volume of 800 μl and BSA concentrations of 0, 1, 2, 5, 7, and 10 μg/ml.

5. Similarly, with the 1/50 dissolved membrane preparation prepare tubes containing 800 μl of final dilutions 1/5000, 1/3333, 1/2500, 1/2000, and 1/1000.

6. Add 200 μl diluted Biorad protein reagent to each tube, and immediately mix by vortexing. Leave for 15 min.

7. Measure the light absorbance of each tube at 595 nm using a spectrophotometer.

8. Using the background-subtracted absorbance values up to about 1.5 ODU, plot the absorbance against the standard protein concentrations. This should be a straight line. Use this standard line to estimate the protein in each of the membrane dilutions, and calculate from these data the protein concentration of the undiluted membrane preparation. Omit estimates of concentration that deviate much from the mean—this occurs when the protein concentration is outside the appropriate range for this assay.

9. After determining the binding characteristics of the membrane preparation (see below), dilute the preparation, if required, and distribute it in appropriate volumes into 1.5 ml microfuge tubes. Store at –80°C. The preparation retains its binding and functional activity for at least 1 year under these conditions, and possibly for much longer.

Protocol 1-4: Saturation Analysis With ^{3}H-N-methylscopolamine

1. Thaw the larger aliquot of membranes. Dilute to 5–50 μg/ml with assay buffer (20 mM HEPES + 100 mM NaCl + 10 mM MgCl$_2$). The dilution will depend on the (suspected) receptor density, and the final receptor concentration should be less than 1/5 the Kd of the radioligand if possible. Store on ice.

2. Prepare a stock solution of radioligand in assay buffer. We have established that ^{3}H-N-methylscopolamine (^{3}H-NMS) does not adsorb to the tube under these conditions. The stock solution, at 100-fold dilution, will be the highest concentration in the assay and should yield a final concentration of 10- to 30-fold higher than the expected Kd of the radioligand. For ^{3}H-NMS at muscarinic receptors this requires stock solutions of 1 to 5 nM. Commercially available ^{3}H-NMS has a specific activity of ~84 Ci/mmol and a radioactive concentration of 1 μCi/μl, which equates to 84 × 2220 dpm/pmol (there are 2.22 × 10^{12} dpm/Ci) and a radioligand concentration of about 10 μM, so 3 μl of commercial solution in 150 μl buffer gives a stock solution of about 200 nM.

3. Perform four further threefold serial dilutions by adding 50 μl solution to 100 μl buffer.

4. Prepare a stock solution of unlabeled competitor for the measurement of nonspecific binding. We use QNB at a final concentration of 1 μM;

alternatively atropine can be used at a final concentration of 10 μM. Prepare 200 μl inhibitor at 100 times its final concentration.

5. The saturation curve will contain total binding tubes in duplicate and nonspecific binding tubes in singleton. For radioligands with more nonspecific binding than ^{3}H-NMS, or tissue with fewer receptors, the number of replicates should be greater.

6. Add 10 μl of each radioligand concentration to sets of three 5 ml polystyrene tubes, starting with the lowest concentration.

7. In addition, add 10 μl of each radioligand concentration to scintillation vials, in duplicate.

8. Add 10 μl buffer to two tubes with each radioligand concentration to measure total binding, then add 10 μl unlabeled competitor to the third tube from each set, starting with the lowest radioligand concentration, to measure nonspecific binding.

9. Add 1 ml membranes to the "total binding" tubes, then add 1 ml membranes to the "nonspecific binding" tubes, in both cases starting with the lowest radioligand concentration. Be vigilant about possible contamination of "total binding" tubes with unlabeled inhibitor. If possible, do not shake or vortex the tubes but let the addition of the membranes mix the reagents.

10. Incubate the tubes for 2 h at 30°C.

11. Soak glass fiber filter strips (Whatman GF/B, Semat, U.K.) in 0.1% PEI, to reduce binding to the filter.

12. Filter the tubes using a cell harvester, and wash each filter disk with 2×3 ml cold assay buffer or water (we have established in this particular assay that water and buffer give the same results).

13. Detach each filter disk from the filter strip and insert into a 5 ml scintillation vial. To each vial add 3 ml of a scintillation fluid which can accept at least 10% water (e.g., ReadySafe, Beckman), cap the tubes, and leave overnight. Alternatively, dry the filter strips on a hotplate or in an oven before distributing the filter disks to scintillation vials and adding scintillation fluid.

14. If the filter disks were added wet to the scintillation vials and left overnight, ensure that the water and scintillation fluid are thoroughly mixed by repeatedly inverting the tubes and checking visually that two fluid phases are not present. This procedure is not necessary if the filter disks were dry when added to the scintillation vials. In both cases, the filters should be uniformly translucent if all the water has dissolved in the scintillation fluid.

15. Count each tube for 10 min (or more if there are very few counts) in a scintillation counter. Check that all the samples show about the same efficiency of detecting dpm: if there are signs of spurious counts (caused, for example, by chemiluminescence) then recount the tubes.

16. The analysis of saturation data is described in detail in Section 1.4.3. Briefly, for routine use:

    a. For each radioligand concentration calculate the specific binding by subtracting nonspecific binding (the mean value if there is more than one replicate) from each of the total binding replicates;

    b. For each total binding tube calculate the free dpm as the difference between added and total bound dpm;

    c. Supply these data to a curve fitting program (we use Enzfitter, Erithacus Software), together with other values which may be required, such as incubation volume, radioligand specific activity, and protein concentration;

    d. Prepare plots of bound dpm vs. added radioligand, and Scatchard or Hofstee plots, either with software or "by hand": check that the values of *Kd* and *Bmax* reported by the program are consistent with the data.

Protocol 1-5: Allosteric Interaction With a Labeled and Unlabeled Ligand— The Affinity Ratio Plot

This assay provides an efficient screening tool for detecting effects of an allosteric agent on the binding of ^{3}H-NMS and unlabeled ACh, and measuring them semi-quantitatively. The specific binding data will be transformed into **affinity ratios**,[46,47] each of which is the apparent affinity of the primary ligand (labeled or unlabeled) in the presence of a particular concentration of allosteric agent divided by its apparent affinity in the absence of allosteric agent. The affinity ratios are plotted against the log concentration of the test agent. In theory, the EC_{50} or IC_{50} of this **affinity ratio plot** corresponds to the K_D of the allosteric agent at the free receptor, and the asymptotic value of the plot corresponds to the cooperativity with the primary ligand.[46]

    The K_D of the agent and its cooperative effects on the labeled and unlabeled primary ligands may be estimated by visual inspection of the affinity ratio plot. Depending on the design of the assay, the K_D of the radioligand may be calculated from the data, or its value may be assumed, and we present a separate protocol for each case. The *Bmax* and apparent affinity of the unlabeled ligand are also calculated from the data. We find that internal estimates of *Bmax*, radioligand K_D, and unlabeled ligand affinity provide a valuable check on the validity of the assay.

Protocol 1-5a: The Affinity Ratio Plot, With Estimation of ^{3}H-NMS Kd

1. Prepare a membrane suspension in assay buffer (see Protocol 1-4, step 1). Add GTP to a final concentration of 0.2 mM to prevent high affinity ACh binding.
2. Prepare a high concentration of ^{3}H-NMS, ~500 times its *Kd* (see Protocol 1-4, step 2).
3. Dilute most of this to a concentration ~100 times the *Kd* of ^{3}H-NMS.
4. Prepare QNB or atropine for measuring nonspecific binding (Protocol 1-4, step 4).

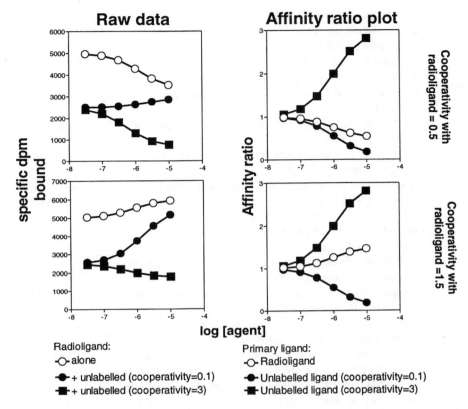

FIGURE 1.5 Allosterism assay—raw data and affinity ratio transformations. The graph shows equilibrium binding of a radioligand, at its K_D concentration, in the presence of an IC_{50} concentration of unlabeled inhibitor and various concentrations of an allosteric agent with a log affinity of 6. Examples of positive and negative cooperativity are shown with both labeled and unlabeled ligands.

5. Prepare a stock solution of unlabeled ligand (ACh in this case) in assay buffer at a concentration 100 times its IC_{50}.
6. Prepare a stock solution of test agent at 100 times the highest concentration in the assay. If the test agent is made up in organic solvent, ensure that
   a. The same concentration of solvent is present in all the tubes in the assay;
   b. The test agent does not precipitate from the solution when diluted;
   c. The final concentration of solvent has little or no effect on binding. This particular assay can tolerate 1% DMSO or 0.5% ethanol.
7. Perform serial dilutions of the test agent in its solvent: we find three concentrations spaced at log intervals to be suitable for initial screening assays, while six concentrations at half log intervals provide a more complete picture and may yield data that are also suitable for nonlinear regression analysis (see Section 1.4.5).

8. The assay will contain tubes to measure:
   a. Total binding of the high [^{3}H-NMS];
   b. Nonspecific binding of the high [^{3}H-NMS];
   c. Total binding of the low [^{3}H-NMS] alone;
   d. Nonspecific binding of the low [^{3}H-NMS];
   e. Binding of the low [^{3}H-NMS] in the presence of the IC_{50} concentration of ACh;
   f. Binding of the low [^{3}H-NMS] in the presence of each concentration of test agent;
   g. Binding of the low [^{3}H-NMS] in the presence of each concentration of test agent and ACh.

9. Add 10 µl of each stock solution to the appropriate tubes in duplicate, in the order ^{3}H-NMS, ACh, test agent, QNB or atropine. In each case start with buffer or the lowest concentration.

10. In addition, add 10 µl of each radioligand concentration to scintillation vials, in duplicate.

11. Add 1 ml membranes to each tube, ending with the tubes measuring nonspecific binding (see Protocol 1-4, step 9).

12. Incubate the tubes at 30°C for 2 h (or at least 5 radioligand dissociation half-lives). If slowing of radioligand association is predicted, then either increase the incubation time or be prepared for kinetic artefacts when interpreting the transformed data.

13. Filter the tubes over PEI-treated filters (see Protocol 1-4, steps 11 and 12). If the assay requires more than one filter, then each set of replicates should be filtered together, and the second replicates should be filtered in reverse order—this is easily achieved by reversing the order in which samples are filtered, and then reversing the filter before distributing the filter disks to scintillation vials. This procedure minimizes possible position effects of the cell harvester.

14. Count the samples (see Protocol 1-4, steps 13 to 15).

15. Perform the calculations below. Subtract the appropriate values of non-specific binding from all points to give specific binding.

   a. The following quantities are known or measured directly in the assay:
      B_L,   binding in the presence of the low [^{3}H-NMS] alone
      B_{LI},   binding in the presence of the high [^{3}H-NMS]
      B_{LA},   binding in the presence of the low [^{3}H-NMS] and ACh
      B_{LX},   binding in the presence of the low [^{3}H-NMS] and a particular concentration of test agent
      B_{LAX}, binding in the presence of the low [^{3}H-NMS], ACh, and the same concentration of test agent
      L,   the low ^{3}H-NMS concentration
      L_I,   the high ^{3}H-NMS concentration
      A,   the concentration of ACh
      q,   the ratio of low and high ^{3}H-NMS concentrations, L/L_I

b. The following quantities are calculated:
   The K_D of ^{3}H-NMS,

$$Kd = \frac{\left(B_{L1} - B_L\right) \cdot L \cdot L_1}{B_L L_1 - B_{L1} \cdot L}$$ (1.36)

The receptor concentration,

$$Bmax = \frac{B_L \cdot B_{L1} \cdot \left(1 - q\right)}{B_L - q \cdot B_{L1}}$$ (1.37)

The apparent affinity of ACh,

$$K_A = \frac{Bmax \cdot \left(B_L - B_{LA}\right)}{A \cdot B_{LA} \cdot \left(Bmax - B_L\right)}$$ (1.38)

The affinity ratio of ^{3}H-NMS in the presence of a single concentration of test agent,

$$r_L = \frac{B_{LX} \cdot \left(B_{L1} - B_L\right)}{B_{L1} \cdot B_L \cdot \left(1 - q\right) - B_{LX} \cdot \left(B_L - q \cdot B_{L1}\right)}$$ (1.39)

The affinity ratio of ACh in the presence of a single concentration of test agent,

$$r_A = \frac{B_L \cdot B_{LA} \cdot \left(B_{L1} - B_L\right) \cdot \left(B_{LX} - B_{LAX}\right)}{B_{LAX} \cdot \left(B_L - B_{LA}\right) \cdot \left[B_{L1} \cdot B_L \cdot \left(1 - q\right) - B_{LX} \cdot \left(B_L - q \cdot B_{L1}\right)\right]}$$ (1.40)

16. Interpretation of the data
    a. Plot the affinity ratios of ^{3}H-NMS and ACh vs. the log concentration of test agent, as shown in Figure 1.5.
    b. If the agent inhibits the affinity ratio of either primary ligand, then the concentration of agent at which the affinity ratio is 0.5 provides an estimate of the K_D of the agent at the unliganded receptor.
    c. The affinity ratio observed with the highest concentration of test agent may be an approximation of the asymptotic affinity ratio, which provides an estimate of the cooperativity with the primary ligand.
    d. If the test agent slows ^{3}H-NMS dissociation, there may be kinetic artefacts: reduction of the affinity ratio of ^{3}H-NMS (which may be misinterpreted as competitive inhibition), and a small increase in the affinity ratio of ACh.

e. If the test agent has strong (more than tenfold) negative cooperativity with ^{3}H-NMS, then affinity ratios of ACh at high concentrations of test agent are likely to be inaccurate.

f. If the test agent has positive cooperativity with ^{3}H-NMS, then affinity ratios of ^{3}H-NMS > 3 are likely to be inaccurate if the concentration of ^{3}H-NMS is more than twofold higher than its Kd.

Protocol 1-5b: The Affinity Ratio Plot, With the Kd of ^{3}H-NMS Already Known

This is the same as Protocol 1-5a, except that only the low concentration of ^{3}H-NMS is prepared and used.

1. Calculations
    a. The following additional quantity is known:
        p the ratio of the low concentration of ^{3}H-NMS and the assumed Kd, L/Kd.
    b. The following quantities are calculated:
        The receptor concentration,

$$Bmax = \frac{B_L \cdot (p+1)}{p} \tag{1.41}$$

The apparent affinity of ACh,

$$K_A = \frac{Bmax \cdot (B_L - B_{LA})}{A \cdot B_{LA} \cdot (Bmax - B_L)} \tag{1.42}$$

The affinity ratio of ^{3}H-NMS in the presence of a single concentration of test agent,

$$r_L = \frac{B_{LX}}{B_L \cdot (p+1) - p \cdot B_{LX}} \tag{1.43}$$

The affinity ratio of ACh in the presence of a single concentration of test agent,

$$r_A = \frac{B_L \cdot B_{LA} \cdot (B_{LX} - B_{LAX})}{B_{LAX} \cdot (B_L - B_{LA}) \cdot [B_L \cdot (p+1) - p \cdot B_{LX}]} \tag{1.44}$$

REFERENCES

1. Bylund, D. B. and Toews, M. L., Radioligand binding methods: practical guide and tips, *Am. J. Physiol.,* 265, L421, 1993.
2. Hulme, E. C., Ed., *Receptor-Ligand Interactions: A Practical Approach,* Oxford University Press, Oxford, 1992.
3. Hulme, E. C. and Birdsall, N. J. M., Strategy and tactics in receptor-binding studies, in *Receptor-Ligand Interactions: A Practical Approach,* Hulme, E. C., Ed., Oxford University Press, Oxford, 1992, 63.
4. Keen, M., The problems and pitfalls of radioligand binding, in *Methods in Molecular Biology, Vol. 41, Signal Transduction Protocols,* Kendall, D. A. and Hill, S. J., Ed., Humana Press, Totawa, NJ, 1995, 1.
5. Lauffenburger, D. A. and Linderman, J. J., *Receptors: Models for Binding, Trafficking, and Signalling,* Oxford University Press, New York, 1993.
6. Limbird, L. E., *Cell Surface Receptors: A Short Course on Theory and Methods,* Kluwer Academic, Boston, 1996.
7. Nahorski, S. R., What are the criteria in ligand binding studies for use in receptor classification?, in *Perspectives on Receptor Classification,* Black, J. W., Jenkinson, D. H., and Gerskowitch, V. P., Eds., Alan R. Liss, New York, 1987, 51.
8. Yamamura, H. I., Enna, S. J., and Kuhar, M. H., Eds., *Methods in Neurotransmitter Receptor Analysis,* Raven Press, New York, 1990.
9. Yamamura, H. I., Enna, S. J., and Kuhar, M. H., Eds., *Neurotransmitter Receptor Binding,* Raven Press, New York, 1978.
10. Hulme, E. C., Analysis of binding data, in *Receptor Binding Techniques,* Keen, M., Ed., Humana Press, Totawa, NJ, 1999, 139.
11. Bennett, J. P. J., Methods in binding studies, in *Neurotransmitter Receptor Binding,* Yamamura, H. I., Enna, S. J., and Kuhar, M. H., Eds., Raven Press, New York, 1978, 57.
12. Weiland, G. A. and Molinoff, P. B., Quantitative analysis of drug-receptor interactions. I. Determination of kinetic and equilibrium properties, *Life Sci.,* 29, 313, 1981.
13. Cheng, Y. and Prusoff, W. H., Relationship between the inhibition constant (K_i) and the concentration of an inhibitor which causes 50 per cent inhibition (I_{50}) of an enzymatic reaction, *Biochem. Pharmacol.,* 22, 3099, 1973.
14. Leeb-Lundberg, L. M. and Mathis, S. A., Guanine nucleotide regulation of B2 kinin receptors. Time-dependent formation of a guanine nucleotide-sensitive receptor state from which [^{3}H]bradykinin dissociates slowly, *J. Biol. Chem.,* 265, 9621, 1990.
15. De Lean, A., Stadel, J. M., and Lefkowitz, R. J., A ternary complex model explains the agonist-specific binding properties of the adenylate cyclase-coupled beta-adrenergic receptor, *J. Biol. Chem.,* 255, 7108, 1980.
16. Tota, M. R. and Schimerlik, M. I., Partial agonist effects on the interaction between the atrial muscarinic receptor and the inhibitory guanine nucleotide-binding protein in a reconstituted system, *Mol. Pharmacol.,* 37, 996, 1990.
17. Lee, T. W., Sole, M. J., and Wells, J. W., Assessment of a ternary model for the binding of agonists to neurohumoral receptors, *Biochemistry,* 25, 7009, 1986.
18. Ehlert, F. J., The relationship between muscarinic receptor occupancy and adenylate cyclase inhibition in the rabbit myocardium, *Mol. Pharmacol.,* 28, 410, 1985.
19. Kenakin, T., The classification of seven transmembrane receptors in recombinant expression systems, *Pharmacol. Rev.,* 48, 413, 1996.

20. Schutz, W. and Freissmuth, M., Reverse intrinsic activity of antagonists on G protein-coupled receptors, *Trends Pharmacol. Sci.,* 13, 376, 1992.

21. Baxter, G. S. and Tilford, N. S., Endogenous ligands and inverse agonism, *Trends Pharmacol. Sci.,* 16, 258, 1995.

22. Costa, T., Ogino, Y., Munson, P. J., Onaran, H. O., and Rodbard, D., Drug efficacy at guanine nucleotide-binding regulatory protein-linked receptors: thermodynamic interpretation of negative antagonism and of receptor activity in the absence of ligand, *Mol. Pharmacol.,* 41, 549, 1992.

23. Samama, P., Cotecchia, S., Costa, T., and Lefkowitz, R. J., A mutation-induced activated state of the beta 2-adrenergic receptor. Extending the ternary complex model, *J. Biol. Chem.,* 268, 4625, 1993.

24. Colquhoun, D., Affinity, efficacy, and receptor classification: is the classical theory still useful?, in *Perspectives on Receptor Classification,* Black, J. W., Jenkinson, D. H., and Gerskowitch, V. P., Eds., Alan R. Liss, New York, 1987, 103.

25. Colquhoun, D., Binding, gating affinity and efficacy, *Br. J. Pharmacol.,* 125, 923, 1998.

26. Samama, P., Pei, G., Costa, T., Cotecchia, S., and Lefkowitz, R. J., Negative antagonists promote an inactive conformation of the beta 2-adrenergic receptor, *Mol. Pharmacol.,* 45, 390, 1994.

27. Kenakin, T., Agonist-receptor efficacy. II. Agonist trafficking of receptor signals, *Trends Pharmacol. Sci.,* 16, 232, 1995.

28. Perez, D. M., Hwa, J., Gaivin, R., Mathur, M., Brown, F., and Graham, R. M., Constitutive activation of a single effector pathway: evidence for multiple activation states of a G protein-coupled receptor, *Mol. Pharmacol.,* 49, 112, 1996.

29. Krumins, A. M. and Barber, R., The stability of the agonist beta2-adrenergic receptor-Gs complex: evidence for agonist-specific states, *Mol. Pharmacol.,* 52, 144, 1997.

30. Kenakin, T., Receptor conformational induction versus selection: all part of the same energy, *Trends Pharmacol. Sci.,* 17, 190, 1996.

31. Leff, P., Scaramellini, C., Law, C., and McKechnie, K., A three-state receptor model of agonist action, *Trends Pharmacol. Sci.,* 18, 355, 1997.

32. Onaran, H. O. and Costa, T., Agonist efficacy and allosteric models of receptor action, *Ann. N.Y. Acad. Sci.,* 812, 98, 1997.

33. Hulme, E. C., Burgen, A. S. V., and Birdsall, N. J. M., Interactions of agonists and antagonists with the muscarinic receptor, *INSERM,* 50, 49, 1975.

34. Freedman, S. B., Harley, E. A., and Iversen, L. L., Relative affinities of drugs acting at cholinoceptors in displacing agonist and antagonist radioligands: the NMS/Oxo-M ratio as an index of efficacy at cortical muscarinic receptors, *Br. J. Pharmacol.,* 93, 437, 1988.

35. Sibley, D. R. and Creese, I., Interactions of ergot alkaloids with anterior pituitary D-2 dopamine receptors, *Mol. Pharmacol.,* 23, 585, 1983.

36. van Rampelbergh, J., Gourlet, P., de Neef, P., Robberecht, P., and Waelbroeck, M., Properties of the pituitary adenylate cyclase-activating polypeptide I and II receptors, vasoactive intestinal peptide1, and chimeric amino-terminal pituitary adenylate cyclase-activating polypeptide/vasoactive intestinal peptide1 receptors: evidence for multiple receptor states, *Mol. Pharmacol.,* 50, 1596, 1996.

37. Tietje, K. M., Goldman, P. S., and Nathanson, N. M., Cloning and functional analysis of a gene encoding a novel muscarinic acetylcholine receptor expressed in chick heart and brain, *J. Biol. Chem.,* 265, 2828, 1990.

38. Zhang, J. and Pratt, R. E., The AT2 receptor selectively associates with Gialpha2 and Gialpha3 in the rat fetus, *J. Biol. Chem.,* 271, 15026, 1996.

39. Page, K. M., Curtis, C. A., Jones, P. G., and Hulme, E. C., The functional role of the binding site aspartate in muscarinic acetylcholine receptors, probed by site-directed mutagenesis, *Eur. J. Pharmacol.,* 289, 429, 1995.
40. Ehlert, F. J., Roeske, W. R., Gee, K. W., and Yamamura, H. I., An allosteric model for benzodiazepine receptor function, *Biochem. Pharmacol.,* 32, 2375, 1983.
41. Macdonald, R. L. and Olsen, R. W., GABA$_A$ receptor channels, *Annu. Rev. Neurosci.,* 17, 569, 1994.
42. Ellis, J., Allosteric binding sites on muscarinic receptors, *Drug Dev. Res.,* 40, 193, 1997.
43. Holzgrabe, U. and Mohr, K., Allosteric modulators of ligand binding to muscarinic acetylcholine receptors, *Drug Discovery Today,* 3, 214, 1998.
44. Christopoulos, A., Lanzafame, A., and Mitchelson, F., Allosteric interactions at muscarinic cholinoceptors, *Clin. Exp. Pharmacol. Physiol.,* 25, 185, 1998.
45. Ehlert, F. J., Estimation of the affinities of allosteric ligands using radioligand binding and pharmacological null methods, *Mol. Pharmacol.,* 33, 187, 1988.
46. Lazareno, S. and Birdsall, N. J. M., Detection, quantitation, and verification of allosteric interactions of agents with labeled and unlabeled ligands at G protein-coupled receptors: interactions of strychnine and acetylcholine at muscarinic receptors, *Mol. Pharmacol.,* 48, 362, 1995.
47. Lazareno, S., Gharagozloo, P., Kuonen, D., Popham, A., and Birdsall, N. J. M., Subtype-selective positive cooperative interactions between brucine analogues and acetylcholine at muscarinic receptors: radioligand binding studies, *Mol. Pharmacol.,* 53, 573, 1998.
48. Murphy, K. M. and Sastre, A., Obligatory role of a Tris/choline allosteric site in guanine nucleotide regulation of [^{3}H]-L-QNB binding to muscarinic acetylcholine receptors, *Biochem. Biophys. Res. Commun.,* 113, 280, 1983.
49. Hosey, M. M., Regulation of ligand binding to cardiac muscarinic receptors by ammonium ion and guanine nucleotides, *Biochim. Biophys. Acta,* 757, 119, 1983.
50. Birdsall, N. J. M., Chan, S. C., Eveleigh, P., Hulme, E. C., and Miller, K. W., The modes of binding of ligands to cardiac muscarinic receptors, *Trends Pharmacol. Sci.,* Suppl., 1989, 31.
51. Pedder, E. K., Eveleigh, P., Poyner, D., Hulme, E. C., and Birdsall, N. J. M., Modulation of the structure-binding relationships of antagonists for muscarinic acetylcholine receptor subtypes, *Br. J. Pharmacol.,* 103, 1561, 1991.
52. Pert, C. B. and Snyder, S. H., Opiate receptor binding of agonists and antagonists affected differentially by sodium, *Mol. Pharmacol.,* 10, 868, 1974.
53. Limbird, L. E., Speck, J. L., and Smith, S. K., Sodium ion modulates agonist and antagonist interactions with the human platelet alpha 2-adrenergic receptor in membrane and solubilized preparations, *Mol. Pharmacol.,* 21, 609, 1982.
54. Hulme, E. C., Berrie, C. P., Birdsall, N. J. M., Jameson, M., and Stockton, J. M., Regulation of muscarinic agonist binding by cations and guanine nucleotides, *Eur. J. Pharmacol.,* 94, 59, 1983.
55. Lazareno, S. and Birdsall, N. J. M., Pharmacological characterization of acetylcholine-stimulated [^{35}S]-GTPγS binding mediated by human muscarinic m1-m4 receptors: antagonist studies, *Br. J. Pharmacol.,* 109, 1120, 1993.
56. Hulme, E. C. and Buckley, N. J., Receptor preparations for binding studies, in *Receptor-Ligand Interactions: A Practical Approach,* Hulme, E. C., Ed., Oxford University Press, Oxford, 1992, 177.
57. Koenig, J. A. and Edwardson, J. M., Endocytosis and recycling of G protein-coupled receptors, *Trends Pharmacol. Sci.,* 18, 276, 1997.

58. Koenig, J. A., Edwardson, J. M., and Humphrey, P. P. A., Somatostatin receptors in Neuro2A neuroblastoma cells: operational characteristics, *Br. J. Pharmacol.,* 120, 45, 1997.

59. Motulsky, H. J. and Mahan, L. C., The kinetics of competitive radioligand binding predicted by the law of mass action, *Mol. Pharmacol.,* 25, 1, 1984.

60. Tränkle, C., Mies-Klomfass, E., Botero Cid, M. H., Holzgrabe, U., and Mohr, K., Identification of a [^{3}H]ligand for the common allosteric site of muscarinic acetylcholine M_2 receptors, *Mol. Pharmacol.,* 54, 139, 1998.

61. Swillens, S., Interpretation of binding curves obtained with high receptor concentrations: practical aid for computer analysis, *Mol. Pharmacol.,* 47, 1197, 1995.

62. Munson, P. J. and Rodbard, D., LIGAND: a versatile computerized approach for characterization of ligand-binding systems, *Anal. Biochem.,* 107, 220, 1980.

63. Wells, J. W., Analysis and interpretation of binding at equilibrium, in *Receptor-Ligand Interactions: A Practical Approach,* Hulme, E. C., Ed., Oxford University Press, Oxford, 1992, 289.

64. Lazareno, S. and Nahorski, S. R., Errors in Kd estimates related to tissue concentration in ligand binding assays using filtration, *Br. J. Pharmacol.,* 77, 571P, 1982.

65. Leppik, R. A., Lazareno, S., Mynett, A., and Birdsall, N. J. M., Characterization of the allosteric interactions between antagonists and amiloride analogues at the human alpha2A-adrenergic receptor, *Mol. Pharmacol.,* 53, 916, 1998.

66. Severne, Y., Ijzerman, A., Nerme, V., Timmerman, H., and Vauquelin, G., Shallow agonist competition binding curves for beta-adrenergic receptors: the role of tight agonist binding, *Mol. Pharmacol.,* 31, 69, 1987.

67. Cohen, F. R., Lazareno, S., and Birdsall, N. J. M., The effects of saponin on the binding and functional properties of the human adenosine A1 receptor, *Br. J. Pharmacol.,* 117, 1521, 1996.

2 Ligand Screening of G Protein-Coupled Receptors in Yeast

John R. Hadcock and Mark H. Pausch

CONTENTS

2.1 INTRODUCTION

2.1.1 G PROTEIN-COUPLED RECEPTORS

The G protein-coupled receptors (GPCRs) comprise a diverse superfamily of integral membrane proteins that communicate changes in the extracellular environment to the cell interior.[1] To do so, members of the GPCR superfamily have the capacity to interact with and bind ligands that range from small biogenic amines and lipids to large complex proteins. Ligand binding induces a conformational change in the GPCR, facilitating activation of the heterotrimeric G proteins, composed of α, β, and γ subunits. The agonist-bound GPCR initiates the guanine-nucleotide cycle by catalyzing guanine nucleotide exchange by the α subunit and dissociation from $\beta\gamma$ complex.[2] Once free, the α subunit and the $\beta\gamma$ subunit complex are able to transduce signals and modulate the activity of a variety of effector proteins, including adenylyl cyclase, phospholipase Cβ, G protein-coupled Ca^{2+} and K^+ channels, phosphatases, sodium/hydrogen exchangers, and the membrane-proximal elements of MAP kinase signal transduction pathways, which produce changes in second messenger molecules or alterations of cell physiology and/or signal transduction that lead to the cellular response.

First recognized in mammalian cells, GPCR-mediated signal transduction pathways have functional homologs in evolutionarily distant organisms like yeast. Indeed, unmistakable similarity has been conserved between the elements of the yeast pheromone response pathway and mammalian GPCR-coupled MAP kinase signaling systems. This similarity suggests that yeast is likely to be a useful model system for study of GPCRs, G proteins, and related signal transduction proteins. Recent advances in GPCR expression in yeast and coupling to the pheromone response pathway suggest that this approach may be particularly useful in examining aspects of structure and function.[3-7] Furthermore, selected adaptations to the haploid yeast cells have made them useful as host cells for the development and implementation of high-throughput screening assays designed to identify novel ligands that modulate the activity of the GPCRs.[8-10]

2.1.2 THE YEAST PHEROMONE RESPONSE PATHWAY

Haploid *Saccharomyces cerevisiae* cells detect the presence of cells of opposite mating type through binding of peptide mating pheromones to G protein-coupled mating pheromone receptors (Ste2 and Ste3), which in turn activate intracellular heterotrimeric G proteins, thus initiating the mating process.[11] A second G protein α subunit, Gpa2, participates in the response of yeast cells to nutrient deprivation and interacts with Gpr1, a putative GPCR.[12,13]

The mating type of haploid yeast cells is determined by secretion of mating-type specific peptide pheromones and by expression of GPCRs that bind and respond to the mating pheromone secreted by cells of the opposite mating type. Thus, **a**-cells secrete **a**-factor and express the α-factor receptor, Ste2, while α-cells secrete α-factor and contain the **a**-factor receptor, Ste3. When cells of opposite mating type come into close proximity, mating pheromone receptors on the cell surface detect the presence of peptide pheromone. Dissociation of Gpa1 from the complex of β (Ste4) and γ (Ste18) subunits induced by pheromone binding to receptor allows the $\beta\gamma$ complex to activate downstream elements of the pheromone response pathway. The pathway is composed of a well-characterized mitogen-activated protein kinase (MAP kinase) cascade which is initiated by Ste20, a p20-activated kinase (PAK) homolog, followed by sequential activation of Ste7 (MAP kinase kinase or

MEK), Ste11 (MEK kinase), and the MAP kinases Fus3 and Kss1.[14] The activated pathway stimulates the cells to undergo the stereotypical responses of the mating haploid yeast cell, i.e., cell cycle arrest and transcriptional induction of pheromone-responsive genes and acquisition of mating-related cell structures.

Elements of the yeast pheromone response pathway are remarkably similar to mammalian GPCR signaling systems, suggesting that yeast is likely to be useful for analysis of GPCRs and G proteins. Since the completion of the genome sequence, the complete complement of yeast GPCRs and G proteins may be replaced with their homologous mammalian counterparts, permitting analysis of these proteins in isolation. Such a condition is not possible in other expression systems.

Efforts to clone new GPCR sequences has led to predictions that 1000 GPCR genes may be present in the human genome and as many as 2000 or more if odorant and pheromone receptors are included. This includes large numbers of orphan GPCRs whose cognate ligands remain unidentified.[15] Advances in functional expression of heterologous G protein-coupled receptors and coupling to the yeast pheromone response pathway suggest that this approach may be particularly useful in examining aspects of structure and function.[3-6] The purpose of this chapter is to introduce the reader to the basis for heterologous GPCR function in yeast, the means by which the technology may be applied to elucidating GPCR structure and activity, and identification of novel agonists and antagonists of GPCRs by high-throughput screening of large chemical and natural products libraries.

2.2   MANIPULATIONS MADE TO YEAST STRAINS

In order to make haploid yeast strains useful as a host for genetic analysis of heterologous GPCR function or in screening for novel agonist or antagonist compounds, the pheromone response pathway must be modified to allow the host cells to exhibit a growth response dependent on agonist activation (Figure 2.1). The usual response of haploid yeast cells to mating pheromone is transcriptional activation of pheromone-responsive genes resulting in cell cycle arrest. In order that this phenotype be removed and cells allowed to continue growing when the pheromone response pathway is activated, the *FAR1* gene, the primary negative regulator of G1 cyclins, was deleted.[16] When activated by the pheromone response pathway, Far1 promotes cell cycle arrest. Thus, elimination of Far1 permits transcriptional induction of pheromone-responsive genes in response to an activated pheromone response signal while allowing cells to continue growing vegetatively even in the presence of an activated pheromone response pathway. When released from cell cycle control, agonist-stimulated *far1* cells are free to proliferate.

In order to detect the yeast cells that respond to agonist a second modification, addition of a pheromone-responsive reporter gene, is required. Pheromone-responsive genes are transcriptionally activated as a result of binding of the transcriptional activating protein, Ste12.[11] Once activated by MAP kinase-catalyzed phosphorylation, Ste12 binds to specific *cis*-acting pheromone responsive elements, resulting in transcription of pheromone-responsive genes. The degree of activation over basal levels varies widely, e.g., the *FUS1*, *FUS2*, and *SST2* genes are induced as much as 700-fold, while the *GPA1* and *STE2* genes are induced to a much lesser degree.[17]

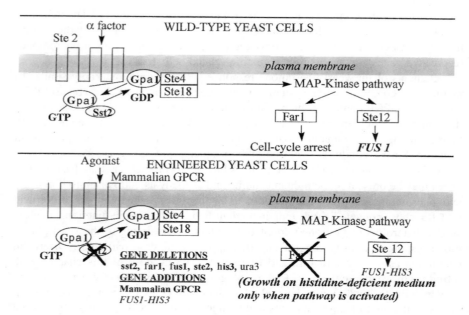

FIGURE 2.1 The wild-type and modified yeast pheromone response pathway. Haploid yeast cells normally undergo cell cycle arrest upon activation of the pheromone response pathway. The *FAR1* gene was deleted. *FAR1* encodes a negative regulator of G_1 cyclins and initiates cell cycle arrest. The endogenous *HIS3* gene was deleted. The yeast cells cannot grow on medium lacking histidine. *FUS1* is a pheromone-responsive gene. The *FUS1* gene was replaced with a reporter gene (*HIS3*) under the control of the *FUS1* promoter. Receptor activation leads to activation of the pheromone response pathway, increased *HIS3* protein expression and allows for growth on medium lacking histidine.

In order to confer a selectable phenotype, the protein coding segment of the *HIS3* gene,[18] which encodes an enzyme required for histidine biosynthesis, was used to replace the *FUS1* protein coding region. Thus, expression of this enzyme necessary for growth on medium lacking histidine is made dependent on activation of the pheromone response pathway. Agonist activation of this *FUS1-HIS3* reporter gene in *his3* cells would be expected to produce a growth response on medium lacking in histidine. In the absence of agonist activation, the *FUS1-HIS3* reporter gene is inactive, little His3 gene product is produced, and the cells make insufficient histidine and therefore cannot grow. As an alternative that permits screening of samples that contain histidine, e.g. natural products extracts, a second reporter gene that confers inducible G418 resistance, is included. As described below, direct screening for antagonists may be accomplished using the stably integrated *FUS2-CAN1* reporter gene.

The yeast cells undergo desensitization or adaptation in response to chronic stimulation of the pheromone response pathway. The primary adaptation mechanism is mediated by the yeast RGS protein, Sst2.[19] Sst2 is the first known member of the RGS family of GTPase-activating proteins demonstrated to play an important role in desensitization of a variety of mammalian GPCR signaling pathways. Yeast cells lacking functional Sst2 are supersensitive to mating pheromone and unable to recover from pheromone-induced cell cycle arrest. Since Sst2 acts at the level of the G protein, its deletion improves assay sensitivity for any expressed GPCR.

The pheromone receptors mediate an additional desensitization function independent of Sst2. The C-terminal tail of the pheromone receptors are hyperphosphorylated in response to pheromone[20] and like mammalian GPCRs, this phosphorylation has been implicated in decreased receptor responsiveness to agonist.[20,21] Receptor phosphorylation may be carried out by a G protein-coupled receptor kinase similar to those that regulate the responsiveness of mammalian GPCRs to agonist stimulation.[22] The C-terminal tail of Ste2 is the site of interaction with components of the endocytic pathway for receptor internalization and the target of the ubiquitin-mediated degradation pathway.[23] Ubiquitination of the mating pheromone receptors appears to be stimulated by membrane-localized casein kinase I isoforms encoded by YCK1 and YCK2.[24] The resulting monoubiquitination leads to internalization of pheromone receptor-bound α-factor.[25] Deletion of the STE2 gene from screening host strains gives rise to improved assay sensitivity, either by removal of the desensitization function exerted by the receptor or due to lack of competition for heterotrimeric G protein complex.

Yeast cells modified as described above are useful as hosts for bioassay of heterologous GPCR function. In order to mobilize a bioassay, the GPCRs must first be expressed in yeast cells at high levels using any of a variety of readily available yeast promoters.[26] Several useful yeast expression plasmids containing strong constitutive promoters have been constructed and are available from the ATCC.[27] The promoters in these plasmids are active in glucose-containing media contributing to high level expression and short (90 to 120 min) generation times. Under the conditions commonly used, the amount of receptor protein expressed is usually not limiting, although vast overexpression of certain GPCRs has occasionally been observed to be toxic to yeast cells.[27a] In this instance, expression of GPCRs under the control of inducible yeast promoters may be useful. The galactose-inducible GAL1,10 promoter confers moderate levels of receptor protein expression and permits a substantial degree of control over expression level. In order to raise GPCR expression to levels sufficient to permit an adequate signal to be transduced, the product of the GAL4 gene, the transcriptional activating protein for the galactose-inducible promoters is often required.[4,28] When modifying the GPCR coding region for insertion into an expression vector, it is worthwhile to eliminate the 5′ and 3′ untranslated regions (UTR) using a PCR-mediated mutagenesis approach. Yeast UTRs are AT rich while those of mammalian cells are often GC rich, leading to occasional difficulties with expression. In addition, the yeast consensus sequence for translational initiation, −4 aAaaATG, is distinct from corresponding mammalian sequences.[29] Modification of protein-coding sequences to include appropriate yeast translational initiation sequences may result in more efficient translation in yeast. GPCR expression plasmids and G protein expression plasmids may be introduced into yeast cells by the convenient lithium acetate methodology described below.

Protocol 2-1: Modified Rapid Protocol for Lithium Acetate Transformation of Yeast

1. Inoculate 2 ml YPD (or selective medium if maintaining a plasmid) with yeast cells to be transformed.
2. Incubate overnight with agitation at 30°C.
3. Inoculate 50 ml YPD with 1 ml of overnight culture.
4. Incubate 3 to 4 h with agitation at 30°C.
5. Harvest cells by centrifugation in 50 ml conical centrifuge tubes at $2000 \times g$, 3 min, room temperature.
6. Wash with 10 ml sterile water.
7. Resuspend with 1 ml sterile 0.1 M LiOAc in TE, prepared fresh from concentrated sterile stocks of 1 M LiOAc, pH 7.5 and 0.1 M Tris, 10 mM EDTA, pH 7.5 by dilution with sterile water.
8. Transfer 0.1 ml samples of resuspended cells to sterile 1.5 ml microfuge tubes containing 5 μl 10 mg/ml denatured salmon sperm DNA and up to 5 μl of DNA sample. Vortex briefly.
9. Add 0.7 ml 40% PEG3350 in 0.1 M LiOAc/TE, prepared fresh from 50% PEG3350 stock by addition of concentrated LiOAc and TE. Vortex until the cells are evenly dispersed.
10. Incubate at 30°C for 30 min.
11. Heat shock at 42°C for 10 to 20 min.
12. Plate 200 to 400 μl on selective agar plates.
13. Incubate 3 d at 30°C.

2.3 PHARMACOLOGICAL ANALYSIS OF GPCRs EXPRESSED IN YEAST

When expressing GPCRs in yeast for the purpose of agonist and antagonist high-throughput screening or structure–function analysis, a pharmacological profile is a critical step for any heterologous expression system. Whether the heterologously expressed receptor is used for screening, structure–function, or protein–protein interactions, the expressed GPCR should display a pharmacological profile equivalent to the eukaryotic system to which it is being compared. For characterization, a side-by-side comparison of the receptor expressed in yeast and mammalian cells is performed. Several different assays can be performed to determine whether the receptor displays equivalent pharmacology in the two systems. Functional signal transduction assays (either a reporter or growth response), saturation radioligand binding, competition radioligand binding, and other assays can all be performed in the yeast expression system.

An advantage of the yeast over other systems is that single components, either transducing or regulating the signal within the pheromone response pathway, can be deleted and replaced within the mammalian equivalent with relative ease. This includes not only receptors, G proteins and MAP kinase cascade components, but also regulatory components such as RGS proteins (regulators of G protein signaling),

G protein-coupled receptor kinases (GRKs), and arrestin homologs. Several of these components have been examined using the rat somatostatin receptor subtype 2 (SST_2),[4] A_{2a}-adenosine receptor,[5] and CCK receptor subtypes.[10]

2.3.1 Radioligand Binding Studies

2.3.1.1 Yeast Crude Membrane Preparation

Protocol 2-2: Yeast Membrane Preparation

1. Expression of GPCR ligand-binding sites in yeast cell membrane fractions can be easily monitored through conventional radioligand binding assays.[30] Yeast cells expressing heterologous GPCRs are cultured overnight at 30°C and 250 rpm in 50 ml of appropriate selective medium.
2. Transfer cells to 50 ml conical tube.
3. Spin down cells at $1500 \times g$ for 10 min at 4°C. Aspirate off supernatant.
4. Add 10 ml cold homogenization buffer (1 mM sodium bicarbonate, 1 mM EDTA, 1 mM EGTA, pH 7.4, with protease inhibitors [Bacitracin, benzamidine, leupeptin, aprotinin]) plus glass beads (Sigma cat. no G-8772, 425-600 microns).
5. Let sit on ice 5 min.
6. Making sure the sample remains cold, vortex three times for 1 min each.
7. Spin down cells $1500 \times g$ for 10 min at 4°C.
8. Transfer supernatant to 50 ml centrifuge tube. Add 10 ml cold homogenization buffer to the cell pellet/glass beads and repeat the vortexing step.
9. Spin down cells $1500 \times g$ for 10 min at 4°C.
10. Transfer supernatant to the first supernatant.
11. Spin 20 min, at $23,000 \times g$. Aspirate off supernatant.
12. Add 20 ml cold homogenization buffer.
13. Spin 20 min, at $23,000 \times g$. Aspirate off supernatant.
14. Add 2 ml cold homogenization buffer.
15. Make 500 µl aliquots. Determine protein concentrations.
16. Freeze down at –80°C in 250 µl aliquots.

2.3.1.2 Saturation Binding Assays

A comparison of the pharmacological characteristics of the rat A_{2a}-adenosine receptor was performed in yeast strains expressing the alpha-mating pheromone receptor, Ste2 (LY594) and in yeast strains where the *STE2* gene was deleted (LY764).[5] Deletion of the *STE2* gene had little effect on either [3H]-NECA saturation binding or competition binding of [3H]-NECA against various adenosine agonists (Figure 2.2). As measured by [3H]-NECA saturation binding, no significant difference was observed in the affinity of [3H]-NECA binding to LY594 and LY764 membranes. The expression level of the A_{2a}-adenosine receptor in strain LY594 (*STE2*) was approximately double that observed in LY764 (*ste2ΔLEU2*) cells (Figure 2.2).

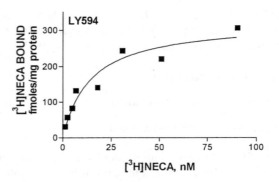

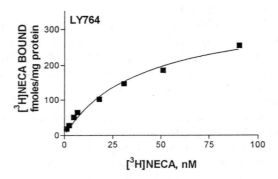

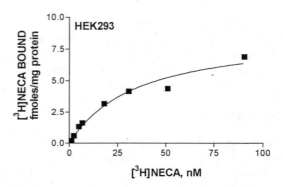

FIGURE 2.2 [³H]-NECA saturation radioligand binding to the rat A_{2a}-adenosine receptor expressed in yeast and HEK 293 cells. (Panels A and B) Liquid cultures of yeast strains LY594 and LY764 bearing the A_{2a}-adenosine receptor expression construct pLP116 and Gpa1 expression plasmid pLP83 in SC glucose (2%)-ura, trp overnight.[5] Membranes from yeast cells expressing the rat A_{2a}-adenosine receptor subtype were prepared. Saturation binding was performed with [³H]-NECA.[5] (Panel C) Membranes from HEK 293 cells expressing the rat A_{2a}-adenosine receptor were prepared and saturation binding was performed.[5] (From Price, L. A. et al., *Mol. Pharmacol.*, 50, 829–837, 1996. Reprinted with permission.)

2.3.1.3   Competition Binding Assays

As with the [^{3}H]-NECA saturation binding, little difference was observed in the K_i values of four prototypical adenosine receptor agonists with affinities ranging almost 3 orders of magnitude (Figure 2.3). The rank order of potency for the agonists was found to be NECA > R-PIA > CHA $\geq$ S-PIA (Table 2.1). For comparison, binding assays were performed with membranes prepared from HEK 293 cells expressing the rat A_{2a}-adenosine receptor. Both the K_d values of [^{3}H]-NECA binding, as well as the K_i values of the four adenosine agonists, were similar in the two systems. In addition, stereoselectivity was found to be conserved in the yeast expression system. R-PIA exhibits a 10-fold greater binding affinity than S-PIA and is a more potent agonist (Figure 2.3).

2.3.1.4   Salt and pH Sensitivity

Two components that can adversely alter receptor function are salt and pH. The effect of different salts on [^{125}I]-somatostatin ([^{125}I]-S-14) binding was examined in membranes prepared from a yeast strain expressing SST_2. [^{125}I]-S-14 binding was decreased by up to 50% in the presence of various salts (Figure 2.4). However, ammonium sulfate is required as a nitrogen source for efficient yeast growth and was used in the growth assays. Yeast cell growth is optimal at low pH (pH 5), whereas radioligand binding and GPCR signal transduction generally require pH approaching neutrality. The effect of varying pH on [^{125}I]-S-14 binding was examined in membranes prepared from a SST_2-expressing yeast strain. As shown in Figure 2.5, radioligand binding is greatly decreased at pH less than 6.8. Yeast cell growth is slowed at pH 6.8 but is maintained to a sufficient degree to allow for overnight in response to agonists. Thus, pH 6.8 was chosen for the whole cell growth assays.

2.3.2   G Protein Coupling

2.3.2.1   GTPγS Sensitivity

The ability of the rat SST_2 and A_{2a}-adenosine receptor to couple to G proteins in yeast was determined by examining the sensitivity of radioligand binding of agonists to guanine nucleotides. For most GPCRs agonists binding is sensitive to nonhydrolyzable guanine nucleotides which uncouple the receptor–G protein complex and shift the receptor from a high affinity state to a low affinity state. Radioligand binding to both SST_2 and the A_{2a}-adenosine receptor was found to be sensitive to incubation of membranes with the nonhydrolyzable GTP analog GTPγS.[4,5] This indicates that agonist binding is sensitive to the association of the receptor with the G protein.

2.3.2.2   GTPγ[^{35}S] Binding

An indicator of efficient coupling of a GPCR and G protein is the ability of an agonist to stimulate GTPγ[^{35}S] binding to membranes expressing the GPCR. Agonist-stimulated GTPγ[^{35}S] binding was examined in yeast strains expressing the human CCK_A and rat CCK_B receptors.[10] As shown in Figure 2.6 the CCK receptor agonist

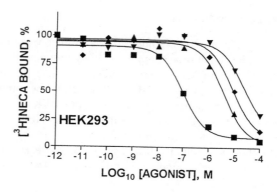

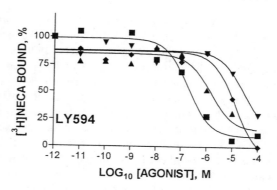

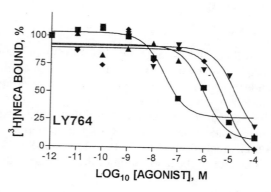

FIGURE 2.3 Competition of [³H]-NECA binding by adenosine receptor agonists. The ability of increasing concentrations of adenosine receptor agonists (10^{-12} M to 10^{-4} M) to displace [³H]-NECA (20 nM) was examined in membranes prepared from yeast strains LY594 and LY764 and HEK 293 cells expressing the rat A_{2a}-adenosine receptor. NECA (■), R-PIA (▲), S-PIA (▼), CHA (◆). (From Price, L. A. et al., *Mol. Pharmacol.*, 50, 829–837, 1996. Reprinted with permission.)

TABLE 2.1
Comparison of Binding Properties of the Rat A_{2a}-Adenosine
Receptor Expressed in Yeast and in Mammalian HEK 293 Cells

Cell Line	System	Saturation Binding (Kd, nM) [³H]-NECA	Competition Binding vs. [³H]-NECA (Ki, nM)			
			NECA	R-PIA	S-PIA	CHA
LY594	Yeast	26	48	469	5,867	4,366
LY764	Yeast	36	12	241	4,857	2,845
HEK 293	Mammalian	42	35	901	5,564	3,397

CCK8-S stimulated GTPγ[³⁵S] binding to membranes expressing either the CCK$_A$ receptor (top panel) or CCK$_B$ receptor (bottom panel). CCK8-S did not stimulate GTPγ[³⁵S] binding to membranes from a yeast strain that was not expressing either subtype (data not shown). Thus, GTPγ[³⁵S] binding can also be used as a measure of agonist efficacy and potency in yeast.

Protocol 2-3: GTPγ[³⁵S] Binding to Yeast Membrane Preparation

1. To each well in a 96-well plate add, in the following order: 10 μl assay buffer (50 mM HEPES, pH 7.4, 5 mM MgCl$_2$, 150 mM NaCl, 2 μM GDP, 1 mM dithiothreitol (DTT), and protease inhibitors [Bacitracin, benzamidine, leupeptin, aprotinin]), or 10 μl 200 μM GTPγS (for nonspecific binding), 20 μl agonist at varying concentrations, 10 μl 2 nM GTPγ[³⁵S] (50 pM final) and 160 μl crude plasma membranes at 0.16 mg/ml in assay buffer.
2. Incubate 60 minutes, shaking at room temperature.
3. Harvest by rapid filtration using a Packard cell harvester followed by five washes with buffer containing ice cold (4°C) 50 mM Hepes, pH 7.4, and 5 mM MgCl$_2$.

2.4　HIGH THROUGHPUT SCREENING FOR AGONISTS AND ANTAGONISTS

2.4.1　AGONIST BIOASSAYS OF FUNCTIONAL EXPRESSION OF HETEROLOGOUS GPCRs IN YEAST

2.4.1.1　Agar-Based Bioassays

Agonist bioassays developed using the modified yeast cells described above are based on the ability of agonists to promote growth on selective media via activation of various reporter genes. For example, yeast cells containing the rat somatostatin SST$_2$ receptor expression plasmid, pJH2,[4] were plated in selective agar media and tested for their ability to grow in response to a variety of somatostatin receptor

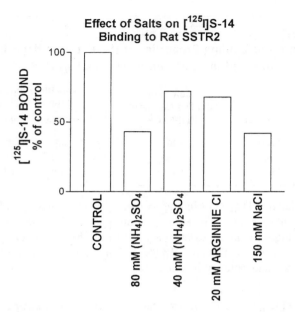

Effect of Salts on [^{125}I]S-14 Binding to Rat SSTR2

FIGURE 2.4 Effect of salt on [^{125}I]-S-14 binding to rat SST$_2$ expressed in yeast, measured under various salt conditions. NaCl and ammonium sulfate were chosen because these are required for mammalian cells and yeast cells, respectively.

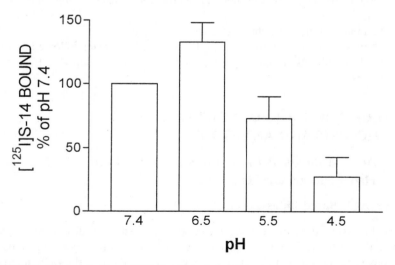

Effect of pH on [^{125}I]S-14 Binding: LY 364 strain

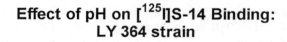

FIGURE 2.5 Effect of pH on [^{125}I]-S-14 binding to rat SST$_2$ receptor expressed in yeast examined at various pH, ranging from optimal yeast cell growth conditions (pH 4.5) to optimal mammalian cell conditions (pH 7.4).

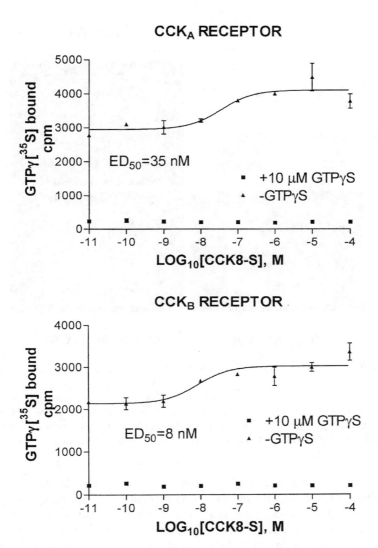

FIGURE 2.6   Cholecystokinin-stimulated GTP$\gamma[^{35}S]$ binding to yeast membranes expressing CCK$_A$ and CCK$_B$ receptor subtypes. The ability of CCK8-S to stimulate GTP$\gamma[^{35}S]$ binding to yeast cell membranes expressing either CCK$_A$ (panel A) or CCK$_B$ receptors (panel B) was examined. No response to CCK8-S was observed in membranes not expressing CCK receptors.

agonists via coupling to the yeast G alpha protein, Gpa1 (Figure 2.7).[10] The rat somatostatin SST$_2$ receptor was shown to couple efficiently to both Gpa1 and to a chimeric G alpha protein composed of an amino terminal $\beta\gamma$ interacting domain from Gpa1 and a receptor interacting domain from rat Gαi2.[4] The dissolved somatostatin agonists applied directly to the surface of the solidified agar diffuse radially from the site of application and bind to the rat somatostatin SST$_2$ receptors

1 mM AT

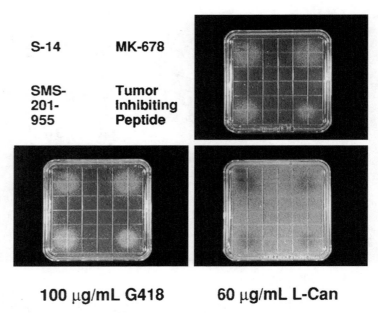

| S-14 | MK-678 |
| SMS-201-955 | Tumor Inhibiting Peptide |

100 µg/mL G418   **60 µg/mL L-Can**

FIGURE 2.7   Agar plate bioassays: MPY647 cells[10] were cultured overnight in SC-glucose-ura-trp medium. Cells (2×10^5) were plated in 30 ml SC-galactose (2%) agar medium lacking trp, ura, and histidine (his) (adjusted to pH 6.8 with concentrated KOH) and equilibrated to 50°C. The plates further contained 3-aminotriazole, the antibiotic G418, and L-canavanine at the indicated concentrations. Selected somatostatin agonists (500 pmol) dissolved in DMSO were applied to the surface of the agar and the plates were incubated at 30°C for 48 to 72 h.[10]

expressed on the surface of cells embedded in the agar. The resulting induction of *FUS1-HIS3* (Figure 2.7, top) and *SST2-G418^r* (Figure 2.7, left) expression permitted growth on histidine-deficient and G418-containing medium, respectively. The responding cells form a dense zone of growth easily detectable over the low level growth of the cells produced by basal reporter gene expression. Background was reduced by addition of the inhibitor of the *HIS3* gene product, 3-aminotriazole (AT), sufficient to reduce growth due to basal His3 expression. Screening of biological samples that may contain substantial amounts of histidine, e.g., natural products extracts, may be facilitated by use of the G418 resistance reporter gene. The size of the induced growth zone is proportional to the efficacy and potency of active compounds, particularly for compounds with similar aqueous solubility and diffusion characteristics. Since concentration varies with the square of the radius of the growth zone, small changes in the zone diameter reflect large differences in apparent affinity and efficacy. Under these conditions, agonist compounds having affinities that vary as little as twofold or by as much as three orders of magnitude may be distinguished in this type of assay.

Protocol 2-4: Agar-Based Bioassay for GPCRs Expressed in Yeast

1. This type of bioassay is mobilized by monitoring the growth of yeast cells expressing GPCRs suspended in selective agar medium. In the example described above, yeast cells expressing the rat somatostatin SST_2 receptor under the control of the galactose-inducible *GAL1* promoter in pJH2 (4) plasmid were cultured overnight at 30°C in SC-glucose-ura-trp medium.[10]

2. In order to induce SST_2 receptor expression, the cells were washed to remove residual glucose and were cultured overnight at 30°C in SC-galactose-ura-trp medium.

3. Cells (1×10^5 to 1×10^6) were added to 30 ml SC-galactose (2%) agar medium lacking trp, ura, and histidine (his) (adjusted to pH 6.7 with concentrated KOH and equilibrated to 50°C) and poured into 100 cm^2 petri plates. The plates were further supplemented with 3-aminotriazole, the antibiotic G418, or L-canavanine at concentrations that were determined empirically to give the best signal-to-noise ratio.

4. After application of dissolved samples (10 µl) of potential agonists and antagonists to the surface of the solidified agar, the plates are incubated at 30°C for 24 to 48 h. Induced yeast cell growth is readily visible as a dense circular zone over a clear background.

2.4.1.2   Liquid-Based Bioassays

Agar-based assays are convenient and can be adapted to high-throughput format, but provide qualitative assessment of the potency and efficacy of test compounds. Quantitative measurements of receptor activation of the pheromone response signal transduction pathway may be obtained using an agonist-induced growth bioassay implemented in liquid culture. Traditionally, these assays have been conducted by measuring the activity of β-galactosidase expressed from reporter gene constructs fused to a pheromone responsive promoter, e.g., *FUS1-LacZ*. The measurement of increase in turbidity of agonist-induced yeast cell cultures in liquid medium is a convenient alternative method for obtaining quantitative measurements of the potency of agonist compounds (Figure 2.8). The response of LY764 cells containing the rat adenosine A_{2a}-receptor to a variety of agonists was assayed in cultures grown in 96-well microtiter dishes.[5] The turbidity that results from yeast cell growth was measured using a 96 well microtiter dish reader. Adenosine A_{2a}-receptor agonists activate yeast cell growth at concentrations equivalent to their potency in activating their cognate effector enzyme, adenylyl cyclase.[5] These results suggest that the yeast-based liquid bioassay system is a suitable model for characterization of ligand interactions of the GPCRs.

Protocol 2-5: Liquid-Based Bioassays for GPCRs Expressed in Yeast

1. This type of bioassay is most conveniently mobilized in 200 µl cultures in 96-well microtiter dishes. In the above example, synthetic SC-glucose (2%) lacking tryptophan (trp) and uracil (ura) (SCD-ura-trp) liquid cultures of

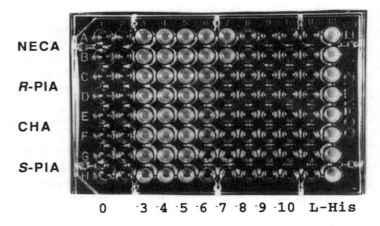

Log_{10} [Agonist], M

FIGURE 2.8 Liquid bioassays: liquid cultures of LY764 cells in SC-glucose (2%) lacking tryptophan (trp) and uracil (ura) (SCD-ura-trp) were grown overnight at 30°C and 250 rpm. The samples were then centrifuged and resuspended in 1 ml sterile H_2O. The cells were diluted 1000-fold in SC-glucose-ura-trp-his, pH 6.8 medium containing adenosine deaminase (0.13 units/ml) and 10 mM 3-aminotriazole (AT) to decrease basal growth rate. Samples of the cell suspension (180 μl) were dispensed into wells of sterile 96-well microtiter dishes containing 20 μl of serially diluted samples (10^{-4} to 10^{-11} M) of known adenosine receptor agonists. The plates were incubated at 30°C for 18 h with agitation (600 rpm). Growth was monitored by recording increases in OD_{620} using a microplate reader. Assays were conducted in duplicate and growth rate measurements obtained during the logarithmic phase of yeast cell growth.[5] (Panel A) Microtiter plate. Column 12 is growth of yeast cells in the presence of histidine. (Panel B) OD_{620} measurements from the microtiter plate. (From Price, L. A. et al., *Mol. Pharmacol.*, 50, 829–837, 1996. Reprinted with permission.)

yeast strain LY764[5] containing expression plasmids encoding the A_{2a}-adenosine receptor and yeast G alpha protein, Gpa1, were grown overnight with agitation at 30°C.[5]

2. The cells were then harvested by centrifugation and resuspended in 1 ml sterile H_2O.

3. The resulting cell suspension was diluted 1000-fold in SC-glucose-ura-trp-his, pH 6.8 medium, and 10 mM 3-aminotriazole (AT) to decrease basal growth rate. Since yeast cells appear to secrete adenosine which interferes in the measurement of growth due to the adenosine agonists, the medium was further supplemented with adenosine deaminase (0.13 units/ml) to degrade the adenosine.[5] Samples of the cell suspension (180 μl) were dispensed into wells of sterile 96-well microtiter dishes containing 20 μl of serially diluted samples of known adenosine receptor agonists.

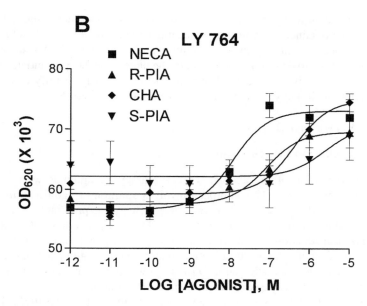

FIGURE 2.8 Continued.

4. The plates may be incubated without agitation at 30°C for 18 h. The cells tend to grow somewhat more quickly with agitation on a plate shaker (600 rpm). The rotary motion of the plate shaker tends to result in an accumulation of the cells in the center of the wells that must be dispersed in order to assure accurate measurement of cell growth.
5. Growth was monitored by recording increases in optical density (590 to 620 nm) using a microplate reader. Assays were conducted in duplicate and growth rate measurements obtained during the logarithmic phase of yeast cell growth.

Both agar and liquid bioassays are well suited to screening compound libraries for GPCR agonists in high-throughput format. These assays may be useful in identifying ligands that interact with orphan GPCRs as well. Orphan GPCRs, those with cognate ligands that have not been identified, may be expressed in selected yeast cells containing a variety of coexpressed heterologous G alpha proteins to facilitate coupling to downstream signal transduction and reporter output. In this regard, the yeast G protein alpha subunit, Gpa1, has been shown to interact productively with a broad range of GPCRs, suggesting that it may be useful for coupling to orphan receptors as well.[4–7,10] A diverse panel of selected agonists is then applied in an effort to identify the cognate ligand. In addition, the induced growth phenotype permits a development of genetic selection schemes useful for identifying gain-of-function mutations in GPCRs, e.g., induction of constitutive activity, response to noncognate agonist, etc. For this application, the DNA sequence of the subject GPCR is mutagenized in a yeast expression vector. The mutagenized GPCR library is used to transform appropriate yeast cells and the resulting transformants plated on selective medium. Agonist-dependent growth responses similar to those described above may be obtained by uniformly spreading cells from an induced culture on the surface of selective agar medium-containing plate. Surviving colonies are subjected to further analysis to determine the nature of the phenotype, i.e. plasmid dependence, DNA sequence analysis. A method of this kind was used to identify constitutively active mutations in the yeast alpha mating pheromone receptor Ste2.[31]

2.4.2   ANTAGONIST BIOASSAYS

2.4.2.1   Inhibition of Cell Growth

Using an adaptation of the agonist screening assay, compounds with antagonist activity may be assayed. This approach involves addition of agonist to the molten agar along with the yeast cells. The presence of agonist, preferably at or slightly higher than the concentration required to reach EC_{50} for activation of GPCR-dependent signal transduction, results in a uniform growth response by all cells in the plate. Application of antagonists or compound libraries containing active compounds to the surface of the plate will cause zones of growth inhibition in a lawn of growing cells. This method was used to identify rat somatostatin SST_2 receptor-selective antagonists.[32] The limitation of this approach lies in the false positives that arise from compounds with fungicidal activity.

2.4.2.2   Canavanine Resistance

A useful alternative involves introduction of a reporter gene that permits induction of a deleterious phenotype, e.g., sensitivity to canavanine. MPY578fc cells contain a mutation (*can1*) that renders them resistant to the toxic arginine analog, L-canavanine, due to the absence of the high-affinity basic amino acid permease that permits its uptake.[10] In addition, these cells contain a *FUS2-CAN1* reporter gene that confers agonist-inducible expression of a wild-type Can1 protein under the control of the pheromone responsive *FUS2* promoter which results in agonist-dependent canavanine sensitivity. Thus, MPY647 cells containing the rat somatostatin SST_2-receptor may be made sensitive to canavanine via somatostatin agonist activation of the *FUS2-CAN1* reporter gene (Figure 2.7, bottom right). Use of this phenotype in drug screening applications facilitates identification of antagonist compounds via rescue of agonist-induced canavanine toxicity. Application of antagonists block agonist-induced canavanine toxicity, allowing cells to grow in the presence of the poison.

A practical use of the yeast expression system is for the discovery of novel agonists and antagonists by high-throughput screening of chemical and natural products libraries. Fifteen thousand compounds can be screened in a single day with relative ease. A further application, though, can be for the structure–function analysis of a series of agonists and antagonists. This is exemplified by the comparison of novel somatostatin agonists and antagonists that were discovered, in part, with the yeast expression system (Figure 2.9). The antagonists discovered, in part, using the yeast bioassay are the first true somatostatin antagonists that have been described.[32] The data obtained from the yeast bioassays offer a quick, simple, and reliable method for examining the agonist and antagonist properties of novel compounds.

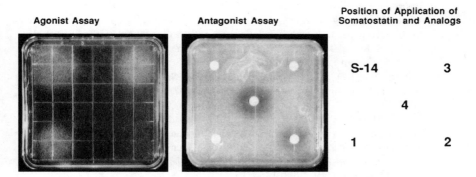

FIGURE 2.9 Identification of novel somatostatin receptor agonists and antagonist. LY364 cells were cultured overnight in SC-glucose-ura-trp medium. Cells (2×10^5) were plated in 30 ml SC-galactose (2%) agar medium lacking trp, ura, and histidine (his) (adjusted to pH 6.7 with concentrated KOH and equilibrated to 50°C). For the antagonist assay plate, somatostatin S-14 was added to a final concentration of 100 nM. The indicated amounts of selected somatostatin receptor agonists and antagonists were applied to the surface of the agar and the plates were incubated at 30°C for 48 to 72 h.[32] (From Bass, R. T. et al., *Mol. Pharmacol.*, 50, 709–715, 1996. Reprinted with permission.)

ACKNOWLEDGMENTS

The authors wish to thank Laura Price and Joann Strnad for their helpful insights and invaluable contributions to the preparation of this chapter and technology, and Don Kirsch and Kurt Steiner for their continued support.

REFERENCES

1. Hadcock, J. R. and Malbon, C. C., Agonist regulation of gene expression of G-proteins and G protein-coupled receptors, *J. Neurochem.*, 60, 1–9, 1993.
2. Neer, E. J., Heterotrimeric G proteins: organizers of transmembrane signals, *Cell*, 80, 249–257, 1995.
3. King, K., Dohlman, H. G., Thorner, J., Caron, M. G., and Lefkowitz, R. J., Control of yeast mating signal transduction by a mammalian β_2-adrenergic receptor and G_s α subunit, *Science*, 250, 121–123, 1990.
4. Price, L. A., Kajkowski, E. M., Hadcock, J. R., Ozenberger, B. A., and Pausch, M. H., Functional coupling of a mammalian somatostatin receptor to the yeast pheromone response pathway, *Mol. Cell. Biol.*, 15, 6188–6195, 1995.
5. Price, L. A., Strnad, J., Pausch, M. H., and Hadcock, J. R., Pharmacological characterization of the rat A_{2a} adenosine receptor functionally coupled to the yeast pheromone response pathway, *Mol. Pharmacol.*, 50, 829–837, 1996.
6. Kajkowski, E. M., Price, L. A., Pausch, M. H., Young, K. H., and Ozenberger, B. A., Investigation of growth hormone releasing hormone receptor structure and activity using yeast expression technologies, *J. Receptor Signal Transduction Res.*, 17, 293–303, 1997.

7. Erickson, J. R., Wu, J. J., Goddard, J. G., Tigyi, G., Kawanishi, K., Tomei, L. D., and Kiefer, M. C., Edg-2/Vzg-1 couples to the yeast pheromone response pathway selectively in response to lysophosphatidic acid, *J. Biol. Chem.*, 273, 1506–1510, 1998.

8. Broach, J. R. and Thorner, J., High-throughput screening for drug discovery, *Nature*, 384 (6604 Suppl.), 14–16, 1996.

9. Pausch, M. H., G protein-coupled receptors in Saccharomyces cerevisiae: high-throughput screening assays for drug discovery, *Trends Biotechnol.*, 15, 487–494, 1997.

10. Pausch, M. H., Price, L. A., Kajkowski, E. M., Strnad, J., Delacruz, F., Heinrich, J. A., Ozenberger, B. A., and Hadcock, J. R., Heterologous G protein-coupled receptors in Saccharomyces cerevisiae, methods for genetic analysis and ligand identification, in *Identification and Expression of G Protein-Coupled Receptors,* Lynch, K. R., Ed., Wiley-Liss, New York, 1998, 196–212.

11. Bardwell, L., Cook, J. G., Inouye, C. J., and Thorner, J., Signal propagation and regulation in the mating pheromone response pathway of the yeast *Saccharomyces cerevisiae, Dev. Biol.*, 166, 363–379, 1994.

12. Lorenz, M. C. and Heitman, J., Yeast pseudohyphal growth is regulated by GPA2, a G protein alpha homolog, *EMBO J.*, 16, 7008–7018, 1997.

13. Xue, Y., Batlle, M., and Hirsch, J. P., GPR1 encodes a putative G protein-coupled receptor that associates with the Gpa2p G alpha subunit and functions in a Ras-independent pathway, *EMBO J.*, 17, 1996–2007, 1998.

14. Herskowitz, I., MAP kinase pathways in yeast: for mating and more, *Cell*, 80, 187–197, 1995.

15. Lynch, K. R., G protein-coupled receptor informatics and the orphan problem: methods for genetic analysis and ligand identification, in *Identification and Expression of G Protein-Coupled Receptors,* Lynch, K. R., Ed., Wiley-Liss, New York, 1998, 54–72.

16. Chang, F. and Herskowitz, I., Identification of a gene necessary for cell cycle arrest by a negative growth factor of yeast: FAR1 is an inhibitor of a G1 cyclin, CLN2, *Cell*, 63, 999–1011, 1990.

17. Erdman, S., Lin, L., Malczynski, M., and Snyder, M., Pheromone-regulated genes required for yeast mating differentiation, *J. Cell. Biol.*, 140, 461–483, 1998.

18. Stevenson, B. J., Rhodes, N., Errede, B., and Sprague, G. F., Constitutive mutants of the protein kinase STE11 activate the yeast pheromone response pathway in the absense of G protein, *Genes Dev.*, 6, 1293–1304, 1992.

19. Dohlman, H. G. and Thorner, J., RGS proteins: filling in the GAPs in heterotrimeric G protein signaling, *J. Biol. Chem.*, 272, 3871–3874, 1997.

20. Reneke, J. E., Blumer, K. J., Courschesne, W. E., and Thorner, J., The carboxy-terminal segment of the yeast alpha factor receptor is a regulatory domain, *Cell*, 55, 221-234, 1988.

21. Konopka, J. B., Jenness, D. D., and Hartwell, L. H., The C-terminus of the S. cerevisiae alpha-pheromone receptor mediates an adaptive response to pheromone, *Cell*, 54, 609–620, 1988.

22. Chen, Q. and Konopka, J. B., Regulation of the G protein-coupled alpha-factor pheromone receptor by phosphorylation, *Mol. Cell. Biol.*, 16, 247–257, 1996.

23. Hicke, L. and Riezman, H., Ubiquitination of a yeast plasma membrane receptor signals its ligand-stimulated endocytosis, *Cell*, 84, 277–287, 1996.

24. Hicke, L., Zanolari, B., and Riezman, H., Cytoplasmic tail phosphorylation of the alpha-factor receptor is required for its ubiquitination and internalization, *J. Cell. Biol.*, 141, 349–358, 1998.

25. Terrell, J., Shih, S., Dunn, R., and Hicke, L., A function for monoubiquitination in the internalization of a G protein coupled receptor, *Mol. Cell. Biol.*, 1, 193–202, 1998.
26. Romanos, M. A., Scorer, C. A., and Clare, J. J., Foreign gene expression in yeast: a review, *Yeast*, 8, 423–488, 1992.
27. Mumberg, D., Muller, R., and Funk, M., Yeast vectors for the controlled expression of heterologous proteins in different genetic backgrounds, *Gene*, 156, 119–122, 1995.
27a. Pausch, M., unpublished observations.
28. Schultz, L. D., Hofmann, K. J., Mylin, L. M., Montgomery, D. L., Ellis, R. W., and Hopper, J. E., Regulated overexpression of the GAL4 gene product greatly increases expression from galactose inducible promoters on multicopy expression vectors in yeast, *Gene*, 61, 123–133, 1987.
29. Cigan, A. M. and Donahue, T. F., Sequence and structural features associated with translational initiator regions in yeast—a review, *Gene*, 59, 1–18, 1987.
30. Blumer, K. J., Reneke, J. E., and Thorner, J., The STE2 gene product is the ligand-binding component of the α-factor receptor of Saccharomyces cerevisiae, *J. Biol. Chem.*, 263, 10836–10842, 1988.
31. Konopka, J. B., Margarit, S. M., and Dube, P., Mutation of pro-258 in transmembrane domain 6 constitutively activates the G protein-coupled α-factor receptor, *Proc. Natl. Acad. Sci. U.S.A.*, 93, 6764–6769, 1996.
32. Bass, R. T., Buckwalter, B. L., Patel, B. P., Pausch, M. H., Price, L. A., Strnad, J., and Hadcock, J. R., Identification and characterization of novel somatostatin antagonists, *Mol. Pharmacol.*, 50, 709–715, 1996.

3 Ligand Screening Using Melanophore Cells: Frog Skin to Combinatorial Chemistry

J. Mark Quillan and Mac E. Hadley

CONTENTS

3.1 INTRODUCTION

The history of pharmacology would be incomplete without mention of melanophores and their impact on the fields of drug discovery and development. The striking color changes that occur in various fish, frog, reptile, and invertebrate species stand in contrast to the gradual modulation of skin tone seen in humans. Yet the two phenomena are governed by related sets of chromatophore cells: melanophores, which change the color of skin quickly, and the melanocytes of humans and other mammals, which change skin color more slowly. While melanophores and melanocytes both synthesize the dark biopolymer pigment melanin, and package it into cytoplasmic vesicles called melanosomes, the ability to quickly reposition melanosomes intracellularly distinguishes the melanophore from its nonmotile melanocyte cousin. And though the amount of melanin in either chromatophore subtype can increase or decrease slowly over time—a change associated with tanning in humans—only animals that possess melanophores can rapidly change their color, an adaptation which helps camouflage and protect them from predation.

The science of drug discovery is a field in constant evolution, and specific techniques employed to link chemicals and the receptors with which they interact are bound to undergo change or become obsolete. Although the long and continued use of melanophores as a tool for drug research may seem surprising to some, such longevity has been nurtured by constant adaptation and integration of new technologies. The goal of this chapter is not simply to discuss the use of melanophores as it relates to one screening method, combinatorial or otherwise, but also to present information that will facilitate adaptation and development of new procedures and approaches.

3.1.1 ANTECEDENTS TO MELANOPHORE BIOASSAYS

Investigation of animal color change has a long history intertwined with that of basic neuroscience and endocrinology. Published reports on the "pigment concentrating" effect of adrenaline on amphibian chromatophores precede the chemical hypothesis of nerve transmission proposed by Elliott in 1904.[1,2] Elliott suggested that this same substance, adrenaline, might be "the chemical stimulant liberated on each occasion when the impulse arrives at the periphery." We now know, of course, that noradrenaline—adrenaline's nonmethylated precursor—is the sympathetic catecholamine released from nerve endings; adrenaline, released from the adrenal medulla, serves primarily as an endocrine signaling molecule, but one that is also capable of stimulating the same set of adrenergic receptors from which Elliott made his now famous assertion. It seems that science is replete with such examples in which circumscribed observations, or the constraints of scientific language itself, help generate hypotheses that are partially correct but incomplete; years later it was discovered that not every species of amphibian melanophore is stimulated to aggregate pigment by adrenergic stimulation. In melanophores of *Xenopus laevis*, one notable exception, the converse is in fact true; adrenergic stimulation of *Xenopus* melanophores produces a β-adrenergic receptor-mediated pigment dispersion instead of an α-receptor-mediated aggregation observed in many other species.[3,4] (More on other, related, topics pertaining to melanophores can be found in texts by Hadley and Eberle.[5,6])

As the chemical hypothesis of nerve transmission was solidified into accepted theory by pioneers of autonomic physiology, such as Dale and Loewi,[7,8] the field of endocrinology was experiencing a renaissance—some may even argue its true birth—on the heels of improvements in surgical ablation techniques introduced by Aschner, Cushing, and others.[9,10] By 1916, Allen and Smith[12,13] had, independently, published their works on hypophysectomy in embryos and tadpoles, noting that removal of the embryonic buccal anlage of the epithelial lobe—what must have been a remarkable feat of dexterity at that time—caused loss of pigmentation and a failure to metamorphose. These results suggested that the intermediate lobe of the adult pituitary, the pars intermedia, released some chemical substance that was responsible for maintaining these organisms in their darkened state, an implication strengthened by observations that pituitary extracts and transplants were able to darken amphibian skin and to reverse the lightening effect of hypophysectomy.[14,15] One year later, in 1917, McCord and Allen[16] noted that extracts of the pineal gland caused lightening of amphibian skin. This observation opened up parallel investigations into pineal extracts which would eventually lead to the discovery of melatonin.

Early autonomic pharmacologists had at their disposal a number of good bioassays—blood vessels and heart preparations, for example—which were highly effective in helping them uncover and test chemical agents that affect the nervous system; there seemed, to many at least, little need for skin assays. Observations in 1922 that physiological effects produced by intermediate lobe extracts on melanophores were unrelated to "pressor-like" activity observed in other preparations[17,18]—a major focus of physiological and pharmacological inquiry at the time—may have contributed further to a waning of interest by some in using frog skin as a tool in drug discovery and autonomic research. For endocrinologists, however, findings that pointed to biologically active substances in brain extracts which could not be assayed using heart or spleen led to inauguration of frog skin as one of the *de facto* bioassays in the quest to identify these mysterious compounds.

Despite a growing desire on the part of early endocrinologists to identify and characterize biologically active substances within the pituitary and pineal glands, testing extracts in whole animals proved difficult and unreliable. Active components within extract material are subject to degradation in the blood and by the liver, and may be influenced by other factors which directly or indirectly affect similar physiological endpoints. Attempts at improving the reliability of whole animal testing and at limiting nonspecific interactions—by ligating limbs to prevent redistribution or passage of injected extracts through the liver, for example[5]—met with limited success, but by 1926 it was becoming clear that whole animal testing was unnecessary as similar results could be obtained, less tediously and perhaps more reliably, simply by using isolated pieces of frog skin.[19,20] The need to find improved methods to administer extracts quickly vanished, and frog skin formed the basis of an important new bioassay. So began a long history of using melanophores as biosensors for drug discovery and screening.

3.1.2 QUANTITATIVE MEASUREMENT OF PIGMENT TRANSLOCATION

Interest in quantitatively measuring physiological responses in melanophore preceded Allen's and Smith's reports on hypophysectomy, for it was in that same year, 1916, that Spaeth[21] published a description of a mechanical device to manually register microscopic movement of pigment within single melanophores. Mechanical collection of data was the exception at that time; most scientists continued to rely on qualitative visual assessments to judge the outcome of their experiments. Even though Spaeth's laborious device never did become a best seller, or widely used, it is interesting to note that variations of this scheme are still used today. One example is David Sugden's recent work in screening analogs of melatonin.[22] While modern techniques may employ computers or other sophisticated types of equipment, and involve isolated cells grown in culture,[22–24] the general concept remains remarkably similar.

The melanophore index (MI) was introduced as a nonmechanical way to visually characterize the degree of pigment dispersion in *Xenopus* melanophores,[20] and was the first system widely used to record physiological responses in melanophores. The importance of this systematic method, which helped standardize experimental data collection and allow data to be accurately compared, is reflected in its continued use long after the development of mechanical optical techniques that began only a few years later. The MI was still used for over half a century later after it was first introduced.[25]

Mechanical optical methods, although crude when first introduced by Hill in 1935 as a way to make photoelectric recordings from the skin of the fish, *Fundulus heteroclitus*,[26] eventually simplified and standardized data collection from pigment cells. By 1948, these methods had been adapted to measurement of transmittance through isolated pieces of frog skin *in vitro*.[27–29] Frieden and coworkers[29] used these adaptations to characterize the melanophore-stimulating activity of intermedin—a name used to describe the active component of intermediate lobe extract responsible for pigment expansion, later called melanocyte-stimulating hormone (MSH) by Aaron Lerner and colleagues.[30] In 1954 Shizume and coworkers[31] introduced a modification of the photoelectric photometer bioassay that replaced transmittance with reflection measurements. This adaptation made it much easier to apply photometric measurements to a broad spectrum of different melanophore-bearing skins. It was widely adopted and played a key role in the isolation of MSH and melatonin—an active component of pineal extracts responsible for melanophore aggregation—by Lee, Lerner, and coworkers.[30,32] A detailed history of these and other topics relating to melanophore bioassays, and their use in isolating and testing melanotropic peptides, can be found in comprehensive texts by Eberle[5] and by Hadley.[6]

The move from screening pituitary extracts in whole animals to using frog skin helped reduce sources of interference from surrounding tissue, circulating hormones, and the nervous system, but it did not eliminate them entirely. The skin matrix contains many cells that can influence bioassay measurements. For example, besides melanophores there are two other types of chromatophores present in frog skin: iridophores and xanthophores. Iridophores contain reflecting platelets comprised of purine crystals, and are able to contract their cell body and dendrites in response to MSH, the active ingredient of intermediate lobe extracts early researchers hoped to identify and characterize using frog skins. Xanthophores do not respond to MSH, but have pigmented pterinosomes and carotenoid vesicles that can reflect or absorb light;[33,34] these can, along with the other non-melanophore cell types present in skin,[35] interfere with assay measurements.

In 1983, DeGraan and coworkers[25] introduced a method to reduce optical interference from surrounding skin by measuring transmittance through pieces of ventral tailfin from *Xenopus laevis* tadpoles suspended in cell culture. The tailfin matrix, which is transparent, produces less interference with phototransmission than earlier methods that relied on thicker, more opaque sections of amphibian skin. The problem of interference, however, was not completely resolved until 1992 when Lerner and Potenza introduced a technique to measure optical responses from cultures of pure

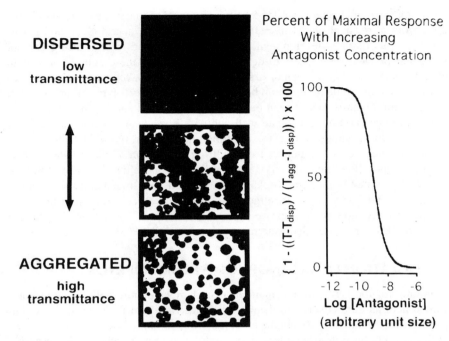

DISPERSED
low transmittance

AGGREGATED
high transmittance

Percent of Maximal Response With Increasing Antagonist Concentration

$\{ 1 - ((T - T_{disp}) / (T_{agg} - T_{disp})) \} \times 100$

Log [Antagonist] (arbitrary unit size)

FIGURE 3.1 Spectrophotometric measurements using melanophores. The center panels illustrate three microscopic images taken from a single field of cultured *Xenopus* melanophore cells in a fully dispersed state (upper image); in an intermediate state (middle image); and in a fully aggregated state (lower image). Images represent a field size of 1 mm^2. (Right) Graph illustrates an idealized concentration-response inhibition curve generated by taking transmittance (T) measurements across the field of cells. T_{disp} = transmittance with cells maximally dispersed. T_{agg} = transmittance with cells maximally aggregated. (From Quillan, J. M., *Development of a Combinatorial Diffusion Assay and Discovery of MSH Receptor Antagonists*, UMI, Ann Arbor, MI, 1995. With permission.)

melanophores using available plate reader spectrophotometers commonly used in enzyme-linked immunosorbent assay (ELISA) systems.[36] This permitted application of more current molecular biological techniques, such as introducing recombinant genes into melanophores, and provided a framework from which to develop combinatorial approaches to ligand screening.[37-39] An example of how measurements are quantified photometrically and used to generate response curves using cultured melanophores is shown in Figure 3.1; procedural details can be found in a number of articles.[37,39-41]

3.1.3 MELANOPHORE CELL CULTURE

Crucial to bioassays employing melanophores, of course, was the development of techniques making it possible to isolate and grow these cells in culture; of particular note is the work of Ide, currently at Tohoku University in Japan.[34,42-45] The ability to culture melanophores was born from the field of developmental biology and the pioneering work of Dushane in 1935[46] on the origins of pigment cells. Melanophores derive from a transient embryonic structure known as the neural crest, and it was studies on this tissue's development, how cells detach and migrate using explant tissue, that first led researchers to devise ways to culture these cells *in vitro*.[47] In 1963, Elsdale and Jones[48] demonstrated that cultures of *Xenopus* melanophores derived from embryonic tissue could be kept and studied in cell-type culture medium instead of the organ-type salt solutions that had been used earlier. In salt solution, melanophore cells do not propagate well and within weeks degenerate and die. By improving upon techniques used for isolating and growing tissue in culture, Ide[42] was able to prepare melanophores from bullfrog (*Rana catesbeiana*) tadpole melanophores, and allow them to survive and propagate in culture.[42,49] The ability to sustain melanophores indefinitely was demonstrated with melanophores of *Rana catesbeiana* and *Xenopus laevis,* after obtaining pure cell cultures using gradients of Ficoll and by employing improved growing conditions that included mitogens such as α-MSH or cyclic adenosine monophosphate.[44,45,50–52]

3.2   ISOLATION OF MELANOPHORES FROM EMBRYOS

Procedures for the isolation and maintenance of melanophore cells in culture are outlined here and are covered in detail elsewhere. Tabti and Poo[53] have written an excellent review on how to obtain embryos from *Xenopus laevis* using *in vitro* fertilization and have described techniques employed in preparing sterile tissue for primary cultures of nerve and muscle cells. These and similar techniques can be used in preparation of sterile melanophore cell cultures as well. Indeed, Tabti and Poo note 6-week-old nerve and muscle primary cell preparations in which only fibroblasts and melanophores survive.[53] These fibroblast/melanophore cultures were obtained by dissection of the neural tube—neural crest tissue from which melanophores differentiate—and surrounding myotomal tissue, using embryos around stage 22 of Nieuwkopp and Faber.[54] Ide and coworkers, on the other hand, have obtained pure melanophore cultures using embryos in stages of development as late as 46–53 by mechanically excising tail skin.[42,44,50]

The particular strategy employed to obtain melanophore cells is dictated to some degree by the stage of development. Neural crest cells, once detached from the neural tube, can take one of at least two pathways; the ventral pathway where cells give rise to neurons and glial cells of the peripheral nervous system, or the dorsolateral pathway where they differentiate into pigment cells of the skin. It is not yet clear exactly at which point pigment cells become fated, but dissections of the neural tube collected from very early embryos do not differentiate in culture into melanophores. Strategies for purifying melanophores, therefore, need to account for the developmental stage of embryonic tissue from which they are isolated. Neural tube explants from embryos in stages 22 to 28, for example, will grow out and differentiate in culture into melanophores. In later stages of development, however, melanophores will have begun or completed differentiation and migration, and should be collected from other areas. In these cases melanophores can be collected from whole embryos or from the dermal layers.

Protocol 3-1: Preparation and Maintenance of Melanophore Cells

Melanophore cell cultures can be obtained from *Xenopus* embryos following methods developed by Ide and coworkers:[42,44]

1. Tail skin from tadpoles in stages 46–53 of Nieuwkopp and Faber[54] is mechanically excised and dissociated for 1 h in Steinberg's solution containing 2% trypsin and 5% fetal bovine serum at 25°C.

2. Dissociated tail skin tissue is then passed through a fine platinum mesh (~150 μm) to help break the tissue apart and remove undissociated clumped tissue.

3. Dissociated tissue is then layered over a discontinuous Ficoll density gradient and centrifuged at low speed (relative centrifugal force of approximately 200 to 400 g) for approximately 20 min. (Melanophores, which contain densely pigmented melanosomes, form a visible band in the lower parts of the gradient.)

4. Melanophore cells are then extracted, washed and plated in tissue culture plates (Falcon) using conditioned medium (described below).

5. Conditioned medium was prepared by Ide[42] by growing neuroretina cells obtained from 11-d-old chick embryos for 4 d in Leibovitz's L-15 medium containing 10% fetal bovine serum. 5.5 parts of filtered neuroretina medium was then diluted with 3 parts water, 1.5 parts fresh fetal bovine serum, and supplemented with 100 units/ml penicillin "G" and 100 μg/ml streptomycin. Melanophore conditioned medium can also be obtained from *Xenopus* fibroblast cultures. Permanent cultures of fibroblasts can be prepared from nerve/muscle preparations by replacing medium used for the optimal growth of nerve and muscle with medium described by Ide for the growth of melanophores.[42] This medium consists of 5 parts Leibovitz's L-15, 2 parts fetal bovine serum, 3 parts double distilled pyrogen-free water, adjusted to pH 7.4, with 100 units/ml penicillin "G" and 100 μg/ml streptomycin added. Fibroblasts from neural tube explants proliferate rapidly and are easily isolated after several months. Filtered fibroblast medium can be used directly as a source of conditioned medium, in place of chick neuroretina conditioned medium, for the isolation, purification, and maintenance of melanophore cells in culture.

6. Continued growth of melanophores is supported by addition of mitogens, such as dibutyryl cyclic adenosine monophosphate (1 mM), α-MSH (60 nM), or insulin,[55] to the conditioned medium. Kondo and Ide[50] noted that at some point during their long-term cultivation melanophores appear to experience a growth crisis, from which they recover. The cause of this sudden die off is not clear but may be related to cell-fate determinants of embryonic tissue at different stages of development.

7. Melanophores are maintained at approximately 25°C and fed fresh conditioned medium every 3 to 4 d and split after 6 to 8 d. Cells can be removed from culture flasks or plates with a 0.25% trypsin solution (5.6 g/l NaCl, 0.7 g/l glucose, 0.2.8 g/l ethylene diamine tetraacetate [EDTA],

2.5 g/l trypsin, 0.41 g/l NaHCO$_3$). Trypsin is quenched with fresh medium and removed by pelleting cells. New cultures are replated at 1/2 to 1/6 confluency for optimal growth. Melanophores can be selected for properties or characteristics of interest—such as high melanin concentration or propensity to grow in a monolayer rather than on top of one another—by selecting individual colonies or applying a finer degree of density gradient separation. Darker cells provide superior signal-to-noise ratios in some photometric assays.

3.3   LIGAND SCREENING WITH MELANOPHORES

Two methods for the testing and screening of ligands using melanophores are described below: one describes protocols used for whole skin assays and the second describes methods that can be used with isolated cultured melanophores.

3.3.1   SKIN BIOASSAYS

The skin of many poikilothermic ("cold-blooded") vertebrates can change color rather rapidly in response to pituitary hormones. Generally, the skin can change from a light to a darker color in response to α-MSH. As discussed earlier, such responses of the skin to hormones in the intact animal can be duplicated *in vitro*. Shizume et al.[31] have described a sensitive frog skin bioassay capable of detecting the presence of α-MSH at a concentration as low as 10 p*M*. This photoreflectance method has been instrumental in development of superpotent melanocortin agonists, and an adaption of this method is described here for use with amphibian, lizard, and fish skins. The flexibility of the skin bioassay provides certain advantages over assays that rely on isolated cultured melanophore cells. Physiological responses of fully developed melanophores in adult or juvenile skin can be different from melanophores isolated from embryonic tissue, and skin assays are easily adapted to the different species of animals from which melanophores have never been isolated or grown in culture. Skin bioassays, therefore, provide a way to utilize the wide array of endogenous receptors found in nature, such as the melanin-concentrating hormone receptor that is found in the skin of the South American swamp eel, *Synbranchus marmoratus*.[56]

Protocol 3-2: Photometric Reflectance Skin/Melanophore Bioassays

1. *Frogs.* Almost any frog species will work, but skins from the leopard frog, *Rana pipiens*, have been most often used. These and other species of frog are seasonally available from the eastern United States and Canada. Although not generally used, skins from the bullfrog, *Rana catesbeiona*, are just as sensitive to α-MSH as are other ranid species. One advantage of *R. catesbeiona* is that the skin is of uniform coloration (no spots) in juveniles (3 to 5 in. snout–vent length). Ranid species are usually available from commercial sources.

2. *Frog skin.* The skin is made up of an epidermis and a dermis. The melanocytes of the epidermis, like human skin, is made up of a so-called "epidermal melanin unit." Epidermal pigment cells, although responsive to α-MSH, do not contribute significantly to color change in the frog skin bioassay. The dermis contains pigment cell of three varieties: melanophores, xanthophores, and geranophores (reflecting cells). These cells comprise the "dermal chomatophore unit" and are responsible for skin color changes as affected by hormones.

3. *Frog skin preparation.* Animals are sacrificed with a guillotine (or razor blade). Animal committee rules at universities and pharmaceutical companies will probably require that the frogs be anesthetized. This can be accomplished by an intraperitoneal injection of tricaine (MS-222; ethyl *m*-aminobenzoate methanesulfonate) which does not itself affect the response of skins to hormones or other drugs. Following decapitation, the spinal column is pithed with a long needle to immobilize the animal. Skins can be removed easily as described in the original publication on the frog skin bioassay.[31] Lower and upper leg skins from *R. pipiens* are used, whereas leg and back skin can be used from *R. catesbeiona*. Skins must be maintained in a physiological saline solution (e.g., Ringer's solution: 111 mM NaCl; 2.6 mM KCl; 3.6 mM NaHCO$_3$; and 1.1 mM CaCl$_2$). Skins are placed over metal rings and held in place by an outer plastic ring as described for the bioassay.[31] One can use 30- or 50-ml graduated beakers containing 10 or 20 ml of Ringer's, respectively.

4. *Photometer recordings.* The reflectometer light source is standardized through a green filter against a white (MgO) plate to some desired number, 82.5, for example. Using *R. pipiens*, normally five frogs are used, but any multiple of that number can also be employed. Four experimental groups (A–D) containing one skin from each of the frogs (n = 5) are used; five generally being adequate for statistical analysis. A reflectance reading is taken from each frog to a total group value (e.g., $5 \times 80 = 400$). Individual skins can be moved within each group so that the basal/resting reflectance value is approximately the same for each group (A to D).

5. At time zero, a test hormone, such as α-MSH, is added and the changes in reflectance measured at various time intervals (15, 30, 60, and 90 min, etc.). The response of skin, which darkens in response to α-MSH and a number of other compounds, is measured as the ratio of the new reflectance value divided by the original time zero value. A dose response curve can be generated using dilutions of an initial stock solution of the hormone. Based on a reflectance change of about 60% for each group with α-MSH, it is possible to obtain complete dose-response curves. Skins can be reused if carefully washed for 60 to 90 min with four to six changes of solution. In this manner parallel concentration–response curves can be generated. Antagonists can be assayed by first pretreating with antagonist and applying the agonist.

3.3.2 LIGAND SCREENING WITH CULTURED MELANOPHORES

Cultured melanophores provide a rapid and inexpensive way to test compounds for functional interactions with endogenous and recombinant seven-transmembrane G protein-linked receptors. The pigment translocation apparatus in *Xenopus* melanophores has been shown to couple well to receptors that activate the Gi/o, Gq, Gs families of G protein α subunits;[36,38,57] as suggested by the recently cloned photoreceptor, melanopsin,[58] it may also couple to Gα subunits associated with the opsin family of G protein-coupled receptors as well. Endogenous receptors for melatonin,[32] melanocortins,[30,59,60] noradrenaline,[3,4,36,60] serotonin,[61–63] endothelin,[64] histamine,[41] vasotocin,[65] vasoactive intestinal peptide (VIP)/pituitary adenylate cyclase-activating polypeptide (PACAP),[40] and light[55,58,66–69] have been identified in melanophores.

Screening ligands against receptors endogenous to frog melanophores may not always give reliable indications of how the same compounds will behave at human receptors. It is possible that dissimilarity between a target receptor—the receptor of interest to the drug developer—and that which may be present in frog cells, could be large enough as to make the frog receptor an unattractive substitute. For example, the nonamer peptide, MPdFRdWFKPV-amide (uppercase letters represent amino acids, and lower case d = dextrorotory conformation) identified first as an antagonist of the *Xenopus* melanophore MC receptor,[40] was later found to have an opposite, agonist effect at the human MC1 receptor (unpublished observation). This unusual finding—of a drug that acts as an antagonist in one species and an agonist in another—illustrates the difficulty in trying to extrapolate results from frog to human. Nevertheless, frog receptors still might be considered for use in the generation of new drug leads, which can be one of the more difficult aspects of drug discovery.

The presence of endogenous receptors in melanophores does not always provide an advantage to the screening process. Endogenous receptors can limit or complicate tests that employ recombinant receptors of similar type. In an elegant study, Potenza and colleagues[37] demonstrated, however, that it was possible to conduct tests in melanophores on a transfected β2-adrenergic receptor, despite the presence of an endogenous β1-like receptor. With adrenergic receptors this approach was simplified because these receptors are very well characterized and because of the large number of ligands available. In screening novel ligands at new or poorly understood receptors such an approach can be significantly more problematic.

Ligand screening in melanophores is greatly enhanced by the ability to express recombinant cDNA for receptors not normally expressed. First introduced by Potenza and coworkers using the β2-adrenergic receptor,[37] expression of recombinant receptors not endogenous to melanophores and which have been demonstrated to couple effectively to the pigment translocation apparatus include the Gq-linked bombesin,[70] substance P, and cholecystokinin subtype A (CCK-A) receptors (Figure 3.2), Gs-linked receptors,[37] and Gi-linked dopamine D2 and μ-opioid[57,71] receptors, among others.

3.3.3 PHOTOMETRIC MEASUREMENTS USING CULTURED MELANOPHORES

Cultured melanophores can be used to analyze receptors linked to Gs, Gq, or Gi/o heterotrimeric G proteins with the aid of a number of commonly available devices.

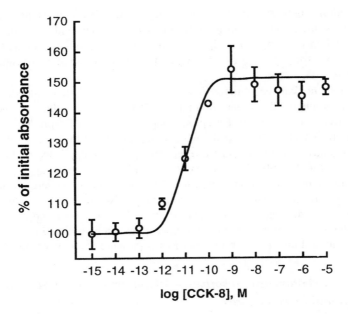

FIGURE 3.2 Testing a recombinant cholecystokinin receptor in melanophores. Melano-phores expressing a recombinant human CCK-A receptor (pcDNA3/CCK-A) respond to stimulation with CCK-8 with an EC_{50} of 11 pM. Nontransfected cells showed no response to equivalent concentrations of CCK-8 (not shown). Measurements were taken from 96-well plates using a multiwell spectrophotometer essentially as described by Quillan and Sadee[41] and Kuhlemann.[68] "Initial absorbance" was measured after a 90 min pretreatment with mela-tonin (10 nM; used to initiate tests from an aggregated or lightened state) and "% of initial absorbance" was calculated from readings taken 60 min after exposure to CCK-8. Each point represents the mean and standard deviation of four measurements. (The author wishes to thank Drs. T. Haga, S. Takeda, and A. Fukuzaki of the University of Tokyo for assistance in preparing plasmids and data, and Dr. A. Funakoshi of the National Kyushu Cancer Center, Fukuoka, Japan, and Dr. K. Miyasaka of the Tokyo Metropolitan Institute of Gerontology, for CCK-A receptor cDNA.)

This can be accomplished using spectrophotometers designed to take measurements from single or multiwell tissue culture plates, as well as a charge-coupled device (CCD) or video camera. Cameras can be used for microscopic measurements from single cells as well as macroscopic measurements from tissue culture plates. Cameras linked to computers and commonly available imaging software (e.g., TLC-Image software from Biological Detection Systems in Pittsburgh, Pennsylvania) can be used to analyze changes in pixel strength using a variety of different formats.

For photometric measurements taken from areas encompassing numerous cells, signal-to-noise can be enhanced by plating melanophores as a confluent monolayer and allowing 2 to 5 d for cells to fill in empty spaces and equilibrate. Afterwards, cells can be assayed by aspirating the culture medium and replacing it with Kreb's HEPES buffer or Leibovitz's L-15 medium with 0.5% bovine serum albumin containing test compounds. Assays can also be performed using melanophore culture medium, but cells appear to give larger and more consistent responses if medium does not contain fresh fetal bovine serum which tends to cause pigment dispersion when first applied to the cells. Bovine serum albumin or melanophore medium drawn from 2- to 5-d-old melanophore cultures seems to produce more consistent results.

Measurements can be initiated from either a lightened (aggregated), darkened (dispersed), or intermediate state of pigment dispersion depending on the type of receptor activity to be assayed. Receptors coupled to Gs and Gq produce a powerful dispersing effect that can reverse the aggregation produced by melatonin; therefore, screening ligands for activity at these receptors can be accomplished by first treating melanophores for 30 to 90 min with 10 nM melatonin. The pigment aggregating effect of Gi/o coupled receptors does not appear to be strong enough to overcome the effects of Gs or Gq activation, so the weaker photoinduced dispersion produced with light is used to initiate screening from a dispersed state. Time-resolved and concentration-response studies on receptors that do not desensitize can be initiated from either state as equilibrium will take place from experiments initiated from either state.

Protocol 3-3: Photometric Absorbance Measurements from Multiwell Plates Using Cultured Melanophores

Cultured melanophores can be assayed from 96-well culture plates using spectro-photometers with ELISA formated plate readers as follows:

1. Melanophore cells are plated to near confluency into a 96-well tissue culture dish and allowed to attach and grow for 2 to 5 d.
2. For compounds that are expected to cause pigment dispersion, the medium is aspirated and replaced with Kreb's-HEPES (diluted to 70%) containing 10 nM melatonin.
3. After 90 min, test compounds are added to the plate. This can be accomplished in less than 30 s by transferring test compounds from a 96-well template with a multichannel pipeting devices.
4. The plate is inserted into a multiwell plate spectrophotometer and the absorbance at 630 nm recorded. (Other wavelengths can also be used, but it should be kept in mind that photoreceptors present in melanophores are responsive to wavelengths in the lower end of the visible spectrum.) This first or initial absorbance reading is denoted Ai.
5. Subsequent readings can be taken at various time intervals (e.g., every 5 min if kinetic curves are desired), but at least one more reading taken at or near equilibrium (generally after 90 to 120 min) is required to complete the assay. This final absorbance measurement is denoted Af.
6. For each well the percent change in initial absorbance, %Ai, is tabulated using the formula (Af/Ai) × 100. Since each well contains the same number of melanophores and will give, correspondingly, equal minimum and maximum responses between wells, tabulated values for %Ai at each test concentration can be used to generate concentrations-response data.

3.4 COMBINATORIAL SCREENING USING MELANOPHORES

"Combinatorial synthesis" describes methods for simultaneously generating large numbers of related molecules; the term "combinatorial library" describes the group of molecules created by a given synthetic strategy. These concepts grew from the field of solid-phase synthetic peptide synthesis (SPPS) pioneered by Merrifield in the early 1960s.[72] SPPS is a method to synthesize peptides using techniques developed in solution for the step-by-step extension of nascent peptide chains that are anchored to an insoluble polymeric support; each reaction is driven to completion by using an excess of soluble reagents which, in turn, can be removed after each step by filtration and washing without loss of peptide. This solid-phase synthetic approach was extended by Geysen and coworkers[73] to what is widely recognized today as the first combinatorial synthesis. The authors described their technique as concurrent synthesis of multiple peptides bound to solid support, and they used these short peptides, which were permanently attached to radiation-grafted polyethylene pins, to examine interactions with viral antigens.

Concurrent synthesis was followed shortly afterwards by another protocol, called simultaneous multiple peptide synthesis (SMPS), introduced by Houghten in 1985.[74] SMPS is a way to construct small combinatorial libraries using standard solid-phase resins and synthesis protocols, and it allows peptides to be released so that they can fold freely and interact untethered with their intended targets. Equally important, SMPS was to form the basis of a simple yet stunning conceptual advance years later, by Furka and coworkers,[75] in which the authors presented "a general method for the synthesis of multicomponent peptide mixtures." This approach, which is now called "split synthesis," can be used to generate enormous numbers of peptide molecules, each attached to its own microscopic solid-phase resin particle. If coupling reactions are driven to completion, synthesis results in a mixture of microscopic resin particles, or beads, each containing a single homogeneous set of peptide molecules.

The advantage of split synthesis is the large number of discrete peptide molecules that can be generated easily; the disadvantages are many. There are problems, for example, associated with trying to attribute an interaction to one of a large pool of different molecules—of "addressing" a specific molecule to a specific biological effect. In addition, the small particle size of synthetic resins, while beneficial in the number of library molecules that can be synthesized, is restrictive in the amount of material per particle available with which to work. The average bead diameter of a typical resin is 110 μm (Figure 3.3), and each bead is able to hold about 300 pmol of peptide.[39]

Strengths and weaknesses of the split synthesis approach are illustrated in two bioassays developed by Lam and coworkers[76] and by Houghten and coworkers[77] in 1991. Lam's group was able to rapidly screen a combinatorial library for antibody interactions by isolating labeled beads from under a microscope, and sequencing the attached peptide. This is an example of a fully addressable assay, an assay in which a biological response can be traced directly to a particular ligand. Thus, each active molecule is identified directly by its association with a combinatorial bead,

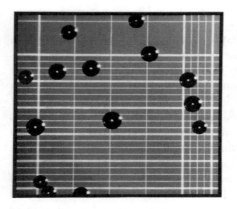

Frequency Distribution of Resin Bead Size

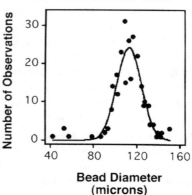

Bead Diameter (microns)

FIGURE 3.3 Frequency distribution of resin bead size: (left) a microscopic image of dry MBHA resin (from Calbiochem-NovaBiochem, La Jolla, CA). (Right) Measured bead diameters ranged in size from 42 to 151 μm with a mean 113 ± 14 μm (n = 282). In larger samples typically used in screening combinatorial libraries, bead diameters outside this range are seen occasionally but these can be easily distinguished from other beads upon microscopic examination. (Reproduced in part from Quillan, J. M., *Development of a Combinatorial Diffusion Assay and Discovery of MSH Receptor Antagonists*, UMI, Ann Arbor, MI, 1995. With permission.)

which in turn, is isolated directly from the biological assay. Deconvolution—the retesting of smaller and smaller pools—is not required in order to identify biologically active sequences. This greatly simplifies screening of chemical libraries, but from a pharmacological point of view, it is important to be able to study receptor interactions with unbound ligand molecules. Once liberated from their covalent supports, however, library molecules are free to mix, and trying to determine the source of a given biological response becomes challenging. Gone is the "one bead–one reaction" level of sensitivity, the high addressability, of Lam's immunoassay. Houghten and coworkers,[77] on the other hand, described an assay in which peptides were released from solid supports and purified of toxic cleavage materials before screening. Liberated peptide, which solvates and folds differently than when attached to resin, is free to mimic hormone and drug interactions so often found in nature. Although the technique is somewhat laborious, Houghten's group was able to deconvolute a large chemical library for antimicrobial sequences using this "solution-based mixture" method.

Solution-based procedures can work well when library molecules are soluble and do not interfere with each other. Interference can result when library molecules bind to each other, precluding interaction with the target; conversely, synergistic or additive interactions can yield false positives in the assay. Such issues may be of little consequence in screens for antimicrobials, but can be significant in assays where hormone-gated receptors are used. The likelihood of success is also compromised by the array of effects that molecules within a given library may exert on a particular receptor; antagonist molecules can block the response to agonists, and the presence of agonists can obscure the activity of competitive antagonists.

3.4.1 COMBINATORIAL DIFFUSION ASSAYS

A number of approaches have attempted to improve addressability in split synthesis free-ligand assays, including double-label and double-linker models, whereby combinatorial beads are given nonlabile tags, or attached with molecules that can be differentially released.[78–80] Presented below is a highly addressable way to use a split synthesis approach to screen ligands for G protein-coupled receptors using melanophores. This system relies on three basic elements in addition to melanophores: (1) multi-use peptide libraries; (2) a gas-phase cleavage and neutralization/detoxification process; and (3) a gel separation system. Together these have been called a melanophore-based combinatorial diffusion assay.

3.4.2 MULTI-USE PEPTIDE LIBRARIES

The term multi-use peptide library (MUPL) describes a process as much as a physical entity. It is a process by which existing solid-phase resins and peptide synthesis protocols are used to increase the addressability of combinatorial beads in split synthesis assays where unbound chemicals are required for pharmacological effect; it is a physical entity in that it actually consists of chemicals attached to microcarrier beads of a solid-phase resin. Obviating the need to employ double-linker or double-label strategies, this strategy allows the identity of the peptide associated with each MUPL microcarrier bead to be determined by residual peptide left after each screening step. Furthermore, unlike most other strategies, MUPL beads can be used repeatedly in successive rounds of screening to narrow choices, and to help in isolation of beads releasing biologically active material. Since it is possible to regulate the amount of material released during each round of the screening process, greater flexibility and control is afforded over the design and execution of screening assays. Once a MUPL bead has been linked to a given biological response, or "purified," the peptide responsible for the response can be identified by sequencing the material remaining on the bead.

   MUPLs were first constructed using 4-methyl-benzhydrylamine (MBHA) linked resin, in combination with standard procedures of 9-fluorenylmethoxycarbonyl (Fmoc) solid-phase peptide synthesis (SPPS) for extension of nascent peptide chains.[81] MBHA resin, originally designed for use with *tert*-butyloxycarbonyl (Boc) chemistry, employs chemical linkers that are fairly stable in strong acids like trifluoroacetic acid (TFA) because this acid is used to remove the N-α-Boc protecting group after each amino acid coupling step. Linkers used with Boc chemistry need to be sufficiently inert to remain unaffected by chemical steps used in elongation of peptide chains, yet labile enough to allow the release of an undamaged product. In Boc synthesis, acids stronger than TFA, like anhydrous hydrogen fluoride or trifluoromethanesulfonic acid, are used to cleave peptide from its solid support. Side-chain protecting groups used to protect certain amino acids from unwanted side reactions during Boc synthesis also need to be stable in TFA and removable at completion. These protecting groups are usually designed to be released during final cleavage. A more detailed history and description of methodologies underlying Boc SPPS can be found in a number of good reviews.[83–85]

In addition to attaching peptides to resin, SPPS linkers serve to functionalize the carboxyl-terminal. Benzhydrylamine (BHA), the first linker developed for the synthesis of α-carboxamides, was followed years later by a variation, p-methyl-benzhydrylamine (MBHA), introduced to help increase Boc SPPS yields.[82,86] MBHA resin was chosen as a foundation on which to construct and test the first MUPLs. Although designed to be stable in TFA for the purposes of Boc synthesis, MBHA is in fact partially labile in this acid, and slowly releases enough peptide to be of potential use in a biological assay. The slow kinetics can be used to regulate the amount of peptide released into assay, and allows beads to be reused in successive rounds of screening, with enough material retained upon completion of screening to serve as sequence identification tags.[39,65,81]

Orthogonal approaches to SPPS—which allow selective removal of protecting groups—were developed out of concern over the repetitive acidolysis procedure used in Boc synthesis.[84,87,88] As an alternative to Boc, Fmoc permits use of milder reaction conditions; side chain protecting groups and chemical linkers, typically labile in TFA, obviate the need for super acids like hydrogen fluoride.[89,90] Protocols for Fmoc synthesis are reviewed elsewhere.[84,91] Since Fmoc chemistry is used in construction of MUPLs, side chain protecting groups can be removed before beads are processed for screening.

In a free release assay, uncleaved residual peptide used to identify the peptide sequence is associated with a given MUPL microcarrier bead. But this sequence identification scheme is of little value if beads cannot be linked directly to the responses they evoke. The goal of tracing bioactive peptides released in a MUPL split synthesis assay back to their beads of origin is accomplished in two parts: first, by developing a cleavage method where released peptide remains noncovalently associated with its bead of origin and, second, by using conditions allowing such peptide to diffuse slowly through an agarose gel, from its point of origin—the MUPL microcarrier bead—to a point in the bioassay where a response is observed. The biological response—color change in melanophore cells—can, therefore, be traced back to the area where the active peptide was released and the bead of origin is located. All the beads in this area can be collected, reprocessed, and retested, until it is determined which of the beads is responsible for the response. The identity of the ligand is determined by sequencing the bead of origin.

Protocol 3-4: Synthesis of MUPLs

MUPLs are constructed on MBHA, or some other resin with favorable release kinetics, using a standard simultaneous multiple peptide synthesis (SMPS)/slit synthesis protocol in conjunction with Fmoc chemistry for extension of peptide chains.[39,74,76,77,81] An example of a library synthesis coupling Fmoc-amino acids to MBHA-linked polystyrene resin is as follows:

1. 0.5 mmol of derivatized amino acid is dissolved in 1.25 ml of N-methylpyr-rolidone, 1.0 ml of N,N-dimethylformamide containing 0.45 M 2-[1H-benzotriazolyl)-1,1,3,3-tetramethyluronium hexafluorophosphate (HBTU) and 0.45 M 1-hydroxybenzotriazole (HOBt) is added to the amino acid solution, and the amino acid/HBTU/HOBt solution is mixed for 10 min.
2. The amino acid/HBTU/HOBt solution is then transferred to and mixed with the MBHA-linked polystyrene resin (substitution level, 0.95 mmol/g), 1.0 mmol of N,N-diisopropylethylamine is added, and the resin solution allowed to react for 30 min at room temperature while mixing.
3. The resin is then filtered and rinsed six times with N-methylpyrrolidone and 2 mg of beads is removed for quantitative ninhydrin testing.[92] If incomplete, the coupling reaction time is extended.
4. Resins from each separate synthesis are then mixed according to standard split synthesis/SMPS protocol for the particular library under construc-tion, and the next amino acid coupled accordingly.
5. Upon completion, the resin is then deprotected with 20% piperidine in N-methylpyrrolidone for 5 min and again for an additional 15 min, filtered, and rinsed six times with N-methylpyrrolidone.
6. Side chain deprotection is achieved by stirring 10 to 100 mg of peptide bearing resin with 2 ml of cleaving reagent (90% TFA, 5% thioanisole, 2.5% ethanedithiol, 2.5% water) at room temperature for 1 h.
7. The cleaving reagent is then filtered and the beads thoroughly washed with dichloromethane, followed by N-methylpyrrolidone, and then meth-anol, and then the resin is dried under vacuum for 6 h.

Protocol 3-5: Gas-Phase Cleavage and Detoxification

To prepare MUPL beads for screening, side chain protecting groups are first removed with the Fmoc cleavage mixture: TFA (90%), thioanisole (5%), ethanedithiol (2.5%), and water (2.5%). Since MUPLs are constructed on MBHA resin, however, this cleavage mixture removes all of the Fmoc amino acid side chain protecting groups but only a small portion of the peptide. MUPL beads, once washed and dried, are then prepared as follows:

1. Beads are resuspended in dichloromethane and spread over a thin layer of transparent, chemically inert plastic film (e.g., polyethylene), that is held in place with chemically inert Teflon rings.
2. Beads are allowed to dry and then are exposed to gaseous TFA for a length of time sufficient to release a desired amount of peptide (typically 0.1 to 5% of the 300 pmol per bead total). This can be accomplished by evacuating a sealed container that contains both MUPL beads and liquid TFA in separate compartments.
3. After a timed exposure to TFA (typically 0.5 to 12 h, depending on the amount of peptide to be released), MUPL beads are removed and dried under vacuum for 12 h.

4. The gaseous TFA cleavage process causes formation of acid salts which are not apparent with conventional liquid methods. These salts remain associated with MUPL beads after cleavage and can result in acid shock and cell death when used directly in biological assays. These toxic salts can be neutralized by exposing beads to a gaseous mixture of ammonia and water (ammonium hydroxide) for approximately 20 min using a vacuum chamber as described for gaseous TFA cleavage. Afterwards the beads are again dried, and are then ready to be applied to a bioassay.

5. MUPL beads can then be applied by cutting the plastic film from the Teflon support rings, inverting, and gently applying it to the surface of the gel layer (described below). Principles involved in gas-phase preparation of MUPLs are illustrated in Figure 3.4, and are discussed in greater detail elsewhere.[39,81]

Protocol 3-6: Gel Assays

Gel assays can be prepared as follows:

1. A tissue culture plate containing a layer of confluent melanophore cells is overlaid with a low-melting-point agarose gel (e.g., 1% SeaPlaque from FMC Corporation, dissolved in a physiological buffer or culture medium) by melting the gel, cooling it to 37°C, and pouring the liquid gel over the cells to a desired thickness (typically 1 to 3 mm), and allowing the gel to solidify at room temperature. Drugs necessary to begin the assay are added to the cooling liquefied gel just before it is applied to the cells. Melatonin is added alone to screen for agonists that cause dispersion, and in screens for antagonists, the corresponding agonist should be added (near its EC_{50}) in addition to the melatonin.

2. After solidifying, the gel is overlaid with plastic film containing prepared MUPL combinatorial beads, face down so the beads contact the gel. Beads remain separated from the melanophore cell layer by the gel layer, and dissociated free peptide diffuses through the gel to interact with cells (Figure 3.5). Responses are recorded by capturing video images on computer; these images are used to analyze signals and localize beads at the gel surface. This combinatorial diffusion assay is illustrated in Figure 3.6.

3. MUPL beads, which are positioned between the gel layer and the transparent plastic film, are then extracted from areas of the gel where reponses are noted by first freezing the gel to remove the plastic film (beads remain embedded in the frozen gel surface) and using positional information captured in video images.

4. Beads are then washed and reprocessed for subsequent rounds of screening, or sequenced to determine the identity of a peptide. Further details of these techniques may be obtained elsewhere.[39,40,65]

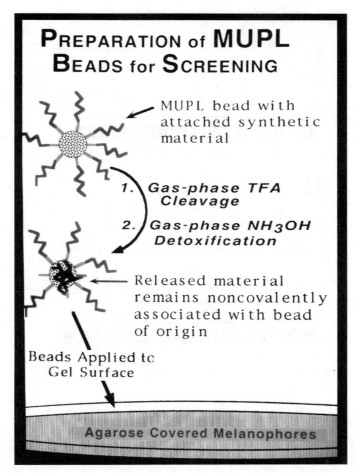

PREPARATION of MUPL BEADS for SCREENING

MUPL bead with attached synthetic material

1. Gas-phase TFA Cleavage

2. Gas-phase NH₃OH Detoxification

Released material remains noncovalently associated with bead of origin

Beads Applied to Gel Surface

Agarose Covered Melanophores

FIGURE 3.4 Combinatorial beads containing free peptide release their contents when they come in contact with the surface of a moist gel. Peptide then begins to diffuse through the gel in a predictable manner. Information on how responses in melanophores vary with time can be used to help determine properties and potency of the relevant ligand molecule causing it, and the location of the bead releasing it. More discussion on mathematical definitions and the physics of diffusion through agarose gels may be found in Quillan.[39]

3.5 CONCLUDING STATEMENTS

Split synthesis has made construction of large chemical libraries more facile, but with such libraries comes the challenge of finding efficient ways to manipulate and screen so many new compounds. Pharmaceutical companies with plentiful resources sometimes favor testing each compound separately; to split library components into multiwell plates and process them robotically can be a capital intensive but effective approach. This "serial processing" strategy is particularly suited to the nonpeptide

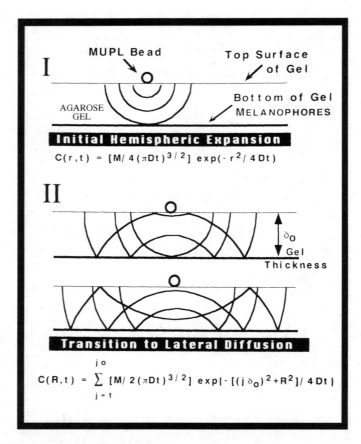

FIGURE 3.5 The physics of assay diffusion.

combinatorial libraries prepared directly in multiwell plates. Serial testing can, however, sap much of the strength inherent in techniques like split synthesis.

The issue of how to "address" a response to a particular ligand is a central obstacle to the efficient use of split synthesis in assays requiring free ligand. Lam and coworkers[76] demonstrated an addressable immunoassay system where peptides remain covalently attached to their bead of origin; untethered library molecules, in contrast, mix and are no longer easily identified. Methods, such as those described by Houghten and coworkers,[77] that rely on mixtures of free peptide molecules require extensive deconvolution. For receptors requiring free ligand, new techniques must be devised. MUPLs and multiple-linker strategies represent the first steps toward improving addressability in split synthesis assays, but themselves are inadequate. Methods to link responses with individual beads must be devised; assays must be sensitive enough to detect minute quantities of material released from each bead, must permit ligands to be separated adequately to minimize interference, and must provide control over ligand concentrations so that conditions may be tailored to address specific problems. Diffusion assays and MUPLs performing in conjunction

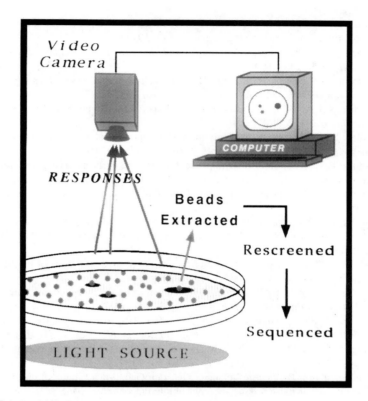

FIGURE 3.6 Diffusion assay response measurements.

meet these objectives, and that venerable chromatophore from frog skin, the melanophore, is once again providing sensitive visual clues which are helping to unlock the secrets of nature.

REFERENCES

1. Burgers, A. C. J. and Van Oordt, G. J., Regulation of pigment migration in the amphibian melanophore, *Gen. Comp. Endocrinol. Suppl.*, 1, 99, 1962.
2. Elliott, T. R., The action of adrenaline, *J. Physiol.*, 32, 401, 1904.
3. Burgers, A. C. J., Boschman, T. A. C., and Van de Kamer, J. C., Excitement darkening and the effect of adrenaline on the melanophores of *Xenopus laevis*, *Acta Endocrinol.*, 14, 72, 1953.
4. Graham, J. D. P., The response to catecholamines of the melanophores of *Xenopus laevis*, *J. Physiol. (Lond.)*, 158, 5P, 1961.
5. Eberle, A., *The Melanotropins: Chemistry, Physiology, and Mechanisms of Action*, Karger, Basel, 1988.
6. Hadley, M. E., *The Melanotropic Peptides. Source, Synthesis, Chemistry, Biological Roles, Mechanisms of Action and Biomedical Applications*, CRC Press, Boca Raton, FL, 1988.

7. Dale, H. H., The action of certain esters and ethers of choline, and their relation to muscarine, *J. Pharmacol. Exp. Ther.*, 6, 147, 1914.
8. Loewi, I., Über humorale Übertragbarkeit der Herznervenwirkung, *Pflügers Arch. Physiol.*, 189, 239, 1921.
9. Aschner, B., Demonstration von Hunden nach Exstirpation der Hypophyse, *Wien. Klin. Wochenschr.*, 22, 1730, 1909.
10. Crowe, S. J., Cushing, H., and Homans, J., Experimental hypophysectomy, *Bull. Johns Hopkins Hosp.*, 21, 127, 1910.
11. Allen, B. M., The results of extirpation of the anterior lobe of the hypophysis and of the thyroid of Rana pipiens larvae, *Science*, 44, 755, 1916.
12. Smith, P. E., The effect of hypophysectomy in the early embryo upon growth and development of the frog, *Anat. Rec.*, 11, 57, 1916.
13. Smith, P. E., Experimental ablation of the hypophysis in the frog, *Science*, 44, 280, 1916.
14. Fuchs, R. F., Die physiologische funktion des chromatophorsystems als organ der physikalischen warmeregulierung der poikilothermen, *Sitzungsber. Phys. Med. Soz. Erlangen*, 44, 134, 1912.
15. Allen, B. M., Experiments in the transplantation of the hypophysis of the adult Rana pipiens to tadpoles, *Science*, 52, 274, 1920.
16. McCord, C. P. and Allen, F. P., Evidence associating pineal gland function with alteration in pigmentation, *J. Exp. Zool.*, 23, 207, 1917.
17. Hogben, L. T. and Winton, F. R., The pigmentary effector system. I. Reaction of frog's melanophores to pituitary extracts, *Proc. R. Soc. London Ser. B*, 93, 318, 1922.
18. Hogben, L. T. and Winton, F. R., Studies on the pituitary. I. The melanophore stimulant in posterior lobe extracts, *Biochem. J.*, 16, 619, 1922.
19. Trendelenburg, P., Weitere Versuche über den Gehalt des Liquor cerebrospinalis an wirksamen Substanzen des hypophysenhinterlappens, *Naunyn-Schmiedebergs Arch. Exp. Pathol. Pharmacol.*, 114, 255, 1926.
20. Hogben, L. T. and Slome, D., The pigmentary effector system. VI. The dual character of endocrine coordination in amphibian colour change, *Proc. R. Soc. London Ser. B*, 108, 10, 1931.
21. Spaeth, R. A., A device for recording the physiological response of single melanophores, *Am. J. Physiol.*, 41, 1916.
22. Sugden, D., Aggregation of pigment granules in single cultured Xenopus laevis melanophores by melatonin analogues, *Br. J. Pharmacol.*, 104, 922, 1991.
23. Messenger, E. A. and Warner, A. E., The action of melatonin on single amphibian pigment cells in tissue culture, *Br. J. Pharmacol.*, 61, 607, 1977.
24. Sugden, D., Melatonin analogues induce pigment granule condensation in isolated Xenopus laevis melanophores in tissue culture, *J. Endocrinol.*, 120, R1, 1989.
25. de Graan, P. N., Molenaar, R., and van de Veerdonk, F. C., A new in vitro melanophore bioassay for MSH using tail-fins of Xenopus tadpoles, *Mol. Cell. Endocrinol.*, 32, 271, 1983.
26. Hill, A. V., Parkinson, J. L., and Solandt, D. Y., Photo-electric records of the colour change in Fundulus heteroclitus, *J. Exp. Biol.*, 12, 397, 1935.
27. Wright, P. A., Response of the melanocytes of frog skin to pituitary extracts in vitro, *Anat. Rec.*, 96, 540, 1946.
28. Wright, P. A., Photoelectric measurement of melanophoral activity of frog skin induced *in vitro*, *J. Cell. Comp. Physiol.*, 31, 111, 1948.
29. Frieden, E. H., Fishbein, J. W., and Hisaw, F. L., An *in vitro* bioassay for intermedin, *Arch. Biochem.*, 17, 183, 1948.

30. Lee, T. H. and Lerner, A. B., Isolation of melanocyte-stimulating hormone from hog pituitary gland, *J. Biol. Chem.*, 221, 943, 1956.
31. Shizume, K., Lerner, A. B., and Fitzpatrick, T. B., In vitro bioassay for melanocyte stimulating hormone, *Endocrinology*, 54, 553, 1954.
32. Lerner, A. B., Case, J. D., Takahashi, Y., Lee, T. H., and Mori, W., Isolation of melatonin, the pineal gland factor that lightens melanocytes, *J. Am. Chem. Soc.*, 80, 2587, 1958.
33. Bagnara, J. T. and Hadley, M. E., *Chomatophores and Color Change*, Prentice Hall, New York, 1973.
34. Ide, H., Transdifferentiation of amphibian chromatophores, *Curr. Top. Dev. Biol.*, 20, 79, 1986.
35. Fukuzawa, T., Hamer, S. J., and Bagnara, J. T., Extracellular matrix constituents and pigment cell expression in primary cell culture, *Pigment Cell Res.*, 5, 224, 1992.
36. Potenza, M. N. and Lerner, M. R., A rapid quantitative bioassay for evaluating the effects of ligands upon receptors that modulate cAMP levels in a melanophore cell line, *Pigment Cell Res.*, 5, 372, 1992.
37. Potenza, M. N., Graminski, G. F., and Lerner, M. R., A method for evaluating the effects of ligands upon Gs protein-coupled receptors using a recombinant melano-phore-based bioassay, *Anal. Biochem.*, 206, 315, 1992.
38. Lerner, M. R., Tools for investigating functional interactions between ligands and G protein-coupled receptors, *Trends Neurosci.*, 17, 142, 1994.
39. Quillan, J. M., *Development of a Combinatorial Diffusion Assay and Discovery of MSH Receptor Anatagonists*, UMI, Ann Arbor, MI, 1995.
40. Jayawickreme, C. K., Quillan, J. M., Graminski, G. F., and Lerner, M. R., Discovery and structure-function analysis of alpha-melanocyte-stimulating hormone antagonists, *J. Biol. Chem.*, 269, 29846, 1994.
41. Quillan, J. M. and Sadee, W., Structure-based search for peptide ligands that cross-react with melanocortin receptors, *Pharm. Res.*, 13, 1624, 1996.
42. Ide, H., Proliferation of amphibian melanophores *in vitro*, *Dev. Biol.*, 41, 380, 1974.
43. Ide, H. and Bagnara, J. T., The differentiation of leaf frog melanophores in culture, *Cell Differ.*, 9, 51, 1980.
44. Fukuzawa, T. and Ide, H., Proliferation *in vitro* of melanophores from Xenopus laevis, *J. Exp. Zool.*, 226, 239, 1983.
45. Ide, H. and Akira, E., Differentiation and transdifferentiation of amphibian chromato-phores, *Prog. Clin. Biol. Res.*, 256, 35, 1988.
46. DuShane, G. P., An experimental study of the origin of pigment cells in amphibia, *J. Exp. Zool.*, 72, 1, 1935.
47. Stevens, L. C., The origin and development of chromatophores of *Xenopus laevis* and other anurans, *J. Exp. Zool.*, 125, 221, 1954.
48. Elsdale, T. R. and Jones, K. W., The culture of small aggregates of amphibian embryonic cells *in vitro*, *J. Embryol. Exp. Morphol.*, 11, 135, 1963.
49. Ide, H., Effects of ACTH on melanophores and iridophores isolated from bullfrog tadpoles, *Gen. Comp. Endocrinol.*, 21, 390, 1973.
50. Kondo, H. and Ide, H., Long-term cultivation of amphibian melanophores. *In vitro* ageing and spontaneous transformation to a continuous cell line, *Exp. Cell. Res.*, 149, 247, 1983.
51. Fukuzawa, T. and Ide, H., Further studies on the melanophores of periodic albino mutant of Xenopus laevis, *J. Embryol. Exp. Morphol.*, 91, 65, 1986.
52. Akira, E. and Ide, H., Differentiation of neural crest cells of Xenopus laevis in clonal culture, *Pigment Cell Res.*, 1, 28, 1987.

53. Tabti, N. and Poo, M.-M., Culturing spinal neurons and muscle cells from *Xenopus* embryos, in *Culturing Nerve Cells*, Banker, G. and Goslin, K., Eds., MIT Press, Cambridge, MA, 137, 1991.

54. Nieuwkopp, P. D. and Faber, J., *Normal Table of* Xenopus laevis *(Daudin). A Systematic and Chronological Survey of the Development From Fertilized Egg Till the End of Metamorphosis*, North Holland, Amsterdam, 1967.

55. Daniolos, A., Lerner, A. B., and Lerner, M. R., Action on light on frog pigment cells in culture, *Pigment Cell Res.*, 3, 38, 1990.

56. Castrucci, A. M., Hadley, M. E., and Hruby, V. J., A teleost skin bioassay for melanotropic peptides, *Gen. Comp. Endocrinol.*, 66, 374, 1987.

57. Potenza, M. N., Graminski, G. F., Schmauss, C., and Lerner, M. R., Functional expression and characterization of human D2 and D3 dopamine receptors, *J. Neurosci.*, 14, 1463, 1994.

58. Provencio, I., Jiang, G., De Grip, W. J., Hayes, W. P., and Rollag, M. D., Melanopsin: an opsin in melanophores, brain, and eye, *Proc. Natl. Acad. Sci. U.S.A.*, 95, 340, 1998.

59. Thing, E., Melanophore reaction and adrenocorticotrophic hormone, *Acta Endocrinol.*, 11, 363, 1952.

60. Novales, R. R., Responses of cultured melanophores to the synthetic hormones α-MSH, melatonin, and epinephrine, *Ann. N.Y. Acad. Sci.*, 100, 1035, 1963.

61. Cerletti, A. and Berde, B., Die Wirkung von D-lysergsayre-diathylamide (LSD-25) und 5-oxytryptamine auf die Chromatophoren von *Poecilia reticulatus*, *Experientia*, 11, 312, 1955.

62. Kahr, H. and Fischer, W., Die Wirkung des 5-oxytrypamins auf das Pigmentsystem der Haut, *Klin. Wochenschr.*, 35, 41, 1957.

63. Potenza, M. N. and Lerner, M. R., Characterization of a serotonin receptor endogenous to frog melanophores, *Naunyn-Schmiedebergs Arch. Pharmacol.*, 349, 11, 1994.

64. Karne, S., Jayawickreme, C. K., and Lerner, M. R., Cloning and characterization of an endothelin-3 specific receptor (ETC receptor) from Xenopus laevis dermal melanophores, *J. Biol. Chem.*, 268, 19126, 1993.

65. Quillan, J. M., Jayawickreme, C. K., and Lerner, M. R., Combinatorial diffusion assay used to identify topically active melanocyte-stimulating hormone receptor antagonists, *Proc. Natl. Acad. Sci. U.S.A.*, 92, 2894, 1995.

66. Bles, E. J., The life history of *Xenopus laevis* Daud, *Trans R. Soc. Edinburgh*, 41, 789, 1905.

67. Van der Lek, B., DeHeer, J., Burgers, A. C. J., and Van Oordt, G. J., The direct reaction of the tailfin-melanophores of *Xenopus* tadpoles to light, *Acta Physiol. Pharmacol. Neerl.*, 7, 409, 1958.

68. Kuhlemann, H., Untersuchungen der Pigmentbewegungen in Embryonalen Melanophoren von *Xenopus laevis* in Gewebekulturen, *Zool. Jahrb. Abt. Allg. Zool. Physiol. Tiere*, 69, 169, 1960.

69. Bagnara, J. T. and Obika, M., Light sensitivity of melanophores in neural crest explants, *Experientia*, 23, 155, 1967.

70. Graminski, G. F., Jayawickreme, C. K., Potenza, M. N., and Lerner, M. R., Pigment dispersion in frog melanophores can be induced by a phorbol ester or stimulation of a recombinant receptor that activates phospholipase C, *J. Biol. Chem.*, 268, 5957, 1993.

71. Quillan, J. M. and Sadee, W., Dynorphin peptides: antagonists of melanocortin receptors, *Pharm. Res.*, 14, 713, 1997.

72. Merrifield, B., Solid phase peptide synthesis. I. The synthesis of a tetrapeptide, *J. Am. Chem. Soc.*, 85, 2149, 1963.

73. Geysen, H. M., Meloen, R. H., and Barteling, S. J., Use of peptide synthesis to probe viral antigens for epitopes to a resolution of a single amino acid, *Proc. Natl. Acad. Sci. U.S.A.*, 81, 3998, 1984.

74. Houghten, R. A., General method for the rapid solid-phase synthesis of large numbers of peptides: specificity of antigen-antibody interaction at the level of individual amino acids, *Proc. Natl. Acad. Sci. U.S.A.*, 82, 5131, 1985.

75. Furka, A., Sebestyen, F., Asgedom, M., and Dibo, G., General method for rapid synthesis of multicomponent peptide mixtures, *Int. J. Peptide Protein Res.*, 37, 487, 1991.

76. Lam, K. S., Salmon, S. E., Hersh, E. M., Hruby, V. J., Kazmierski, W. M., and Knapp, R. J., A new type of synthetic peptide library for identifying ligand-binding activity, *Nature*, 354, 82, 1991. [Published errata appear in *Nature* 358(6385), 434, 1992; and 360(6406), 768, 1992.]

77. Houghten, R. A., Pinilla, C., Blondelle, S. E., Appel, J. R., Dooley, C. T., and Cuervo, J. H., Generation and use of synthetic peptide combinatorial libraries for basic research and drug discovery, *Nature*, 354, 84, 1991.

78. Lebl, M., Patek, M., Kocis, P., Krchnak, V., Hruby, V. J., Salmon, S. E., and Lam, K. S., Multiple release of equimolar amounts of peptides from a polymeric carrier using orthogonal linkage-cleavage chemistry, *Int. J. Peptide Protein Res.*, 41, 201, 1993.

79. Salmon, S. E., Lam, K. S., Lebl, M., et al., Discovery of biologically active peptides in random libraries: solution-phase testing after staged orthogonal release from resin beads, *Proc. Natl. Acad. Sci. U.S.A.*, 90, 11708, 1993.

80. Ohlmeyer, M. H., Swanson, R. N., Dillard, L. W., Reader, J. C., Asouline, G., Kobayashi, R., Wigler, M., and Still, W. C., Complex synthetic chemical libraries indexed with molecular tags, *Proc. Natl. Acad. Sci. U.S.A.*, 90, 10922, 1993.

81. Jayawickreme, C. K., Graminski, G. F., Quillan, J. M., and Lerner, M. R., Creation and functional screening of a multi-use peptide library, *Proc. Natl. Acad. Sci. U.S.A.*, 91, 1614, 1994.

82. Matsueda, G. R. and Stewart, J. M., A p-methylbenzhydrylamine resin for improved solid-phase synthesis of peptide amides, *Peptides*, 2, 45, 1981.

83. Merrifield, B., Solid phase synthesis, *Science*, 232, 341, 1986.

84. Barany, G., Kneib-Cordonier, N., and Mullen, D. G., Solid-phase peptide synthesis: a silver anniversary report, *Int. J. Peptide Protein Res.*, 30, 705, 1987.

85. Grant, G. A., *Synthetic Peptides: A User's Guide*, W.H. Freeman, New York, 1992.

86. Pieta, P. G. and Marshall, G. R., Amide protection and amide supports in solid-phase peptide synthesis, *Chem. Commun.*, 11, 650, 1970.

87. Barany, G. and Merrifield, R. B., A new amino protecting group removable by reduction. Chemistry of the dithiasuccinoyl (Dts) function [letter], *J. Am. Chem. Soc.*, 99, 7363, 1977.

88. Albericio, F. and Barany, G., Mild, orthogonal solid-phase peptide synthesis: use of N alpha-dithiasuccinoyl (Dts) amino acids and N-(iso-propyldithio)carbonylproline, together with p-alkoxybenzyl ester anchoring linkages, *Int. J. Peptide Protein Res.*, 30, 177, 1987.

89. Chang, C. D. and Meienhofer, J., Solid-phase peptide synthesis using mild base cleavage of N alpha-fluorenylmethyloxycarbonylamino acids, exemplified by a synthesis of dihydrosomatostatin, *Int. J. Peptide Protein Res.*, 11, 246, 1978.

90. Atherton, E., Fox, H., Harkiss, D., Logan, C. J., Sheppard, R. C., and Williams, B. J., A mild procedure for solid phase peptide synthesis: use of fluorenylmethyloxycarbonylamino-acids, *J. Chem. Soc. Chem. Commun.*, 13, 537, 1978.

91. Fields, G. B. and Noble, R. L., Solid phase peptide synthesis utilizing 9-fluorenylmethoxycarbonyl amino acids, *Int. J. Peptide Protein Res.*, 35, 161, 1990.

92. Sarin, V. K., Kent, S. B., Tam, J. P., and Merrifield, R. B., Quantitative monitoring of solid-phase peptide synthesis by the ninhydrin reaction, *Anal. Biochem.*, 117, 147, 1981.

4 Screening Orphan Receptors for Native Ligands

Shelagh Wilson, J. Randall Slemmon, Jeffrey S. Culp, Brian D. Hellmig, Dean E. McNulty, Robert S. Ames, Henry M. Sarau, James J. Foley, Janet E. Park, Jon K. Chambers, Alison I. Muir, Jeffrey M. Stadel, and Derk J. Bergsma

CONTENTS

4.1 INTRODUCTION

Rapid advances in DNA sequencing technologies in recent years have led to a huge explosion in the amount of genomic information available to researchers and resulted in the identification of large numbers of novel genes. The challenge now lies in being able to characterize these novel genes in biological systems as rapidly and efficiently as possible in order to determine their function and gain an understanding of their physiological and pathophysiological roles. This chapter focuses on one family of novel genes identified via the use of genomic techniques, the family of orphan G protein-coupled receptors (GPCRs), and describes the methods one can use to try and characterize these and identify their native ligands.

The superfamily of GPCRs is one of the largest families of genes identified to date, with over 800 members cloned from a wide range of species. The characteristic motif of the superfamily is the seven distinct hydrophobic regions, thought to represent the transmembrane domains of these integral membrane proteins. There is little conservation of amino acid sequence across the entire superfamily of receptors, but key sequence motifs can be found within phylogenetically related subfamilies, and these motifs can be used to help classify new members.

Since the first cloning of GPCR cDNAs more than a decade ago, new genes have continued to emerge with sequences that place them firmly within the GPCR superfamily, but whose ligands remain to be identified. These "orphan" receptors show levels of homology with known GPCRs (typically less than 40%) that are too low to classify them with any confidence into a specific receptor subfamily. Many orphan receptors in fact show closer homology to each other than to known GPCRs, suggesting that they represent new subfamilies of receptors with distinct, possibly novel, ligands. These sub-families are distributed throughout the GPCR superfamily tree, suggesting that they will have a diverse range of functions and physiological roles. Figure 4.1 shows the phylogenetic relationship of a representative set of known and orphan receptors for illustration.

4.2 REVERSE PHARMACOLOGY

The overall strategy for characterizing orphan receptors has often been termed a "reverse pharmacology" approach,[1] since it involves using a receptor of unknown function to identify an activating ligand, rather than the more conventional approach of using the biological activity of a known ligand to identify its receptor. Figure 4.2 illustrates a generic "reverse pharmacology" strategy that can be used to screen orphan receptors for activating ligands. Briefly, following cloning of the receptor it is expressed in appropriate cell lines and subjected to functional analysis by screening in a range of assay systems against libraries of known and putative GPCR ligands, as well as against biological extracts of tissues, to search for novel ligands. Another option is to screen against peptide or compound libraries to identify "surrogate" agonists that can be used as tool compounds to explore the biology of the receptor. There are a number of technical issues that must be addressed at each of the various stages in order to successfully identify cognate ligands of orphan receptors, and these will be discussed in the relevant sections below.

4.3 EXPRESSION

The choice of expression system is critical to the success of ligand screening. Cell lines with a good history of GPCR expression which also contain a wide range of

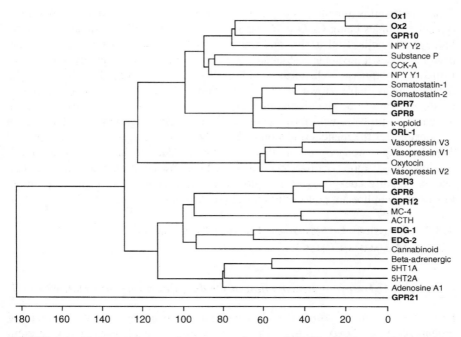

FIGURE 4.1 Phylogenetic tree of a representative set of GPCRs. Bold text indicates receptors that are (or were) orphans.

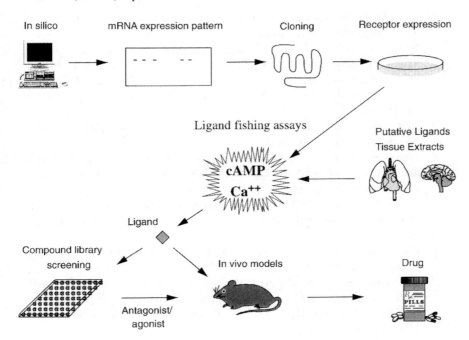

FIGURE 4.2 Reverse pharmacology approach for screening orphan receptors.

G proteins to allow functional coupling to downstream effectors is of key importance, typical examples being CHO (Chinese hamster ovary) or HEK (human embryonic kidney) 293 cells. Both stable and transient expression systems can be used and each has its own advantages. Stable expression of the receptor readily allows confirmation of expression via northern blotting, or western blotting if antibodies to the receptor are available. However, clonal variation in the pattern of endogenous receptor expression can be a confounding factor in the use of stable cell lines, leading to a significant problem with false positives. Such clonal variation has probably been responsible for the misidentification of a number of orphan receptors in the past.[2-4] Transient expression systems are therefore often used as an alternative, as these can be subjected to more rigorous control procedures. An additional advantage is that one can readily coexpress the receptor with promiscuous G proteins to promote functional coupling to an assay of choice (see Section 4.4.2.1).

4.3.1 Transient Expression of Orphan Receptors in Mammalian Cells

Protocol 4-1: Transient Transfection of Orphan Receptors into HEK 293 Cells

HEK 293 cells are transfected with a mammalian expression plasmid containing orphan receptor cDNA, with or without co-transfection of additional G protein cDNA, using LipofectAMINE PLUS reagent (Life Technologies).

1. Seed HEK 293 cells at ~10^7 per 150 to 175 cm^2 flask and incubate at 37°C overnight to allow to adhere.
2. In a 15 ml conical tube add 15 µg plasmid DNA (mammalian expression vector containing the orphan GPCR) and 40 µl PLUS reagent to 1.6 ml serum-free Earl's minimum essential medium (EMEM, Life Technologies Ltd.), vortex gently and incubate the mixture at room temperature for 15 min.
3. In a separate tube add 60 µl LipofectAMINE to 1.6 ml EMEM, gently vortex and incubate as above.
4. Following the 15 min incubation, mix the solution containing the DNA and PLUS reagent with the LipofectAMINE EMEM mixture, gently vortex and incubate at room temperature for 15 min.
5. While DNA LipofectAMINE solution is incubating, aspirate culture medium from the HEK 293 cells and replace with 12.5 ml serum-free EMEM.
6. Following the 15 min incubation, add the DNA LipofectAMINE mixture to the cells and incubate for 3 h at 37°C. The amount of DNA utilized per transfection is usually limited to 30 µg. When co-transfecting cells with receptor DNA and an expression plasmid encoding a G protein such as Gα_{16} (see Section 4.4.2.1), 15 µg receptor DNA is added together with 15 µg G protein DNA.

7. Add 15.5 ml EMEM supplemented with 20% fetal bovine serum to the cells and incubate at 37°C for 24 h. Cells can then either be plated into appropriate plates and incubated for a further 24 h before use, or left in the flask for a further 24 h (see subsequent assay protocols).

4.4 FUNCTIONAL ASSAYS

The choice of functional assays used for screening orphan receptors is crucial to success. The assays should be as generic as possible in order to pick up a wide range of coupling mechanisms. Measurement of metabolic activation of cells using the Cytosensor microphysiometer is probably the most generic assay available, but is hampered by its low throughput. Alternative assay systems in mammalian cells focus largely on measuring changes in intracellular cAMP or Ca^{2+} levels, either directly using standard methods or via the use of reporter gene assays. It is becoming increasingly important to use high throughput assay systems to allow screening of relatively large libraries of compounds and peptides in microtiter plate format; 96-well or 384-well format assays can readily be developed for measurement of cAMP levels.

For fluorescence-based assays such as Ca^{2+} mobilization, 96-well fluorescent plate readers offer greater throughput over conventional fluorimeters, but standard plate-readers only allow sequential read-out from individual wells, which limits the rate of data capture when measuring rapid responses such as Ca^{2+} mobilization. The use of a charge-coupled device (CCD) imaging camera to allow simultaneous 96-well fluorescent readout in real time—the FLIPR system developed by Molecular Devices—can dramatically increase throughput. Since the coupling pathway for a particular receptor is difficult to predict *de novo* from sequence information, a variety of assay systems should be used when screening orphan receptors.

4.4.1 THE CYTOSENSOR MICROPHYSIOMETER

The principles of microphysiometry as employed by the Cytosensor microphysiometer (Molecular Devices) have been fully described elsewhere,[5] but essentially the approach utilizes the detection of changes in extracellular acidification rates as a generic measurement of cellular metabolic activation when receptors on the cell surface are stimulated.

Briefly, cells are immobilized in close proximity to a light addressable potentiometric sensor (LAPS) which acts as a highly sensitive pH detector. A low buffer capacity "running medium" is perfused across the cells, and at regular time intervals the flow of the running medium is temporarily stopped. As a result, the medium surrounding the cells becomes progressively more acidic and this change in pH is detected by the LAPS. The rate of acidification during the stop-flow phase is the measured parameter. Typically, sampling of the extracellular acidification rate in this way occurs once every 1 to 2 min, depending upon the exact configuration of the assay. Most cell types can be maintained in the microphysiometer for several hours, allowing considerable time for data to be accumulated. Extracellular acidification rates are generally expressed as a percentage of basal rates immediately prior to the

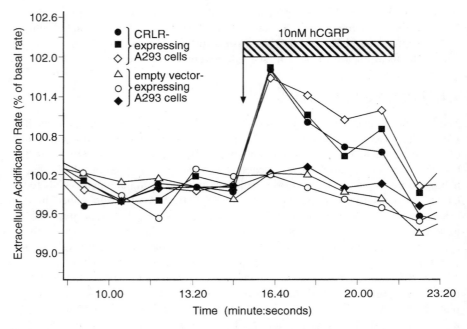

FIGURE 4.3   The orphan receptor CRLR (calcitonin receptor-like receptor) expressed transiently in HEK 293 cells, mediates a specific extracellular acidification response when challenged with 10 n*M* human CGRP. Each point represents the acidification response from an individual chamber of cells.

addition of test drug, and plotted against time (as in Figure 4.3). An elegant fluidics system enables the operator to control the exposure of cells to test drugs without flow artefacts.

Protocol 4-2: Use of Cytosensor Microphysiometer to Screen Orphan Receptors

1. 48 h after receptor transfection, gently lift adherent cells from the flask by addition of 5 ml ethylene diamine tetraacetate (EDTA, 200 mg/l in phosphate-buffered saline without Ca^{2+} and Mg^{2+}) for 1 to 2 min. Gently tap flask if necessary to detach cells. Aspirate the EDTA and add to 20 ml microphysiometer running medium (minimum essential medium, MEM, with Earle's salts without $NaHCO_3$, supplemented with 6.6 ml/l 4 *M* NaCl). Repeat the EDTA incubation until all the cells have detached.
2. Count cells, centrifuge for 5 min at $300 \times g$, then resuspend in running medium to a concentration of 2 to 3×10^6 cells/100 µl.
3. Immobilize the suspension of transfected cells in agarose on a cell capsule cup assembly as follows: melt a 200 µl aliquot of agarose (cell entrapment kit, Molecular Devices, cat no. R8023) in boiling water and then bring to 37°C in a heating block. Mix one volume of agarose with three volumes

of cell suspension, in a prewarmed tube, by pipetting the mixture up and down several times, taking care not to introduce air bubbles, and not to allow the mixture to cool below 30°C (below which it will solidify).

4. Transfer the cells into the microphysiometer by placing a blue spacer ring into the capsule cup, carefully spotting 10 μl of cell/agarose mixture into the center of the ring (taking care not to allow the mixture to touch the ring otherwise it will wick away from the center of the capsule), and allowing the mixture to solidify (2 to 3 min).

5. Carefully add 1 ml of running medium to the well holding the cup and 0.2 ml to the cup itself. Add an insert, and then 0.2 ml of running medium to the insert. When the insert has sunk to the bottom of the cup, the whole assembly may then be transferred to the cytosensor.

6. Set up the cytosensor microphysiometer as per manufacturer's instructions, with parameters set as follows: total pump cycle time = 1 min, 30 s; 1 minute at 50% full speed, then 30 s pump off; "Get rate" from 1 min 8 s to 1 min 28 s. Chamber temperatures are set to 37°C. Allow at least 1 h for rates to stabilize before adding test ligands at 30 s into pump cycle.

The orphan G protein-coupled receptor CRLR (calcitonin receptor-like receptor) was identified as a calcitonin gene-related peptide (CGRP) receptor by Aiyar et al.[6] following analysis of HEK 293 cells stably expressing the receptor in both functional (cAMP) and radioligand binding assays. However, other groups were unable to repeat this finding and there has been some speculation that CRLR is not the authentic CGRP receptor.[7,8] The cytosensor microphysiometer is an alternative assay system that can be used to demonstrate that CRLR will produce a robust, specific response to CGRP when expressed transiently in HEK 293 cells (Figure 4.3), confirming the findings of Aiyar et al.[6] More recent publications are also consistent with this receptor functioning as a CGRP receptor when expressed under appropriate conditions.[9]

4.4.2 Ca²⁺ MOBILIZATION

Changes in intracellular calcium levels can be routinely detected using fluorescent calcium indicator dyes such as Fluo-3 and Fluo-4. Cells are incubated with the acetoxymethyl ester form of the dye which permeates the cell membrane. Once inside the cell, the dye is converted to its free acid form by cleavage of the ester portion by intracellular esterases. This conversion renders the dye cell-impermeant, thus trapping it inside the cell. Only the free acid form and not the esterified counterpart will fluoresce on binding calcium ions and this property forms the basis of the assay.

Two methods are described for analyzing Ca²⁺ mobilization responses; the first involves detection via a standard fluorescent plate reader (Fluoskan Ascent, Labsystems) and is suitable for cells in suspension; the second involves detection via FLIPR (Fluorometric Imaging Plate Reader; Molecular Devices), which has the ability to read all 96 wells of a microtiter plate simultaneously with data readouts every second, and is hence higher throughput. The FLIPR imaging system is configured in a semiconfocal manner, which renders it more suitable for use with adherent cells in monolayer. Sensitivity is also enhanced by eliminating background fluorescence from the fluid layer above the cells.

Protocol 4-3: Ca²⁺ Mobilization Measurements Using Fluoskan Ascent Fluorescent Plate Reader (Labsystems)

1. Prepare assay buffer: Hanks balanced salts solution (HBSS) containing 10 mM HEPES, 200 μM CaCl$_2$ and 0.1% bovine serum albumin (BSA), pH 7.4 at 37°C.
2. Reconstitute Fluo-3/AM (Molecular Probes) in dimethyl sulfoxide (DMSO). A stock concentration of 1 mM is appropriate.
3. 48 h after transfection lift adherent cells from flask by adding 2 ml versene (Life Technologies). Rock flask gently to ensure versene completely covers monolayer, then incubate flask for 5 to 10 min at 37°C.
4. Add 8 ml growth medium to flask (to neutralize ethylene bis(oxyethylenenitrilo)tetraacetic acid [EGTA]). Pipette solution up and down several times to prepare a homogeneous cell suspension.
5. Transfer cell suspension to a 50-ml conical tube. Pellet cells by centrifuging at 1500 rpm for 5 min. Aspirate growth medium then resuspend cell pellet in 10 ml assay buffer and repeat centrifugation process.
6. Aspirate buffer and resuspend cells in 12 ml fresh assay buffer, ensuring that all clumps of cells are well dispersed. Wrap tube in foil to protect contents from light and then add Fluo-3/AM such that final concentration of dye is 1 μM. Mix thoroughly. For certain cell types (e.g., CHO), it is necessary to include 2.5 mM probenecid in the loading buffer to prevent leakage of dye from the cells.
7. Incubate at 37°C for 20 min, giving tube a swirl occasionally to prevent cells settling and clumping.
8. Add 36 ml assay buffer, mix well, and incubate at 37°C for a further 25 min. This should ensure complete hydrolysis of the dye.
9. Centrifuge cells as in step 6. Resuspend cell pellet in 20 ml fresh assay buffer.
10. Perform cell count, centrifuge cells as in step 6 then resuspend in assay buffer to a density of 2×10^6 cells/ml. Cells should be used within 2 h of loading with Fluo-3.
11. Program Fluoskan reader to read a single column of eight wells repeatedly. Repeat reads on each well can be made every 3.4 s and this time frame is sufficient to allow detection of Ca²⁺ mobilization responses. Set filters to 485 nm excitation and 538 nm emission.
12. Immediately before use, transfer 1.7 ml cell suspension to a 2-ml Eppendorf tube and microfuge briefly to pellet cells. Aspirate buffer and resuspend cell pellet in 1.7 ml fresh assay buffer.
13. Pipette 190 μl cell suspension into each of the 8 wells in the first column of a 96-well, clear bottom, black-walled plate (Becton Dickinson Labware) and start assay. Take six baseline readings on each well at 3.4 s intervals.

14. Using a multichannel pipette, add test ligands to all eight wells in the column, mixing three or four times, then continue reading plate for 30 readings, again at 3.4 s intervals. Restart readings as quickly as possible following addition of agonist.

15. Repeat steps 13 to 15 for remaining columns on plate. Reader can be set up such that machine will automatically assay column 2 after column 1, etc.

The plate reader will automatically produce a results sheet at the end of the plate run. Fluorescence values are listed for each well at each time point, with plate well reference across the rows (A1, A2, A3, etc.) and time points down the columns. (Note: Fluoroskan automatically inserts empty cells around each value, these cells must be taken into account when designing macros for data analysis.)

Readings are taken every 3.4 s and a 10-s delay is assumed between the sixth and seventh reading per well (i.e., last baseline reading and first reading following agonist addition). Results can be analyzed as follows:

1. Average first six readings to calculate mean baseline value.
2. Determine increase in fluorescence above baseline for each time point following agonist addition and express as a percentage of baseline value.
3. Plot data as change in fluorescence against time.

Protocol 4-4: Ca²⁺ Mobilization Measurements Using FLIPR (Fluorometric Imaging Plate Reader)

1. 24 h after receptor transfection harvest and plate cells into poly-D-lysine-coated 96-well, black-walled/clear bottom microtiter plates (Becton Dickinson Labware) at 3×10^4 cells per well.

2. After 18 to 24 h aspirate off medium and add 100 µl of fresh EMEM containing 4 µM Fluo-3/AM (Molecular Probes; prepare stock solution at 2 mM in DMSO containing 20% pluronic acid), 0.1% BSA, and 2.5 µM probenecid to each well and incubate for 1 h at 37°C.

3. Aspirate medium and replace with 100 µl of the same medium, but without Fluo-3/AM, and incubate for 10 min at 37°C. Wash cells three times with a Denley cell washer (Labsystems) with Krebs Ringer Henseleit (KRH; 118 mM NaCl, 4.6 mM KCl, 25 mM NaHCO$_3$, 1 mM KH$_2$PO$_4$, 11 mM glucose, 1.1 mM MgCl$_2$) containing 0.1% BSA, 2.5 µM probenecid, and 20 mM HEPES, pH 7.4 (buffer A). After the last wash aspirate down to a final volume of 100 µl.

4. Prepare plates of potential ligands, tissue extracts, or compound libraries in 96-well polypropylene microplates at three times the final concentration in buffer A and warm to 37°C.

5. Place microplates containing Fluo-3 loaded cells and the ligand plates in FLIPR and monitor.[10] At initiation of the fluorescence recording read measurements every 1 s for 60 s and then every 3 s for the following 60 s. Add agonist (50 µl) to the cell plate at 10 s and use the maximal fluorescent counts above background after addition of agonist to define maximal

activity for that ligand. FLIPR software normalizes fluorescent readings to give equivalent readings for all wells at zero time.

4.4.2.1   Use of Promiscuous G Proteins with Orphan Receptors

More recently, it has become possible to funnel heterologous signal transduction of GPCRs through a common pathway involving phospholipase C and Ca^{2+} mobilization by co-expression of the receptor with the promiscuous G proteins $G\alpha_{15/16}$, or by co-expression with chimeric G proteins based on Gq.[11,12] For example, the C3a receptor,[13,14] originally described as an orphan receptor,[15] produces weak or no Ca^{2+} mobilization responses to C3a when expressed in HEK 293 cells. However, when co-expressed with $G\alpha_{16}$, robust responses are seen (Figures 4.4 and 4.5). Although this cotransfection approach may not be universally successful, it can be a useful technique to promote functional coupling of orphan receptors and to streamline their screening by focusing predominantly on one signal transduction system.

4.4.3   Determination of cAMP Levels via Use of Flashplate Technology

Protocol 4-5: cAMP Determination Using Flashplates

1. Seed HEK 293 cells into 96-well microtiter plates 24 h prior to assay at a density of 10^5 cells/well in 100 µl growth medium. Incubate overnight at 37°C. The wells should reach at least 95% confluency on the following day.
2. Add 10 µl IBMX (3-isobutyl-1-methylxanthine; final concentration 0.5 mM) and incubate cells for 30 min at 37°C.
3. Add 10 µl putative ligand immediately followed by 10 µl growth medium (for determination of cAMP stimulatory responses) or 10 µl forskolin (for determination of cAMP inhibitory responses) at 30 µM final concentration. Incubate for 10 min at 37°C.
4. Terminate reaction with the addition of 10 µl ice cold 5% perchloric acid (PCA).
5. Leave for 1 h at 4°C to allow equilibrium to be reached before neutralizing samples with 100 µl 75 mM KOH in 50 mM sodium acetate.
6. Incubate at 4°C for at least 1.5 h to ensure neutralization is complete and allow sedimentation of the precipitate before transfering samples into cAMP [^{125}I] FlashPlates (NEN Life Sciences; Catalogue no. SMP001), taking care not to disturb the sediment.
7. Typically for HEK 293 cells, transfer 10 to 40 µl of neutralized sample to FlashPlates for assay. Reconstitute kit reagents as per the manufacturer's instructions and make sample volume up to 100 µl in FlashPlates with assay buffer.
8. Add 100 µl [^{125}I]-cAMP working tracer, seal plate, and incubate overnight at 4°C prior to counting.

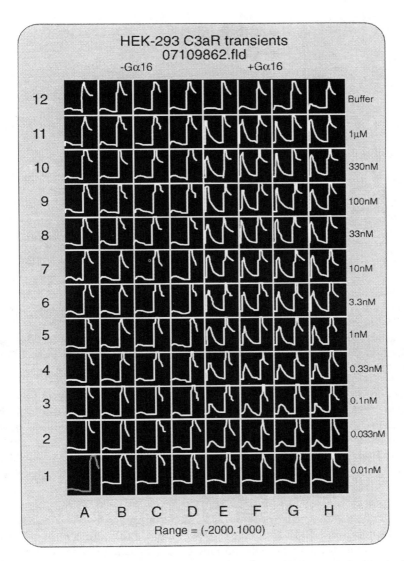

FIGURE 4.4 Ca^{2+} mobilization responses of the C3a receptor expressed with and without the promiscuous G protein $G\alpha_{16}$. Each square represents responses as determined simultaneously via FLIPR from an individual microtiter plate well of cells. The y axis in each square corresponds to intracellular calcium levels and the x axis corresponds to time. The left-hand response of each trace corresponds to challenge with C3a (or buffer) as indicated; the right-hand response of each trace corresponds to challenge with 200 μM muscarine as an internal control for each well. Columns A–D were generated using HEK 293 cells transiently transfected with C3a receptor and columns E–H were generated using cells transfected with both C3a receptor and $G\alpha_{16}$. Quantitation of these data is presented in Figure 4.5.

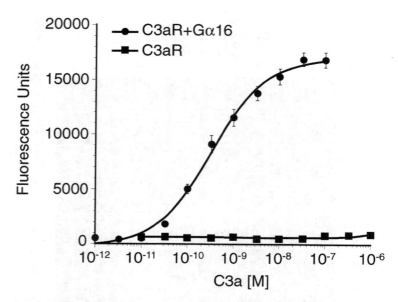

FIGURE 4.5 Calcium mobilization induced by C3a in HEK 293 cells transiently transfected with C3a receptor alone (■) or C3a receptor with $G\alpha_{16}$ (●). The maximum Ca^{2+} response (peak height) following C3a addition to each well, as illustrated in Figure 4.4, was used to construct the dose-response curves. Each point in Figure 4.5 represents the mean ± SEM of eight determinations.

The assay can be used not only to screen for ligands that activate Gs- or Gi-coupled orphan receptors, but also to look for receptors that might be constitutively active. For example, Figure 4.6 illustrates the elevated basal cAMP levels that can be detected in HEK 293 cells expressing the orphan receptor GPR3, consistent with the constitutive activity reported for this receptor.[16] Such receptors are amenable to screening for compounds with inverse agonist activity which reduce the elevated basal levels of activity, which can then be used as tools to explore the biology of the receptor *in vivo*.

Recently, a "homogeneous" version of the flashplate assay has been developed by the manufacturer (SMP004, NEN Life Sciences). In this format the cells and test substance are added directly to the flashplate, the incubation terminated by a single addition (which includes the radiolabel) and the plate then counted 2 to 24 h later at room temperature. The reduced number of pipetting steps results in a higher throughput. However, a caveat when screening tissue extracts concerns the lack of an acid precipitation step to remove excess proteinaceous material, which may thus interfere with the assay.

4.5 SOURCES OF ACTIVATING LIGANDS FOR SCREENING ORPHAN RECEPTORS

Although the sequence homology with known receptors is low, orphan receptors would first be screened against a bank of known or putative GPCR ligands when

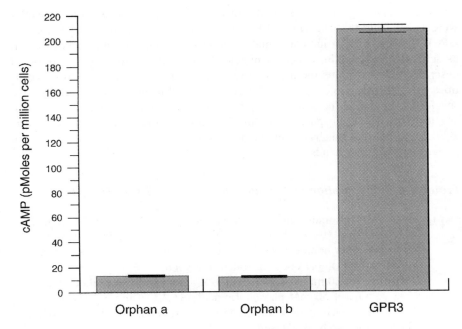

FIGURE 4.6 Constitutive activity of the orphan receptor GPR3 expressed in HEK 293 cells. Bars show basal levels of cAMP in cells expressing either two unrelated orphan receptors (a and b) or GPR3, as determined by Protocol 4-5.

"fishing" for the cognate ligand. There is some precedent for the existence of very low homology between subtypes of receptors (e.g., less than 35% identity between Y_1, Y_2, and Y_5 subtypes of NPY receptors), so it is possible that some orphan receptors represent new paralogues of known GPCRs. However, an important activity is to search for novel ligands by screening receptors against biological extracts of tissues, fluids, or cell supernatants. Precedent for success with this approach comes from the identification of nociceptin as a novel peptide ligand for the orphan receptor ORL-1[17] and more recently from the identification of the orexins, two related peptides purified from hypothalamic tissue, as endogenous ligands for two related orphan receptors.[18]

4.5.1 PREPARATION OF PEPTIDE EXTRACTS OF TISSUES

FLIPR technology has provided a major improvement in throughput and sensitivity when screening orphan receptors but, as with many new technologies, additional technical problems arise, particularly when using the system to screen extracts of biological tissues. Many of the fractions obtained after typical reverse-phase high performance liquid chromatography (HPLC) fractionation of acid extracts of frozen tissues (e.g., bovine brain) produce large, nonspecific responses when screened against orphan receptor-expressing cell lines using FLIPR. However, we found that after removal of high-molecular-weight peptides (>15,000 Da), very few of the resultant fractions produced nonspecific responses. This offered a major improvement in signal to noise, but still preserved the peptides with molecular weights below 15,000 Da. This is likely to be an important pool for screening for GPCR ligands, since it includes small peptides (e.g., somatostatin, GRP) as well as acid stable chemokines such as platelet factor-4 and RANTES. The protocol described below exploits the utility of removing high-molecular-weight polypeptides prior to assay and provides for further enrichment of potential bioactive species by separating the resultant peptide pool into cationic, anionic, and neutral species. The use of phosphoric acid, pH 2.5, in the initial extraction procedure is also employed since this has given the best recovery of total peptide material from frozen tissues in our hands.

Protocol 4-6: Preparation of Peptide Libraries from Tissue Extracts

1. Make a 20% homogenate (weight/volume in 50 mM phosphoric acid, pH 2.1) from 400 g fresh frozen tissue (e.g., bovine brain) in a Waring blender. Further homogenize the disrupted tissue with a Tekmar Tissumizer (Tekmar Company) or equivalent for 3 min.
2. Centrifuge for 30 min at 4°C at 28,000 × g. Re-extract the pellet with the same volume of 50 mM phosphoric acid, pH 2.1, using the Tissumizer. Repeat the centrifugation and combine the supernatants. Filter through 35 μm nylon mesh (BioDesign).
3. Prepare 250 g of C18-silica (Waters), grade WAT 010001 (or equivalent) in a sintered-glass funnel by washing with 1 l HPLC-grade methanol followed by 2 l 50 mM phosphoric acid, pH 2.1. Add this directly to the peptide supernatant and stir for 1 h at ambient temperature.
4. Pack a 5 cm × 25 cm solvent-resistant column with the slurry and wash with 1 l 0.1% trifluoroacetic acid (TFA) in water (vol/vol). Elute the bound peptides with 800 ml 60% acetonitrile in 0.1% TFA (vol/vol).
5. Concentrate the peptide elution on a Rotovap concentrator to about 125 ml. Filter the concentrate with a Sterivex-HV 0.2 μm filter or equivalent (Millipore).
6. Fractionate the concentrated peptides on a 5 cm × 94 cm column of Sephadex G-50 (medium) equilibrated in 50 mM sodium phosphate, pH 2.5. Develop the column for the first 810 ml and then begin collecting the elution through the next 1080 ml (peptides <15,000 Da).
7. Prepare 75 g of C18-silica as described above and pack into a 5 cm × 25 cm solvent-resistant column. Load the pooled peptides from G-50 onto the column and wash with 300 ml 0.1% TFA in water (vol/vol) followed by elution of the peptides in 200 ml of 60% acetonitrile:0.1% trifluoroacetic acid in water (vol/vol). Concentrate the elution to about 100 ml using the Rotovap concentrator.
8. Dilute the concentrate to 500 ml with 10 mM MES (12-(N-morphilino)-ethanesulfonic acid, sodium salt). Assure that the pH of the sample is 6.0 by the addition of 6 N sodium hydroxide. Centrifuge to remove any flocculent material if necessary.

9. Prepare one each of a 10 g Accell Plus QMA and CM cartridge (Waters) by wetting in 30 ml methanol, washing in 30 ml water, charging in 50 ml of 500 mM sodium MES, pH 6.0, followed by equilibration in 10 mM MES, pH 6.0.

10. Pass the neutralized peptide sample over the QMA cartridge (use mild vacuum if necessary) and wash completely through the column with 40 ml of 10 mM MES, pH 6.0. Pass the combined unbound and wash fractions from the QMA cartridge over the CM cartridge and wash with another 40 ml of 10 mM sodium MES, pH 6.0. This flow through is the "neutral peptide" fraction.

11. Elute the bound peptides from both the QMA and CM cartridges with 40 ml of 10 mM MES, pH 6.0, containing 0.75 M sodium chloride. The peptides from the QMA cartridge are the "anionic peptides" and those from the CM cartridge are the "cationic peptides."

12. The three peptide pools are each adjusted to pH 2.5 by the addition of concentrated TFA, and the neutral peptide fraction is captured on C18-silica as described for the capture of the peptides from Sephadex G-50 chromatography. Eluted material, containing the neutral peptide fraction, is concentrated on a Rotovap concentrator to approximately 50 ml.

13. Each peptide pool is chromatographed on reverse-phase HPLC using a water:acetonitrile:0.1% TFA system on a Vydac C18 (2.2 cm × 25 cm, 6 ml/min flow rate, detection at A_{280}) and 1 min fractions are collected across a 85 min gradient of 12 to 36% acetonitrile and a 72% acetonitrile strip, both in 0.1% TFA. The fractions are dried in a SpeedVac concentrator (Savant Instruments) and then resuspended in an appropriate assay buffer and dispensed into a 96-well polypropylene plate for assay.

4.5.2 PURIFICATION OF AN ORPHAN RECEPTOR PEPTIDE LIGAND FROM BOVINE HYPOTHALAMUS

One demonstration of the utility of peptide libraries prepared as described above was their use in identifying a novel peptide, Lig10 (also termed PrRP31[19]) for the orphan receptor GPR10.[20] A cationic peptide from a sample of bovine hypothalamus library prepared as per protocol 4.6 gave robust, specific calcium mobilization responses in HEK 293 cells over two sets of consecutive HPLC fractions (Figure 4.7). Fractions containing the largest responses, fractions 45 to 49, were pooled and the activity further purified as follows.

Protocol 4-7: Purification and Identification of a Novel Peptide Ligand for the Orphan Receptor GPR10

1. The active fractions from reverse phase-HPLC (about 30 ml) were pooled and applied directly to a 7.5 × 75 mm TosoHaas SP-5PW cation exchange column equilibrated at 3 ml/min in 10 mM sodium phosphate in 25% acetonitrile in water, pH 2.7, and developed in a linear gradient from

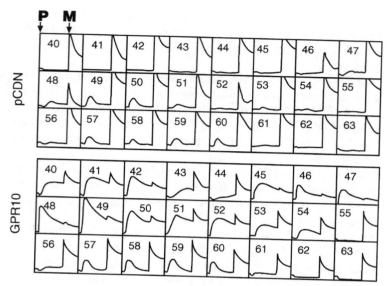

FIGURE 4.7 Fractions from a bovine hypothalamus extract eliciting specific activity with GPR10. The bovine hypothalamus cationic peptide library RP-HPLC fractions, prepared as described in Protocol 4-6, were screened for Ca^{2+} responses in FLIPR against HEK 293 cells transfected with GPR10 or with empty expression vector (pCDN) as control. Each square represents responses as determined simultaneously via FLIPR from each individual well of cells challenged with sequential peptide library fractions. The fraction number is indicated in the upper left corner of each square. The left-hand response of each trace corresponds to challenge with the peptide library fraction (indicated by **P** arrow at top). The second response corresponds to challenge with 200 μM muscarine (indicated by **M** arrow at top) as an internal positive control to assess the quality of the cells in each well. Note the strong specific responses in GPR10 transfected cells with fractions 40–43, 45–47, and 51–54, accompanied by desensitization of the response to muscarine. Also note GPR10 activity above background responses in fractions 48–50. Background responses are observed in both cell lines in fractions 57–60.

0 to 0.5 M of sodium chloride in equilibration buffer. Three-milliliter fractions were collected and 150 μl aliquots were diluted for assay. The active peptide was eluted in four fractions, which were combined. Acetonitrile was removed using a SpeedVac concentrator.

2. Further fractionation was then accomplished on a Vydac pH stable (polymeric) C8 column (0.46 × 25 cm, using a 0.1% TFA in water buffer system and acetonitrile mobile phase at 1 ml/min). The column was developed over 55 min with a gradient of 16 to 28% acetonitrile. The activity eluted in three fractions which were prepared for the next chromatography by removal of acetonitrile as before.

3. The concentrated material was diluted times 2 with one volume of buffer A (equilibration buffer) and chromatographed on a Vydac C18 narrow-bore column (0.21 × 25 cm, 0.25 ml/min; buffer A: 20 mM phosphoric acid in water, buffer B: 80% acetonitrile in water). The column was developed over 55 min in a 20% to 30% gradient of buffer B and 125 μl

fractions were collected. Two consecutive fractions containing the activity were pooled and diluted to 0.8 ml in 0.06% TFA.

4. The sample was fractionated on a Vydac Diphenyl column (0.21 × 25 cm, 0.15 ml/min), using a 0.1% TFA in water buffer system and acetonitrile (mobile phase). The column was developed over 55 min with a gradient of 21% to 26% acetonitrile. The active fraction from this separation was submitted for sequence determination. Preliminary automated Edman sequencing yielded the sequence XRAXQXXMEIXTPDINPAXY and MALDI-MS determined the mass to be 3576 Da, indicating the active peptide to be the full-length form of PrRP31. Scale-up of the purification procedure outlined above is required to obtain the complete sequence of the peptide, which as reported by Hinuma et al.[19] is SRAHQHSMEIRT-PDINPAWYAGRGIRPVGRFa.

ACKNOWLEDGMENT

We would like to thank Dr. P. Szekeres for generation of the phylogenetic tree in Figure 4.1.

REFERENCES

1. Libert, F., Vassart, G., and Parmentier, M., Current developments in G protein-coupled receptors, *Current Opinion Cell Biol.*, 8, 218, 1991.
2. Cook, J. S., Wolsing, D. H., Lamech, J., Olson, C. A., Correa, P. E., Sadee, W., Blumenthal, E. M., and Rosenbaum, J. S., Characterization of the RDC1 gene which encodes the canine homolog of a proposed human VIP receptor, *FEBS Lett.*, 300, 149, 1992.
3. Jazin, E. E., Yoo, H., Blomqvist, A. G., Yee, F., Weng, G., Walker, M. W., Salon, J., Larhammar, D., and Wahlestedt, C., A proposed bovine neuropeptide Y (NPY) receptor cDNA clone, or its human homologue, confers neither NPY binding sites nor NPY responsiveness on transfected cells, *Regul. Peptides*, 47, 247, 1993.
4. Yokomizo, T., Izumi, T., Chang, K., Takusa, Y., and Shimizu, T., A G protein coupled receptor for leukotriene B4 that mediates chemotaxis, *Nature*, 387, 620, 1997.
5. McConnell, H. M., Owicki, J. C., Parce, J. W., Miller, D. L., Baxter, G. T., Wada, H. G., and Pitchford, S., The Cytosensor microphysiometer: biological applications of silicon technology, *Science*, 257, 1906, 1992.
6. Aiyar, N., Rand, K., Elshourbagy, N. A., Zeng, Z., Adamou, J. E., Bergsma, D. J., and Li, Y., A cDNA encoding the calcitonin gene-related peptide type 1 receptor, *J. Biol. Chem.*, 271, 11325, 1996.
7. Fluhmann, B., Muff, R., Hunziker, W., Fischer, J. A., and Born, W., A human orphan calcitonin receptor-like structure, *Biochem. Biophys. Res. Commun.*, 206, 341, 1995.
8. Wimalawansa, S. J., CGRP and its receptors, *Endocr. Rev.*, 17, 533, 1996.
9. Mclatchie, L. M., Fraser, N. J., Main, M. J., Wise, A., Brown, J., Thompson, N., Solari, R., Lee, M. G., and Foord, S. M., RAMPs regulate the transport and ligand specificity of the calcitonin-receptor-like receptor, *Nature*, 393, 333, 1998.
10. Schroeder, K. S. and Neagle, B. D., FLIPR: a new instrument for accurate high throughput optical screening, *J. Biomol. Screening*, 1, 75, 1996.

11. Offermans, S. and Simon, M. I., Ga$_{15}$ and Ga$_{16}$ couple a wide variety of receptors to phospholipase C, *J. Biol. Chem.*, 270, 15175, 1995.

12. Conklin, B. R., Herzmark, P., Ishida, S., Voyno-Yasenetskaya, T. A., Sun, Y., Farfel, Z., and Bourne, H. R., Carboxyl-terminal mutation of G$_{q\alpha}$ and G$_{s\alpha}$ that alter the fidelity of receptor activation, *Mol. Pharmacol.*, 50, 885, 1996.

13. Ames, R. S., Li, Y., Sarau, H. M., Nuthulaganti, P., Foley, J. J., Ellis, C., Zeng, Z. Z., Su, K., Jurewicz, A. J., Hertzberg, R. P., Bergsma, D. J., and Kumar, C., Molecular-cloning and characterization of the human anaphylatoxin C3a receptor, *J. Biol. Chem.*, 271, 20231, 1996.

14. Crass, T., Raffetseder, U., Martin, U., Grove, M., Klos, A., Köhl, J., and Bautsch, W., Expression cloning of the human C3a anaphylatoxin receptor (C3aR) from differentiated U-937 cells, *Eur. J. Immunol.*, 26, 1944, 1996.

15. Roglic, A., Prossnitz, E. R., Cavanagh, S. L., Pan, Z., Zou, A., and Ye, R. D., cDNA cloning of a novel G protein-coupled receptor with a large extracellular loop structure, *Biochim. Biophys. Acta*, 1305, 39, 1996.

16. Eggerick, D., Denef, J.-F., Labbe, O., Hayashi, Y., Refetoff, S., Vassart, G., Parmentier, M., and Libert, F., Molecular cloning of an orphan G protein-coupled receptor that constitutively activates adenylate cyclase, *Biochem. J.*, 309, 837, 1995.

17. Meunier, J. C., Molleraue, C., Toll, L., Suaudeau, C., Moisand, C., Alvinerie, P., Butour, J. L., Giullimot, J. C., Ferrara, P., Monsarrat, B., Mazarguil, H., Vassart, G., Parmentier, M., and Costentin, J., Isolation and structure of the endogenous agonist of opioid receptor-like ORL1 receptor, *Nature*, 377, 532, 1995.

18. Sakurai, T., Amemiya, A., Ishii, M., Matsuzaki, I., Chemelli, R. M., Tanaka, H., Williams, S. C., Richardson, J. A., Kozlowski, G. P., Wilson, S., Arch, J. R. S., Buckingham, R. E., Haynes, A. C., Carr, S. A., Annan, R. S., McNulty, D. E., Liu, W. S., Terret, J. A., Elshourbagy, N. A., Bergsma, D. J., and Yanagisawa, M., Orexins and orexin receptors: a family of hypothalamic neuropeptide and G protein-coupled receptors that regulates feeding behaviour, *Cell*, 92, 573, 1998.

19. Hinuma, S., Habata, Y., Fuji, R., Kawamata, Y., Hosoya, M., Fukusumi, S., Kitada, C., Masuo, Y., Asano, T., Matsumoto, H., Sekiguchi, M., Kurokawa, T., Nishimura, O., Onda, H., and Fujino, M., A prolactin-releasing peptide in the brain, *Nature*, 393, 272, 1998.

20. Marchese, A., Heiber, M., Nguyen, T., Heng, H. H. Q., Saldivia, V. R., Cheng, R., Murphy, P. M., Tsui, L. C., Shi, X., Gregor, P., George, S. R., O'Dowd, B. F., and Docherty, J. M., Cloning and chromosomal mapping of three novel genes, GPR9, GPR10, and GPR14, encoding receptors related to interleukin 8, neuropeptide Y and somatostatin receptors, *Genomics*, 29, 335, 1995.

Section II

Function and Regulation:
Receptor–G Protein Coupling
and Desensitization

5 Reconstitution of Receptor-Promoted Signal Transduction Using Purified Receptor, G Protein, and Effector

Gabriel Berstein and Gloria H. Biddlecome

CONTENTS

0-8493-3384-9/00/$0.00+$.50
© 2000 by CRC Press LLC

5.1 INTRODUCTION

5.1.1 OVERVIEW OF G PROTEIN-COUPLED RECEPTOR SIGNAL TRANSDUCTION

Cells use heterotrimeric G proteins to differentially modulate the speed and through-put of various signaling pathways by coupling one or more class of G proteins to different receptors and effectors. Heterotrimeric G proteins control the kinetics and amplitude of signal transduction via a regulated cycle of guanosine triphosphate (GTP) binding and hydrolysis. When inactive, the G_α subunits contain guanosine diphosphate (GDP) in their active site and are bound to the $G_{\beta\gamma}$ subunits. Upon encounter with a ligand-activated receptor, the subunits dissociate, GDP is released, and GTP binds to and activates G_α. Both GTP-activated G_α subunits and the liberated $G_{\beta\gamma}$ subunits regulate effector molecules until GTP is hydrolyzed and G_α and $G_{\beta\gamma}$ subunits reassociate. The duration of the active state depends on the rate at which a particular G_α hydrolyzes bound GTP. The intrinsic rate of GTP hydrolysis by G_α is accelerated by GTPase-activating proteins (GAPs) such as phospholipase C-β (PLC-β), RGS (Regulator of G Protein Signaling) and GAIP (Gα interacting protein) proteins.[1–4]

5.1.2 APPLICATIONS OF *IN VITRO* RECONSTITUTION OF RECEPTOR-PROMOTED SIGNALING

There are distinctive advantages of using the reconstitution approach described here to study these signaling proteins. One, this is a functional assay for G protein-coupled receptor activity, equivalent to an enzymatic assay in which rates of receptor-pro-moted nucleotide exchange, concentration dependencies of cofactors, and effects of receptor ligands on G protein activity can be unambiguously determined. Addition of effector to the reconstituted system increases the experimental complexity some-what, but provides crucial, and otherwise unattainable, mechanistic information about receptor–G protein–effector signaling. Such questions cannot be answered by cell transfection experiments and are nearly impossible in most membrane-based assays. Two, once adequate background knowledge is attained for a particular recon-stituted system, it can then be used to screen for effects of soluble proteins, cellular factors, and therapeutic drugs on the interaction of specific receptors, G proteins, and effectors. Three, the method provides great flexibility in the choice of proteins, scale of preparation, and types of experimental questions one can ask. No other experimental system like this is available. Related articles that review reconstitution methods are given in the Reference section.[5,6]

We used the techniques presented in this chapter primarily to investigate M_1 muscarinic acetylcholine receptor regulation of G_q and PLC-$\beta1$, although they should be applicable to many other systems. In particular, the reconstitution of purified receptors and G proteins has been used extensively for investigating the regulation of various G proteins by receptors. Some examples are the following: β-adrenergic receptor,[7–9] α-adrenergic receptors,[10,11] muscarinic receptors,[12–19] dopamine recep-tors,[20] thromboxane receptors,[21] and pituitary adenylate cyclase-activating polypep-tide receptors.[22]

5.2 METHODS

5.2.1 cDNA Cloning, Plasmid Constructs, and Production of Recombinant Baculoviruses

Description of standard methods for cloning complementary DNAs that encode the proteins used here and for production of baculoviruses for recombinant protein production in Sf9 cells is beyond the scope of this chapter and can be found elsewhere.[23,24] Original references for cloning of human M_1 muscarinic receptor,[25,26] mouse $G_{q\alpha}$,[27] rat $G_{i\alpha1}$,[28] bovine $G_{\beta1}$,[29] bovine $G_{\beta2}$,[30] bovine $G_{\gamma2}$,[31] and rat phospholipase C-$\beta1$ (PLC-$\beta1$)[32] are given in the Reference section.

Sf9 cells are infected at 1.5×10^6 cells/ml at 1 plaque-forming unit/cell for each recombinant virus coding for the following proteins: M_1 receptor; hexahistidine-tagged PLC-$\beta1$; $G_{q\alpha}$, hexahistidine-tagged $G_{\beta2}$, and hexahistidine-tagged $G_{\alpha1}$ to prepare heterotrimeric G_q; and $G_{\beta1}$, $G_{\gamma2}$, and hexahistidine-tagged $G_{i\alpha1}$ to prepare $\beta\gamma$ dimers.

Several of the purification methods described here use hexahistidine motifs as tags for affinity chromatography on Ni-nitrilotriacetate (Ni-NTA) resin. The position of the affinity tag influences both the behavior of the protein during purification and its interaction with other signaling components.[33] The following positions are rec-ommended: for rat $G_{i\alpha1}$, position 121, the cognate position where the α subunit of *Saccharomyces cerevisiae* GPA 1 has a long insert;[34] for bovine G_γ, at the amino terminus;[34] for rat PLC-$\beta1$, replace the first seven amino acids by Met-Gly-His$_6$.[18]

5.2.2 GENERAL CONSIDERATIONS FOR PROTEIN PURIFICATION

1. All the steps must be carried out at 4°C unless indicated otherwise.
2. Water should be doubled-distilled quality.
3. Bovine serum albumin (BSA) should be protease-free grade (e.g., from Calbiochem).
4. Purity of protein preparations is often judged by sodium dodecyl sulfate-polyacrylamide gel electrophoresis (SDS-PAGE) followed by silver staining of the gel. For some of the preparations, immunoblotting is used to detect the presence of the desired protein as well as contaminants in various chromatographic fractions.
5. Protein concentration in membrane preparations and chromatography fractions can be determined by amido black staining of nitrocellulose-bound protein according to Schaffner and Weissmann[35] (see protocol 5-16). This procedure is relatively unaffected by the presence of the various detergents and reducing agents that interfere with many other protein determination methods.
6. Receptors and G proteins are sensitive to the type and purity of detergents used for purification and assay of their activity. It is important to use high grade commercial detergents or buy lesser grades and repurify them.

> *Digitonin:* The product from BioSynth AG (water-soluble grade) consistently renders clear solutions upon boiling at 5 to 10% (w/v). These stock solutions should be prepared shortly before use.
>
> *Cholate:* Sodium cholate from Fluka (MicroSelect Grade) is adequate. High-purity cholic acid can be obtained by purification of any commercial preparation through DE-52 resin according to the method by Ross and Schatz.[36]
>
> $C_{12}E_{10}$ *("Lubrol PX"):* To prepare a 10% (w/v) stock, warm solid polyoxyethylene 10 lauryl ether ($C_{12}E_{10}$, Sigma) in a 55°C water bath until a viscous solution forms, then dissolve 100 g with 800 ml water and adjust to 1 l. Add 5 g mixed bed resin AG 501-X8/Bio-Rex MSZ 501(D) (BioRad) and stir overnight at room temperature. Filter through a 0.45 µm nylon filter and store at 4°C.

7. Because of the propensity of PIP_2 to stick to glass, it is essential to siliconize any sonication tubes, gel filtration columns, storage vials, and Hamilton syringe barrels that will be in contact with this phospholipid. Siliconization can be done in batch according to the manufacturer's instructions (for example, Aquasil, Pierce).

5.2.3 MEMBRANE PREPARATION

All the proteins that are needed for the reconstitution experiments are isolated from membrane preparations of recombinant baculovirus-infected Sf9 cells. A crude particulate fraction is prepared by lysing the cells, followed by centrifugation. Because the preparation contains substantial amounts of DNA which may interfere with protein extraction, deoxyribonuclease I is included. Metal chelators are included in the initial lysing step as metalloprotease inhibitors, but they are omitted in later steps because they interfere with Ni-NTA resin chromatography.[15,18]

Protocol 5-1: Sf9 Cell Membrane Preparation

1. *Solutions*
   a. Solution A (200 ml): 20 mM Tris-Cl (pH 8), 1 mM EDTA, 1 mM EGTA, and PI (protease inhibitors : 0.1 mM PMSF, 2 µg/ml aprotinin, and 10 µg/ml leupeptin).
   b. Solution B (100 ml): 20 mM Tris-Cl (pH 8), 3 mM MgCl$_2$, 10 µg/ml deoxyribonuclease I (Sigma), and PI.
2. The following protocol is designed for 1 l Sf9 cells but can be scaled up or down as needed. Harvest Sf9 cells 48 to 60 h after infection by centrifugation at 1000 × g for 10 min. Discard supernatant after bleaching or autoclaving. The optimal time of harvesting must be determined experimentally by carrying out a time course of protein expression. In some instances, the production of protein does not correlate with the amount that is extractable from the membranes or that is active. Thus, it is advantageous to take samples before and after extraction, and perform a Western blot and an activity assay if possible. In our experience protein expression is maximal between 48 and 60 h after infection, but longer[17] incubations have been reported. Before harvesting, cells should be examined for (a) signs of baculoviral infection, such as cell enlargement, presence of multinucleated cells, and poor growth after infection (<2 × 10^6 cells/ml), and (b) absence of microbial contamination.
3. Resuspend cell pellet with 200 ml (10^7 cells/ml) ice-cold solution A.
4. Homogenize the cell suspension with a Dounce homogenizer (type A, i.e. tight), 10 strokes.
5. Centrifuge homogenate at 30,000 × g (e.g., Beckman JA-14 rotor at 14,000 rpm) for 20 min.
6. Discard supernatant, resuspend pellet with 50 ml solution B, and repeat Dounce homogenization.
7. Centrifuge at 30,000 × g for 20 min.
8. Discard supernatant and resuspend pellet with ~20 ml solution B (this will give a protein concentration of ~15 mg/ml) with a Dounce homogenizer as above.
9. Aliquot as needed and freeze with liquid nitrogen. Freezing can be done in a batch manner by adding the membrane suspension dropwise from a syringe and cannula into a polypropylene beaker containing liquid nitrogen, while constantly stirring with a glass bar. The small pellets that are formed can be easily aliquoted and stored below −70°C. Biological activity of any of the proteins dealt with here is maintained indefinitely when membranes are stored under these conditions.

5.2.4   PREPARATION OF M₁ MUSCARINIC ACETYLCHOLINE RECEPTOR

5.2.4.1   Receptor Purification

M_1 muscarinic acetylcholine receptor (M_1 receptor) is extracted from Sf9 cell membranes with detergent and is purified by ligand affinity chromatography essentially as described by Haga and Haga[37] and Haga et al.[38] In this two-step procedure the membrane extract is initially subjected to aminobenztropine (ABT, a muscarinic receptor antagonist)-Sepharose chromatography. Receptor is eluted with atropine (another muscarinic antagonist), and chromatographed through a hydroxyapatite column which is subsequently eluted with potassium phosphate in step-gradient manner. A summarized protocol is given here. For a detailed protocol see the article by Haga et al.[38] This procedure takes two consecutive working days.

Protocol 5-2: Purification of M₁ Muscarinic Acetylcholine Receptor

---

1. Materials
   a. 2× extraction buffer (40 ml): 40 mM potassium phosphate (KPi) buffer, pH 7.0, 200 mM NaCl, 2 mM EDTA, 0.2% (w/v) sodium cholate, and 2% (w/v) digitonin, PI (protease inhibitors: 0.2 mM PMSF, 4 μg/ml aprotinin, 20 μg/ml leupeptin).
   b. Affinity resin: aminobenztropine (ABT) coupled to Sepharose 4B. ABT is commercially available from Research Biochemicals International (RBI) and is coupled to epoxy-activated Sepharose 4B according to the protocol described by Haga et al.[38]
   c. Chromatography buffers
      R1 (100 ml): 20 mM KPi buffer (pH 7.0), 0.1% (w/v) digitonin, PI
      R2 (100 ml): R1 plus 150 mM NaCl
      R3 (100 ml): R2 plus 0.1 mM atropine sulfate
      R4 (100 ml): 200 mM KPi (pH 7.0), 1 mM carbachol (to stabilize binding sites during storage), and 0.1% (w/v) digitonin
      R5 (2 ml): like R4 except that the concentration of KPi (pH 7.0) is 500 mM
      R6 (2 ml): like R4 except that the concentration of KPi (pH 7.0) is 900 mM
2. Mix membranes from a 1 l Sf9 cell culture with 2× extraction buffer to yield 5 mg/ml membrane protein and 1× extraction buffer.
3. Stir on ice for 1 h.
4. Centrifuge at >100,000 × g for 1 h at 4°C (e.g., Beckman Ti45 rotor at 35,000 rpm).
5. Collect extraction supernatant; hold 50 μl aliquot on ice for assay. Proceed immediately with chromatography.
6. Load extraction supernatant at a flow rate of 1 ml/min onto a 10 ml bed volume ABT column that has been equilibrated with R1.
7. Wash the column with R2 until A_{280} returns to baseline (it requires at least three column volumes).

8. Prepare 0.5 ml hydroxyapatite (BioRad) column by defining thoroughly and pouring into an Isolab polypropylene Mini Column®. Wash with 5 ml 1 M KPi buffer (pH 7.0), 2 ml water, and then equilibrate with R2.

9. Connect the outlet of the ABT column in series to the hydroxyapatite column.

10. Elute the receptor from the ABT column with 50 ml R3. Receptor will almost quantitatively bind to hydroxyapatite. It is important to closely monitor this step for excessive increase in back-pressure in the hydroxyapatite column which leads to overflow and loss of material.

11. Close the outlet of the hydroxyapatite column, and carefully disconnect it from the ABT column.

12. Wash the hydroxyapatite column with 5 ml R4. A small amount of receptor is eluted in this washing step but it is heavily contaminated with other proteins.

13. Elute receptor sequentially with 1.5 ml R5 and 1.5 ml R6. Collect three 0.5 ml fractions for each eluate, and hold 10 µl aliquots for assay. Depending on the preparation, 30 to 70% of the eluted receptor is present in the R5 eluate. Although this eluate is usually more pure than that obtained with R6, biological properties of receptors in both eluates are identical.

14. Perform binding assay (see below) to measure receptor concentration. For fractions that contain less than 0.5 pmol receptor per microliter, concentrate eluates up to 10-fold by centrifugal ultrafiltration with Centricon 50 (Millipore). Measure receptor binding of the concentrated samples; receptor recovery after concentration is >80%.

15. Freeze single-use aliquots with liquid nitrogen, and store below −70°C. You will need approximately 10 pmol per reconstitution experiment. Biological activity is preserved for several years under these storage conditions provided that repeated freezing and thawing is avoided.

16. This purification procedure should yield ~200 pmol purified receptor (8 to 10% of the number of binding sites in the membrane preparation). M_1 receptor appears as a broad band at ~60 kDa in silver-stained polyacrylamide gels. Purity judged by this method is >90%. However, specific activity of the final preparation is usually only 10 to 15% of the theoretical value, which is ~17 nmol/mg protein. This is most likely due to receptor inactivation during chromatography. Rechromatography on ABT resin increases specific activity several-fold.

5.2.4.2 Receptor Quantitation

Receptor is quantified by radioligand binding assay using [³H]-quinuclidinyl benzylate methyl chloride ([³H]-QNB). Receptor-bound radioligand is separated from free radioligand and then measured by scintillation counting. Nonspecific binding is determined by addition of a high concentration of atropine. Separation of receptor-bound from free radioligand is achieved by different methods, depending on whether membrane or soluble preparations are assayed. For membranes, reaction mixtures are filtered through glass fiber disks which retain the membrane fragments but not free radioligand. After extensive washing, filter-associated radioactivity is determined. For soluble preparations, reaction mixtures are subjected to gel filtration. Receptor elutes with the void volume, whereas free radioligand elutes with the total volume. The protocol given here uses centrifugal gel filtration.[39] Alternatively, conventional gel filtration gives identical results.[38]

Protocol 5-3: Assay for Membrane-Bound M_1 Muscarinic Receptor

1. *Solutions*
   a. Receptor binding buffer (2 ml): 20 mM KPi buffer (pH 7.0), and 100 mM NaCl.
   b. QNB solution (1 ml): receptor binding buffer containing 2 nM [^{3}H]-QNB (New England Nuclear, 13 μM, 80 Ci/mmol or ~90 cpm/fmol at ~45% counting efficiency).
   c. QNB+atropine solution (0.5 ml): QNB solution containing 10 μM atropine sulfate.
   d. Washing solution (1 l): 20 mM Tris-Cl (pH 8.0), 10 mM MgCl$_2$, and 100 mM NaCl.
   e. 1% PEI (500 ml): polyethyleneimine (Sigma), 1% (w/v) solution in water.
2. Immerse eight glass fiber filters (Whatman, GF/F) in ~25 ml 1% PEI and incubate for 1 to 2 h at room temperature. This procedure greatly reduces nonspecific binding of radioligand to glass fiber filters.
3. Dilute 10 μl membranes prepared from M_1 recombinant baculovirus-infected Sf9 cells (see Section 5.2.3) with 990 μl receptor binding buffer and keep on ice.
4. Label eight polypropylene test tubes (12 × 75 mm). Add 90 μl QNB solution to tubes 1, 2, 5, and 6 and 90 μl QNB+atropine solution to tubes 3, 4, 7, and 8.
5. Start the reaction by adding diluted membranes to tubes 1 to 4 and receptor binding buffer to tubes 5 to 8.
6. Incubate at 30°C for 1 h and then transfer to an ice bath.
7. Filter reaction mixture through glass fiber filters in a vacuum filtration manifold (for example, Hoeffer SH225V). Wash the filters three times with 4 ml ice-cold washing solution.
8. Dry filters at room temperature, place them in scintillation vials, add 10 ml scintillant, and count for ^{3}H.
9. Values obtained from tubes 5 to 8 represent the blank, that is nonspecific binding of [^{3}H]-QNB to filters, and should be subtracted (usually less than 1% of total [^{3}H]-QNB). The difference between the values from tubes 1 and 2 and 3 and 4 represent specific binding of [^{3}H]-QNB to the membrane preparation. Because the dissociation constant of this radioligand to membrane-bound M_1 receptor is ~50 pM, this assay should measure ~100% of the receptors provided that radioligand is in excess. Check that less than 10% of total [^{3}H]-QNB in the reaction is bound to

the membranes. M_1 receptor expression in Sf9 cell membrane preparation is ~5 to 10 pmol/mg protein. Thus, specific binding should be ~700 to 1400 cpm under the conditions described here.

Protocol 5-4: Assay for Soluble M_1 Muscarinic Receptor

1. *Solutions*
   a. Receptor binding buffer (360 ml): 20 mM KPi buffer (pH 7.0), 100 mM NaCl, 0.1% (w/v) digitonin.
   b. QNB solution (1.2 ml): receptor binding buffer plus 40 nM [³H]-QNB.
   c. QNB+atropine solution (1.2 ml): QNB buffer plus 200 μM atropine sulfate.
   d. Samples:
      1. Extract (2 μl)
      2. ABT column flow-through (10 μl)
      3. ABT column wash (90 μl)
      For samples 4 and 5 atropine is removed by applying 100 μl to Sephadex G-25 columns. For sample 4 load 100 μl each onto four columns, combine the eluates and use 90 μl/reaction. For sample 5, load 100 μl onto one column and use 20 μl/reaction.
      4. ABT column eluate/hydroxyapatite column flow-through (90 μl)
      5. Hydroxyapatite column wash (20 μl)
      Samples 6 to 17 contain 1 mM carbachol which will be diluted to 10 μM or less in the reaction mixture. Because this concentration is more than 10-fold lower than its Kd for soluble receptors, and [³H]-QNB is present at 10 times its Kd, carbachol should not have any appreciable effect in the assay.
      6 to 11.   Hydroxyapatite column eluates (1 μl each)
      12 to 17. Hydroxyapatite column eluates after concentration (0.2 μl each)
2. Hydrate Sephadex G-25 medium (Pharmacia) with 3 bed volumes of water for 1 h at 90°C or >3 h at room temperature.
3. Rinse hydrated Sephadex on a sintered glass funnel with water. Resuspend with 1.5 bed volumes receptor binding buffer.
4. Place polypropylene columns (Mini Columns®, Isolab) in glass test tubes (16 × 100 mm).
5. Pour 4.5 ml gel slurry into each column. A 3 ml bed volume should be formed.
6. Place the columns at 4°C for at least 1 h before use.
7. For each receptor sample, prepare duplicate tubes for QNB and for QNB+atropine. Each tube should contain 30 μl of either QNB solution or QNB+atropine solution and the sample volumes indicated above in 120 μl total reaction volume. The extract and ABT column flow-through should be assayed the same day they are obtained because receptor binding activity declines rapidly with time, presumably due to proteolysis.

Approximately 50% of the binding sites are usually extracted from the membrane preparation.

8. Incubate at 30°C for 1 h then transfer to an ice bath.
9. Immediately before use, centrifuge up to 12 columns at $1400 \times g$ at 4°C in a swinging-bucket rotor for 5 min.
10. Transfer spun columns to scintillation vials (columns fit in Omni Vials®, Wheaton Scientific).
11. Carefully apply 100 µl reaction mixture on the top of the gel. This step should be completed in less than 3 min.
12. Thirty seconds after applying the last sample centrifuge columns at $1600 \times g$ as in step 9.
13. Remove columns from the vials. Add 2.5 ml scintillant and count for radioactivity.

5.2.5   PREPARATION OF $G_{q\alpha}$

5.2.5.1 $G_{q\alpha}$ Purification

The procedure described here is a modification of the method originally developed by Kazasa and Gilman,[34] eliminating one chromatography step while maintaining high purity.[18] Sf9 cells are co-infected with recombinant baculoviruses for $G_{q\alpha}$, hexahistidine tagged bovine $G_{\beta2}$ (His6G$_{\beta2}$), and hexahistidine tagged bovine $G_{\gamma2}$ (His6G$_{\gamma2}$). Trimeric $G_{q\alpha}$/His6G$_{\beta2}$/His6G$_{\gamma2}$, which is formed spontaneously in the Sf9 cells, is extracted with detergent and subjected to affinity chromatography on a Ni-NTA resin. After removing contaminant proteins and other G proteins by a variety of washing steps $G_{q\alpha}$ is dissociated from the resin-bound His6G$_{\beta2}$/His6G$_{\gamma2}$ complex with aluminum tetrafluoride (AlF$_4$). The AlF$_4$ concentration is subsequently lowered to an innocuous level by repeated dilution and concentration. This procedure can be done in one day (~12 working hours).

Protocol 5-5: Purification of $G_{q\alpha}$

1. *Solutions* (all should be ice-cold unless otherwise indicated):
   a. 2X extraction buffer (100 ml): 2% (w/v) sodium cholate, 40 mM Tris-Cl (pH 8.0), 200 mM NaCl, 10 mM 2-mercaptoethanol, 20 µM GDP (to stabilize G_q trimer), PI (protease inhibitors: 0.1 mM PMSF, 2 µg/ml aprotinin, 10 µg/ml leupeptin).
   b. 2X G1 (2 l): 40 mM Na-HEPES (pH 8.0), 200 mM NaCl, 2 mM MgCl$_2$, 10 mM 2-mercaptoethanol, PI.
   c. G2 (1.5 l): 1X G1 plus 0.5% C$_{12}$E$_{10}$ and 10 µM GDP.
   d. G3 (750 ml): G2 plus 0.5 M NaCl.
   e. G4 (750 ml): G2 plus 0.5 M NaCl and 10 mM imidazole (pH 8.0).
   f. G5 (100 ml): G2 plus 0.2% (w/v) sodium cholate and 3 µM GTPγS (lithium salt, Boehringer Mannheim).
   g. G6 (80 ml): 1X G1 plus 0.3% (w/v) sodium cholate.

    h. G7 (100 ml): 1X G1 plus 1% (w/v) sodium cholate, 30 μM AlCl$_3$, 10 mM NaF, and 50 mM total MgCl$_2$. *Important note:* to avoid precipitation, add while stirring in the following order: NaF, MgCl$_2$, AlCl$_3$.

    i. G8 (300 ml): 1X G1 plus 0.4% sodium cholate and 1 mM DTT (instead of 2-mercaptoethanol).

2. Mix Sf9 cell membranes from a 2 l culture with 2X extraction buffer to yield 5 mg/ml membrane protein and 1X extraction buffer final.

3. Stir on ice for 1 h at 4°C; hold 100 μl aliquot on ice for assay and SDS-PAGE.

4. Centrifuge at >100,000 × g for 1 h at 4°C (e.g., Beckman Ti45 rotor at 35,000 rpm).

5. Collect extraction supernatant; hold 100 μl aliquot on ice for assay and SDS-PAGE. Proceed immediately with chromatography.

6. Dilute extraction supernatant five-fold with G2 and mix with 12 ml Ni-NTA-agarose (Qiagen) preequilibrated with G2.

7. Stir with a glass/Teflon overhead stirrer (to avoid crushing the agarose beads) for 1 h at 4°C and collect in a chromatographic column (i.e. 2.5 × 10 cm).

8. Wash the resin sequentially at ~5 ml/min (or maximal gravity flow) with G2, G3, G4, and again with G2. Each step is deemed complete when the A$_{280}$ remains flat for at least 24 ml (2 column volumes).

9. Move the column to room temperature (or use a water-jacketed column).

10. Wash the column at 2 ml/min with 100 ml of ambient G5 to remove endogenous Sf9 G$_{\alpha i/o}$-like proteins. Some G$_{q\alpha}$ unavoidably elutes at this stage also.

11. Wash the column at 5 ml/min with 80 ml ice-cold G6 to remove residual GTPγS and C$_{12}$E$_{10}$. Some G$_{q\alpha}$ is lost during this step, so washing quickly with cold buffer is important.

12. Elute G$_{q\alpha}$ at 5 ml/min with 100 ml ambient G7, collecting the eluate on ice. AlF$_4$ will bind to G$_{q\alpha}$ and promote its dissociation from the G$_{\beta\gamma}$ dimer that remains attached to the resin via its hexahistidine tags.

13. Concentrate the eluate at 4°C to <10 ml by ultrafiltration through an Amicon PM10 membrane (Millipore) in a pressurized vessel with magnetic stirring (200 ml Stirred Cell®, Millipore). Dilute the concentrated eluate to 100 ml with cold G8.

14. Repeat the concentration/dilution procedure three times. These repeated cycles result in reduction of the AlF$_4$ concentration more than 1000-fold and therefore eliminate its activating effect on the G protein.

15. Concentrate finally to ~1 ml; stir 20 min without pressure to increase protein recovery, then transfer sample to a tube on ice. Rinse interior surfaces of the stirred cell and ultrafiltration membrane with ~300 μl of G8, stir several minutes and add this rinse to the concentrated G$_{q\alpha}$.

16. Save a 20 μl aliquot to measure the concentration of G$_{q\alpha}$ based on bound GDP determination (see below).

17. Add GDP (10 μM final concentration to stabilize $G_{q\alpha}$ activity), aliquot as needed, freeze with liquid nitrogen, and store below $-70°C$.

5.2.5.2 $G_{q\alpha}$ Quantitation

The procedure described above should yield between 400 and 600 μg $G_{q\alpha}$ protein. The purity of the final preparation is tested by the following criteria:

1. Silver staining of a 10 or 12% polyacrylamide gel to reveal general protein contamination ($G_{q\alpha}$ should appear as a 42- to 43-kDa band which accounts for >95% of the staining in that lane).
2. Immunoblot with antibody P-960,[40] which does not recognize either Sf9 cell $G_{q\alpha}$ or expressed $G_{q\alpha}$, but detects other G protein α subunits. Sf9 cell $G_{\alpha i/o}$-like proteins should be detected up to the G5 wash (perhaps also some in G6).
3. Alternatively, one can use pertussis-toxin catalyzed ADP-ribosylation to test for contamination with $G_{\alpha i/o}$ class of G proteins. Protocols for this assay can be found elsewhere.[41,42] $G_{\alpha i/o}$-like proteins should be detected in the G5 wash but not in the final eluate.
4. Since Sf9 cell $G_{q\alpha}$ and mouse $G_{q\alpha}$ migrate identically on SDS-PAGE and since we do not know of any antibodies that recognize only Sf9 cell $G_{q\alpha}$, making the assessment of contamination by this G protein is problematic.[43,44] However, in our experience, using this same procedure to purify truncated $G_{q\alpha}$ that migrated apart from full-length Sf9 cell $G_{q\alpha}$, we find that Sf9 cell $G_{q\alpha}$ comprises less than 3% of the purified $G_{q\alpha}$ protein.[45]

Identification of recombinant $G_{q\alpha}$ can be made by immunoblot with antibody Z811, W082, or B825 and by G_q-dependent activation of PLC-β1 (see below).

1. Z811[43,44] recognizes a common 15 amino acid sequence at the C-terminus of both $G_{q\alpha}$ and $G_{11\alpha}$. Calbiochem carries a similar antibody (10 amino acids of the C-terminus, catalog no. 371751-S).
2. W082[46] recognizes amino acids 115–133 which are present in $G_{q\alpha}$ but not in $G_{11\alpha}$ and can be purchased from Calbiochem (no. 371752-S).
3. B825[43,44] recognizes mammalian $G_{q\alpha}$ but not Sf9 cell $G_{q\alpha}$.

The specific activity of the preparation is determined by estimating the proportion of active $G_{q\alpha}$ as a fraction of total protein. Active $G_{q\alpha}$ is estimated by measuring the GDP content of the preparation as explained below. Typical specific activity is 10 to 17 nmol/mg protein which represents 40 to 70% of the theoretical value (24 nmol/mg).

The standard method to measure the amount of active purified G protein α subunits is Mg^{2+}-stimulated [^{35}S]-GTPγS binding assay.[47] Unfortunately purified $G_{q\alpha}$ has unusual guanine nucleotide binding properties which makes it impractical to carry out this assay. Therefore, the proportion of functional $G_{q\alpha}$ in the final preparation is determined by measuring protein-bound GDP.

Protocol 5-6: Measurement of Protein-Bound GDP

This GDP determination assay is based on the observation that GDP binds stoichiometrically to G protein α subunits.[48,49] $G_{q\alpha}$ is heat-denatured to release bound GDP, which is then used as a competitive inhibitor of [^{35}S]-GTPγS binding to $G_{i\alpha1}$. By comparing the inhibition produced by GDP in the boiled $G_{q\alpha}$ sample to the inhibition by known concentrations of GDP, the concentration of active $G_{q\alpha}$ can be estimated.

1. *Solutions*
   a. Binding buffer (600 μl): 140 mM NaHEPES (pH 8.0), 37 mM EDTA, 2 mM DTT, 2 nM [^{35}S]- GTPγS (1250 Ci/mmol, 10 μM, New England Nuclear), 0.1% (w/v) $C_{12}E_{10}$.
   b. Quench solution (3 ml): 20 mM NaHEPES (pH 8.0), 100 mM NaCl, 10 mM MgCl$_2$, 1 mM GTP, 1 mM DTT.
   c. G8: see composition in Protocol 5-5.
   d. GDP standards: 1, 2, 4, 8, 12, 24, 40, 80, 120, 400, 1200 nM GDP diluted from 0.1 mM stock GDP into buffer G8.
   e. Washing solution (1 l): 20 mM Tris-Cl (pH 8.0), 10 mM MgCl$_2$, and 100 mM NaCl. Transfer to a bottle furnished with a bottle-top dispenser. Keep on an ice bath.
   f. $G_{i\alpha1}$ buffer (3 ml): 20 mM NaHEPES (pH 8.0), 1 mM EDTA, 1 mM DTT, and 0.2% (w/v) $C_{12}E_{10}$.
   g. $G_{i\alpha1}$ solution (1 ml): 2 nM solution of rat $G_{i\alpha1}$ in $G_{i\alpha1}$ buffer. $G_{i\alpha1}$ can be prepared from Sf9 cells infected with recombinant baculoviruses encoding for this protein (without hexahistidine tag), bovine G_β, and His6G_γ. The protocol is similar to that described for $G_{q\alpha}$ preparation. Alternatively, this protein can be purchased from Calbiochem (catalog no. 371793).
2. Dilute final $G_{q\alpha}$ preparation 100-, 300-, and 1000-fold with buffer G8. Incubate 10 μl samples in duplicate at 100°C for 5 min in microcentrifuge tubes. Place the tubes in an ice bath for 5 min, then centrifuge them for 5 min in a microcentrifuge at maximal speed. Transfer the supernatants to new tubes; these are the samples to be tested.
3. Soak 50 nitrocellulose filters (BA85, Schleicher and Schuell) in 50 ml washing solution.
4. Add 10 μl binding buffer to 46 polypropylene test tubes (12 × 75 nm).
5. Add 10 μl buffer G8 to tubes 1 to 10, 10 μl GDP standards to tubes 11 to 40 (10 standards in duplicate), and 10 μl samples from step 2 in tubes 41 to 46 (three samples in duplicate).
6. Add 20 μl $G_{i\alpha1}$ buffer to tubes 1 to 4, and $G_{i\alpha1}$ solution to tubes 5 to 46, and incubate at 30°C for 30 min.
7. Add 60 μl ice-cold quench solution, and transfer to an ice bath.
8. Filter quenched reaction mixtures through nitrocellulose filters in a standard filtration manifold, for example Hoeffer SH225V, which holds 10 filters. Filter reaction mixtures one by one by adding 2 ml ice-cold washing solution in the reaction tube and then pouring the contents of the tube on

the filter under vacuum. This procedure ensures a relatively even distribution of the reaction mixture on the surface of the filter and thus avoids
losses of G protein binding to the filter due to high local concentration
of $C_{12}E_{10}$. Rinse the filters with at least 6 ml washing solution.

9. Transfer filters to scintillation vials, add 10 ml scintillation fluid, shake
well, and count for [35]S.

10. Tubes 1 to 4 are blanks (that is they measure nonspecific binding of the
radionucleotide to the filters) and their average value should be subtracted
from each of the other determinations. Use the average value from tubes
5 to 10 to determine the binding of ^{35}S-GTPγS to $G_{i\alpha1}$ in the absence of
competing nucleotide. Use the values from tubes 5 to 40 to construct a
calibration curve of ^{35}S-GTPγS binding as a function of increasing
concentrations of GDP. This curve is plotted in a semilog scale and can
be fitted to a single-site inhibition model assuming a Hill coefficient =
1 ($B_i = B_0/(1 + IC_{50})$, where B_i is the binding in the presence of a concentration of GDP = i, B_0 is the binding in the absence of GDP, and IC_{50}
is the concentration of GDP at which $B_i = 0.5 B_0$). IC_{50} for GDP is ~10 nM,
which allows the calibration curve to be used over a range of 2.5 to 25 nM
GDP.

Protocol 5-7: Assay of PLC-β1 Activation by Purified $G_{q\alpha}$

This assay is based on the ability of GTPγS-bound $G_{q\alpha}$ to stimulate PLC-β1-
catalyzed ^{3}H]-phosphatidylinositol 4,5-bisphosphate (^{3}H]-PIP$_2$) hydrolysis into
diacylglycerol and ^{3}H]-inositol triphosphate (^{3}H]-IP$_3$).[16,18,50] The reaction is terminated by addition of acid, which also precipitates the substrate. [3]H]-IP$_3$ in the
supernatant is quantitated by scintillation counting.

1. *Solutions*
   a. 100 mM Ca-EGTA buffer (500 ml): mix 0.05 mol EGTA (free acid)
   with 20 ml 5 N NaOH, then bring up to ~400 ml with water. Add 45 ml
   1 M CaCl$_2$ and check pH. Add NaOH (dropwise) until pH increases
   to 7.40. Add 500 μl of 1 M CaCl$_2$, note pH. Add NaOH until pH reaches
   7.40 and note the volume added. Repeat dropwise addition of NaOH
   and 500 μl of CaCl$_2$ until the Ca^{2+}-induced drop in pH decreases 5 to
   20%. Bring pH back to 7.40 with NaOH then bring volume up to 500
   ml with water.
   b. 100 mM Na-EGTA buffer (500 ml): mix 0.05 mol EGTA (free acid)
   with 20 ml 5 N NaOH, then bring volume up to ~450 ml with water;
   bring pH to 7.4 with either NaOH or HCl and volume up to 500 ml
   with water.
   c. Lipid stock solutions: bovine liver phosphatidylethanolamine (PE) and
   bovine brain phosphatidylserine (PS) are from Avanti Polar Lipids (10
   mg/ml chloroform solutions, ~15 mM and ~13 mM, respectively). [3]H]-
   PIP$_2$ (NEN, 1 μM, 10 Ci/mmol or ~10,000 cpm/pmol assuming ~45%

counting efficiency). Phosphatidylinositol 4,5-bisphosphate (PIP_2, Sigma or Boehringer Mannheim) is dissolved in its shipping vial with enough 2:1 chloroform:methanol to yield ~2 mM calculated concentration of PIP_2. Mix gently and allow it to sit at room temperature for ~30 min. If it has not gone into solution completely, add 1 drop of diethylamine and mix. Transfer to siliconized vials (#03-339-23B, Fisher), prefilled with Ar or N_2, and seal quickly with screw-cap Teflon-lined caps (#02-88303A, Fisher). Use of these vials and caps are highly recommended. Store at –20°C. Determine the PIP_2 concentration by phosphorus assay (see Protocol 5-17).

d. Lipid rehydration buffer (3 ml): 80 mM NaHEPES (pH 7.5), 160 mM NaCl, and 0.2 mM Na-EGTA.

e. Calcium buffer (2 ml): 4.93 mM Ca-EGTA buffer, 10.7 mM $MgCl_2$, 1.40 mM Na-EGTA buffer.

f. G_q dilution buffer (10 ml): 55.5 mM NaHEPES (pH 7.5), 111 mM NaCl, 1.11 mM DTT, 0.11 mg/ml BSA, 4.44 mM $MgCl_2$, 1.11% sodium cholate.

g. G_q activation buffer (150 µl): add 30 µl of 10 mM GTPγS (Boehringer Mannheim, lithium salt) to 120 µl of G_q dilution buffer.

h. PLC dilution buffer (3 ml): 50 mM NaHEPES (pH 7.5), 100 mM NaCl, 1 mM DTT, 0.1 mg/ml BSA.

i. PLC-β1 (750 µl): membrane preparation from Sf9 cells infected with 6HisPLC-β1 recombinant baculovirus as described in protocol 5-1. For membranes at 15 mg/ml, dilute 10 µl with 740 µl PLC dilution buffer just before assay and store on ice. Usually 1 to 2 µg membranes per assay tube is adequate, but it is advisable to perform a preliminary G_q-dependent PLC assay at several concentrations of PLC membranes. Choose the concentration that gives very low Ca^{2+}-stimulated signal but strong G_q-stimulated activity, where less than 15% of the total cpm are hydrolyzed during the 10 min assay.

j. TCA: 10% (w/v) stock. Store at 4°C.

k. BSA: 3.3 mg/ml BSA stock. Store at 4°C.

2. The assay is conducted by mixing 50 µl of substrate vesicles with 30 µl calcium buffer and 10 µl of activated $G_{q\alpha}$; 10 µl PLC-β is added to initiate the assay and the tubes are transferred to 30°C for 10 min. The reaction is quenched with acid, carrier BSA is added, and [^{3}H]-IP_3 formed is counted from the supernatant. The following protocol is written for 20 tubes.

3. Mix lipid stock solutions in a siliconized glass test tube (Pyrex, 13×100 mm): 480 pmol PE (32 µl of 15 mM stock), 120 pmol PS (9.2 µl of 13 mM stock), 120 pmol PIP_2 (60 µl of 2.0 mM stock), and 24 µl [^{3}H]-PIP_2 (10,000 cpm per assay tube). Dispense lipids in the order listed using a Hamilton syringe. After dispensing [^{3}H]-PIP_2 rinse the syringe barrel with a chloroform:methanol (2:1) solution into the test tube to ensure that most of the PIP_2 and [^{3}H]-PIP_2 are delivered.

4. Dry the mixture with a gentle stream of Ar or N_2 such that a uniform film is formed in the bottom of the tube. Exert care to avoid splashing lipid solution on the walls of the tube.

5. Add 1.2 ml lipid rehydration buffer. Vortex briefly, seal the tube with several layers of parafilm, and sonicate to clarity at room temperature (it usually takes 15 to 20 min) to form lipid vesicles containing the substrate. Keep on ice.

6. Add 720 μl calcium buffer to the lipid vesicles and mix well by vortexing. This is the substrate solution.

7. Dilute $G_{q\alpha}$ samples with G_q dilution buffer as follows, then mix 1:1 with G_q activation buffer. Because $C_{12}E_{10}$ strongly inhibits PLC activity only the fractions indicated below can be assayed. If more than 15% of the total radioactivity is hydrolyzed, repeat assay and increase the dilution of G_q samples as needed.

	Sample (μl)	G_q dilution buffer volume (μl)
Membranes	1	49
Extract	3	47
Eluate (Protocol 5-8, step 10; Ni-NTA chrom.)	1	35 (and add 14 μl 1% (w/v) sodium cholate)
Concentrate (Protocol 5-8, step 12)	1	5,000
No-G_q control	0	50

Add 25 μl of diluted G_q sample to 25 μl of G_q activation buffer. Incubate at 30°C for 1 h then keep activated sample on ice until assay. The concentrations here are 1% cholate, 1 mM GTPγS, 4.0 mM $MgCl_2$.

8. Label 20 polypropylene tubes (12 × 75 mm). Add 80 μl substrate solution to each tube on ice. Add 10 μl activated No-Gq control sample to tubes 1 to 4, membrane sample to tubes 5 to 8, extract sample to tubes 9 to 12, elution sample to tubes 13 to 16, and concentrate sample to tubes 17 to 20.

9. Start the reaction by adding 10 μl PLC-β1 dilution buffer or PLC-β1 to tubes, vortexing, and transferring to a 30°C water bath for 10 min. Add dilution buffer to the first pair in each set (i.e. 1, 2, 5, 6, etc.) and PLC-β1 to the others (i.e. 3, 4, 7, 8, etc.).

10. Terminate reaction by adding 200 μl ice-cold 10% TCA and returning samples to an ice bath.

11. Add 300 μl 3.3 mg/ml BSA to facilitate [³H]-PIP_2 precipitation.

12. Prepare a mock tube with 200 μl 10% TCA, 300 μl BSA, 10 μl G_q dilution buffer, 50 μl lipid rehydration buffer, 30 μl calcium buffer, and 10 μl PLC dilution buffer.

13. Centrifuge at 1300 × g for 15 min at 4°C. A white compact pellet will form.

14. Transfer 450 μl of the supernatant, which contains [³H]-IP_3, into scintillation vials, add 10 ml scintillant, and count for [³H]. Remember to adjust

for counting only 75% of the total supernatant in your calculations. To determine the specific activity, you have to account for the quenching effect of the acid supernatant. Therefore, count an aliquot of lipid substrate vesicles in 450 µl of mock tube supernatant (described above) and divide by the moles of PIP_2.

15. Final concentrations during the assay are 50 mM NaHEPES (pH 7.5), 100 mM NaCl, 0.2 mM DTT, 200 nM free Ca^{2+}, 3.6 mM $MgCl_2$, 100 µM GTPγS, 0.1% sodium cholate, 0.02 mg/ml BSA, 200 µM PE, 50 µM PS, and 50 µM [^{3}H]-PIP_2.

16. Tubes 1 and 2 represent the blank to be subtracted ([^{3}H]-IP_3 produced in the absence of both G_q and PLC). This value should be less than 0.5% of the total amount of [^{3}H]-PIP_2. Tubes 3 and 4 represent PLC activity in the absence of $G_{q\alpha}$. Because of the low free Ca^{2+} concentration in the assay PLC is submaximally stimulated. This value should not exceed 3% of the total amount of [^{3}H]-PIP_2 to allow for a suitable G_q-stimulated signal. The final $G_{q\alpha}$ purified preparation will substantially stimulate PLC activity over this value, but should hydrolyze less than 15% of the total [^{3}H]-PIP_2 in the assay.

This basic protocol can be modified easily to determine the potency of GTPγS-bound $G_{q\alpha}$ to stimulate PLC-β1 activity, as well as its modulation by other factors.

5.2.6 PREPARATION OF $G_{\beta\gamma}$

5.2.6.1 $G_{\beta\gamma}$ Purification

The $G_{\beta 1\gamma 2}$ subunit complex is purified exactly as described by Kazasa and Gilman.[34] Sf9 cells are infected with recombinant baculoviruses for hexahistidine-tagged rat $G_{i\alpha 1}$, bovine $G_{\beta 1}$, and bovine $G_{\gamma 2}$. Trimeric $His6G_{i\alpha 1}/G_{\beta 1}/G_{\gamma 2}$, which is spontaneously formed in Sf9 cells, is extracted with detergent and subjected to affinity chromatography on Ni-NTA resin. $G_{\beta\gamma}$ is dissociated from resin-bound $His6G_{i\alpha 1}$ with AlF_4 and further purified by ion-exchange chromatography on a MonoQ column. These two chromatographies can be completed in one day (~12 working hours).

Protocol 5-8: Purification of $G_{\beta\gamma}$

1. *Solutions* (all should be ice-cold unless otherwise indicated)
   a. G2 (1.2 l): see Protocol 5-5.
   b. B1 (300 ml): G2 plus 300 mM NaCl and 5 mM imidazole (pH 8.0).
   c. B2 (50 ml): 20 mM NaHEPES (pH 8.0), 50 mM NaCl, 10 mM 2-mercaptoethanol, 10 µM GDP, 1% (w/v) sodium cholate, 30 µM $AlCl_3$, 10 mM NaF, 5 mM imidazole (pH 8.0), 50 mM $MgCl_2$, PI (protease inhibitors: 0.1 mM PMSF, 10 µg/ml leupeptin, 2 µg/ml aprotinin).
   d. B3 (250 ml): 20 mM NaHEPES (pH 8.0), 1 mM EDTA, 3 mM DTT, 3 mM $MgCl_2$, 0.7% (w/v) CHAPS ({3-[(3-cholamido propyl) dimethylammonio]-1-propane sulfonate}, Sigma) PI.

     e. B4 (150 ml): B3 plus 1 M NaCl.

2. $G_{\beta\gamma}$ is extracted from the membrane preparation exactly as described for $G_{q\alpha}$ (Protocol 5-5). Proceed immediately with chromatography.

3. Dilute the extract five-fold with G2 and apply it at 5 ml/min to a 4 ml $(1.5 \times 2.3$ cm) Ni-NTA column pre-equilibrated with G2.

4. Wash the column with 300 ml G2.

5. Wash the column with 100 ml B1 until A_{280} returns to baseline.

6. Move the column to room temperature (or use a water-jacketed column).

7. Elute $G_{\beta\gamma}$ complex at 2 ml/min with 40 ml (10 column volumes) of B2.

8. Collect the A_{280} peak fractions.

9. Dilute Ni-NTA eluate threefold with B3 and apply it at 2 ml/min to a MonoQ HR 5/5 column (Pharmacia) pre-equilibrated with B3.

10. Elute the column with a 20 ml linear gradient of 0% to 40% buffer B4 (0 to 400 mM NaCl), collecting 0.5 ml fractions. $G_{\beta\gamma}$ eluates at ~180 mM NaCl in 4 ml.

11. Concentrate the eluate to ~0.5 ml by ultrafiltration through Centricon 10 (Amicon).

12. Dilute the concentrated eluate with B3 to a volume that will give 100 mM final NaCl.

5.2.6.2 $G_{\beta\gamma}$ Quantitation

This procedure should yield ~350 µg/l purified $G_{\beta\gamma}$ subunit. Purity of the preparation is determined by silver staining of polyacrylamide gels to estimate general protein contaminants and by standard [^{35}S]-GTPγS binding assay to test for contaminating G_{α} subunits. The latter is carried out by a procedure essentially identical as described above for determination of protein-bound GDP (Protocol 5-6), except that 100 nM [^{35}S]-GTPγS is used and no competing GDP is added. Total protein concentration is determined by amido black protein assay (see Section 5.4.1). For detection of G_{β} by silver staining, it is usually adequate to load 1 µl of membranes and Ni-NTA flow-through, 5 µl of each wash, 20 µl of the Ni-NTA eluate, and 3 µl of the MonoQ peak elution fractions on a 10 or 12% mini-gel. However, G_{γ} will not be detectable unless a 10 to 20% acrylamide, 4 M urea gel system is used.[51]

5.2.7   PREPARATION OF PLC-β1

5.2.7.1   PLC-β1 Purification

Because it is a peripheral membrane protein, His6PLC-β1 produced in Sf9 cells can be extracted with a high-salt buffer.[52] The extract is subjected to a three-step chromatographic procedure on Ni-NTA, MonoQ, and MonoS resins.[18] Highly purified His6PLC-β1 is obtained in reasonable yields. These three chromatographies can be performed in three days. Higher yield may be obtained by omitting the MonoQ and MonoS chromatographies. Although purity is sacrificed somewhat, this may be adequate if no contaminating GTPase activities are detected.[45]

Protocol 5-9: Purification of PLC-β1

1. *Solutions* (all should be ice-cold unless otherwise indicated)
   a. 2X high-salt buffer (150 ml): 40 mM Tris-Cl (pH 7.5), 2 M NaCl, 10 mM 2-mercaptoethanol, PI (5 μM calpain inhibitor I and II, 20 μg/ml leupeptin, 4 μg/ml aprotinin, 0.2 mM PMSF).
   b. 2X P1 (1 l): 40 mM Tris-Cl, pH 7.5, 10 mM 2-mercaptoethanol, 40% glycerol, PI.
   c. P2 (500 ml): 1X P1 plus 1 M NaCl.
   d. P3 (500 ml): 1X P1 plus 1 M NaCl and 10 mM imidazole (pH 8.0).
   e. P4 (250 ml): 1X P1 plus 100 mM NaCl and 10 mM imidazole (pH 8.0).
   f. P5 (250 ml): 1X P1 plus 100 mM NaCl and 50 mM EDTA (pH 8.0).
   g. Q1 (250 ml): 20 mM Tris-Cl (pH 7.5), 1 mM DTT, 20% glycerol, PI.
   h. S1 (250 ml): 20 mM 4-morpholineethanesulfonic acid (MES, pH 6.0), 1 mM DTT, 1 mM EDTA, 20% glycerol, PI.
   i. Q2 (250 ml): Q1 plus 1 M NaCl.
   j. S2 (250 ml): S1 plus 1 M NaCl.
2. Mix membranes from 2 l Sf9 cell culture with cold 2X high-salt buffer such that the membrane protein concentration in the mixture is 5 mg/ml and the buffer is reduced to 1X.
3. Stir for 1 h at 4°C.
4. Centrifuge at 100,000 × g for 1 h, hold a 100 μl aliquot for SDS-PAGE and activity assay. Proceed immediately with chromatography.
5. Stir the extract for 1 h at 4°C with 8 ml Ni-NTA agarose pre-equilibrated with P2 using a glass/Teflon overhead stirrer.
6. Pour the resin into a chromatographic column.
7. Sequentially wash the column with the following solutions at 1 ml/min: P2 until A_{280} returns to baseline; a 12 ml gradient of P2 to P3; 30 ml P3; a 12 ml gradient of P3 to P4; and 30 ml P4.
8. Elute PLC-β1 with 30 ml P5.
9. Load the Ni-NTA eluate onto a MonoQ column (HR5/5, Pharmacia) that has been equilibrated with 10% Q2 (100 mM NaCl) at 0.5 ml/min.
10. Wash the column with 10% Q2 until A_{280} returns to baseline.
11. Elute PLC-β1 with a 15 ml linear gradient of 10% to 60% Q2 (100 mM to 600 mM NaCl). PLC-β1 elutes between 200 mM and 300 mM NaCl in four 1 ml fractions.
12. Dilute pooled fractions from MonoQ chromatography 10-fold with S1. Load onto a MonoS column (HR 5/5, Pharmacia) that has been equilibrated with 1% S2 (10 mM NaCl) at 0.5 ml/min.
13. Wash the column sequentially with 5 ml of 1% S2 and then with 20% S2 (200 mM NaCl) until A_{280} returns to baseline
14. Elute PLC-β1 with a 20 ml linear gradient of 20% to 70% S2 (200 mM to 700 mM NaCl). PLC-β1 elutes between 350 and 450 mM NaCl in four 1 ml fractions.

15. Pool the active fractions, concentrate them with a Centricon-30 (Amicon) to at least 1 mg/ml protein (~250 μl). Aliquot as necessary, freeze with liquid nitrogen, and store the protein below −70°C.

5.2.7.2 PLC-β1 Quantitation

This procedure yields 500 to 800 μg PLC-β1 per 1 Sf9 cells that is >98% pure by silver stained SDS-PAGE. Occasionally, bands at ~100 and ~50 kDa are detected.[53] These are proteolytic products that may be minimized by inclusion of calpain inhibitors I and II and/or calpeptin (Calbiochem, catalog no. 208719, 208721, 03-34-0051).

Protocol 5-10: Assay of Purified PLC-β1 Activation by $G_{q\alpha}$

This assay is conducted as described in Protocol 5-7, for testing $G_{q\alpha}$ activity during purification, except that Sf9 cell membranes expressing G_q are used to test PLC activity during purification of PLC-β1.

1. G_q (1 ml): dilute 14 μl G_q membranes (~15 mg/ml) with 986 μl G_q dilution buffer. This results in 1 μg per assay tube, which is usually adequate. One to 5 nM purified $G_{q\alpha}$ can be used. Whether membranes or purified $G_{q\alpha}$ is used, we recommend that a preliminary assay be conducted to ensure that an appropriate G_q concentration is added. Check several dilutions of G_q and choose one that allows significant PLC stimulation, but does not result in hydrolysis of more than 15% of total [^{3}H]-PIP$_2$ during the 10 min assay.
2. PLC samples: to follow activity over the course of purification, test the membranes, extract, Ni-NTA eluate, MonoQ peak elution fractions, MonoS peak elution fractions, and the concentrated PLC.

	Sample (μl)	PLC dilution buffer volume (μl)
Membranes	1	750
Extract	1	500
Ni-NTA eluate	1	500
MonoQ fractions	1	3,000
MonoS fractions	1	3,000
Concentrated PLC (if 1 mg/ml)	1	10,000
No-PLC control	0	100

3. Activate G_q (300 μl, see Protocol 5-7).
4. Prepare "mock G_q," a 1:1 mix of G_q dilution buffer + G_q activation buffer (300 μl).
5. Assay: pipette 80 μl substrate solution into tubes on ice. Add 10 μl of either mock G_q or activated G_q to each tube. Mock G_q is added to the first pair of each set (i.e. 1, 2, 5, 6) and activated G_q to the others (i.e. 3, 4, 7, 8). Add 10 μl of no-PLC control to the first four tubes, then 10 μl of each PLC sample to the rest.

6. The first two tubes have no protein and the average of these values should be deducted from all other samples. Tubes 3 and 4 have no PLC and account for the presence of any PLC activity in the G_q membranes. This value should be very low. The remaining sets show the amount of G_q-stimulated activity induced by the PLC present in the purification sample. As before, limit the amount of PLC added so that no more than 15% of the total [^{3}H]-PIP$_2$ is hydrolyzed. The specific activity is ~25 pmol [^{3}H]-PIP$_2$ hydrolyzed/min/ng PLC-β1.

5.3   RECONSTITUTION OF RECEPTOR-PROMOTED SIGNALING

This method models the signal transduction from receptor to effector that occurs in cells.[16–19,45] That is, addition of receptor agonist to membrane bound M_1 and G_q activates PLC-β1 when GTP, Mg^{2+}, Ca^{2+}, and PIP$_2$ are present. This reconstitution system displays appropriate dependencies on agonist, GTP, Ca^{2+}, and PIP$_2$ as well as selectivity for M_1 receptor and G_q (but not M_2 receptor or G_s, for example). The general strategy presented here should be applicable to activation of other effectors, although this has not yet been demonstrated.

5.3.1   PREPARATION OF LIPID VESICLES THAT CONTAIN RECEPTOR AND G PROTEIN

Receptor and G protein are co-reconstituted into lipid vesicles by gel filtering a mixture of phospholipid-detergent micelles and purified proteins.[54,55] The following protocol describes the preparation of [^{3}H]-PIP$_2$-containing vesicles which are amenable for assaying PLC activity upon receptor stimulation (see below).[18] For assays that do not require the measurement of PLC activity, simply omit the additions of PIP$_2$ and [^{3}H]-PIP$_2$.

Protocol 5-11: Preparation of Lipid Vesicles that Contain M$_1$ Muscarinic Receptor and G Protein

1. *Solutions*
    a. Lipid stock solutions: These solutions are described in Protocol 5-7, step 1c. An additional lipid used here is cholesteryl hemisuccinate (CHS, tris salt, Sigma) which is dissolved with chloroform at 1.8 mM.
    b. Deoxycholate stock solution: (sodium salt, Calbiochem, Ultrol Grade), dissolve to 10% (w/v).
    c. Lipid rehydration buffer: 20 mM NaHEPES (pH 7.5), 100 mM NaCl, 1 mM EGTA (pH 8.0), and 0.3% deoxycholate (pH 8.0).
    d. Reconstitution column buffer: 20 mM NaHEPES (pH 7.5), 2 mM MgCl$_2$, 1 mM EGTA (pH 8.0), and 100 mM NaCl.
    e. Proteins: G_q: 35 pmol $G_{q\alpha}$ plus 70 pmol $G_{\beta\gamma}$; prepare trimeric G_q stocks and freeze in single-use aliquots at –70°C. M_1 receptor: 10 pmol.

2. Mix lipid stock solutions in a siliconized glass tube (Pyrex brand, 12×75 mm) as follows: 165 nmol PE, 98 nmol PS, 18 nmol CHS, 10 nmol PIP_2, and 150 µl $[^3H]$-PIP_2 (to achieve ~150 cpm/pmol PIP_2).

3. Dry the mixture with a gentle stream of Ar or N_2 such that a film forms at the bottom of the tube. Exert care to avoid splashing the lipid solution on the walls of the tube.

4. Rehydrate the lipid mixture with 0.1 ml lipid rehydration buffer.

5. Blow a gentle stream of Ar or N_2 inside the tube for approximately 20 s, seal the tube with parafilm, and sonicate for 20 min. Hold two 1 µl aliquots for counting.

6. Centrifuge the cloudy sonicated suspension at $1300 \times g$ for 10 min at 4°C, and transfer the clear supernatant to a new tube.

7. Count two 1 µl aliquots to determine the recovery of $[^3H]$-PIP_2 (which should be >90%). If less than 90%, combine supernatant and pellet, and repeat steps 5 and 6.

8. From this step forward all operations are carried out at 4°C. Mix 75 µl lipid-detergent micelles with G_q by brief vortexing.

9. Hold on ice for 15 min.

10. Add purified M_1 receptor and vortex briefly.

11. Add reconstitution column buffer such that the final volume is 150 µl.

12. Apply the mixture to a 7.5-ml bed volume AcA34 (Sepracor) column (see Section 5.4.3).

13. Collect 9-drop fractions. Protein-lipid vesicles are formed as detergent is removed during the gel filtration process and are recovered in the void volume. Save fractions spanning the void volume (1 ml), determined previously during column calibration (see Section 5.4.3).

14. Inspect the fractions containing vesicles as follows: (1) the meniscus should be flat; (2) the volume should be between 220 and 250 µl; (3) no bubbles should form upon light flicking of the tube contents. All of these criteria are consistent with detergent removal.

15. Pool vesicle-containing fractions (usually 0.75 to 1 ml). Count an aliquot to determine the recovery of $[^3H]$-PIP_2 which is usually 30 to 50%. Save 20 µl for GDP determination. Add BSA to 0.1 mg/ml final and either use immediately or flash-freeze with liquid N_2 and store below −70°C. The activity of receptor and G protein is maintained after a single freeze and thaw.

16. Wash the gel filtration column with at least 20 column volumes of reconstitution column buffer to ensure that all detergent is eluted prior to next use.

5.3.2 QUANTITATION OF VESICLE PROTEINS

5.3.2.1 Quantitation of Receptor

Receptor content in protein/lipid vesicles is quantified by determining $[^3H]$-QNB binding to the vesicles, which is measured by two different methods depending on whether or not vesicles contain $[^3H]$-PIP_2. If vesicles do not contain $[^3H]$-PIP_2 the assay is performed exactly as described for measuring binding to membranes in

Protocol 5-3, with the following exceptions: (1) for receptor binding buffer, use reconstitution column buffer supplemented with 0.1 mg/ml BSA; and (2) test 10 μl of vesicles instead of diluted membranes. Receptor recovery is typically 10%. Thus, expected specific binding is ~1000 cpm or ~11 fmol.

Protocol 5-12: Quantitation of M_1 Muscarinic Receptor in Vesicles that contain $[^3H]$-PIP_2[18]

1. *Materials*
   a. Protein/lipid vesicles prepared as described above.
   b. Sephadex G-25 spin columns (see Protocol 5-4).
   c. QNB solution: 2 μM $[^3H]$-QNB in 50% (v/v) ethanol.
   d. Atropine solution: 1 mM atropine sulfate.
   e. Digitonin solution: 5% (w/v).
   f. QNB dilution buffer (1 ml): 20 mM NaHEPES (pH 7.5), 2 mM $MgCl_2$, 1 mM EGTA, 100 mM NaCl, 0.1 mg/ml BSA, and 2.5 nM $[^3H]$-QNB. Keep on ice.
   g. Atropine dilution buffer: 247 μl QNB dilution buffer plus 2.5 μl 1 mM atropine sulfate. Keep on ice.
   h. TLC plates: polyester-backed silica gel G thin layer chromatography plate (Whatman). Wash plate with 100% methanol. Store the plate in vacuum desiccator.
   i. TLC solvent: ethanol:acetic acid:water, 60:30:10.
2. Mix 40 μl protein/lipid vesicles with 1 μl QNB solution (Q sample), and 20 μl protein/lipid vesicles with 0.5 μl QNB solution and 0.5 μl atropine solution (A sample) in polypropylene (12 × 75 mm) test tubes.
3. Incubate at 30°C for 1 h.
4. Transfer the tubes to an ice bath, and incubate for 10 min.
5. Add 10 μl digitonin stock solution to the Q sample and 5 μl to the A sample. Vortex for about 10 s and place on ice for at least 10 min to solubilize the vesicles.
6. Add 449 μl ice-cold QNB dilution buffer to the Q sample and 224 μl atropine dilution buffer to the A sample.
7. Carry out centrifugal gel filtration (see Protocol 5-4), applying 100 μl of the Q sample to each of four columns and 100 μl of the A sample to each of two columns.
8. Freeze-dry column eluates by speed-vacuum. Resuspend the samples with 30 μl 70% (v/v) ethanol, washing down the sides of the tube thoroughly.
9. Evaporate solvent and resuspend in 10 μl 70% ethanol.
10. Mark 1.25 cm increments with a pencil across the TLC plate horizontally at a height of 2 cm and 14 cm from the bottom.
11. Apply the samples between the pencil marks at 2 cm such that the spots are smaller than 5 mm in diameter. Rinse tubes with 4 μl 70% ethanol and apply to existing spots.

12. Equilibrate a TLC tank with TLC solvent and develop plate up to the 14 cm marks (approximately 3 h). Use [^{3}H]-PIP$_2$ and [^{3}H]-QNB diluted into equivalent assay buffer as standards which migrate with R_f values of 0 to 0.2 and 0.5 to 0.8, respectively.

13. Dry the plate at room temperature. Align the marks at 2 cm and at 14 cm vertically with a straight edge and cut through the plastic with a razor. Cut each lane in 1 cm increments between origin and solvent front.

14. Place the square pieces into scintillation vials, add 10 ml scintillant and count for ^{3}H. Maximal counts are measured after 2 d in scintillant. The areas corresponding to [^{3}H]-QNB account for receptor binding. This method detects ~90% of the receptor content detected by the filter binding assay described above.

5.3.2.2 Quantitation of G$_{q\alpha}$

Total amount of active G$_{q\alpha}$ in the vesicles is determined by measuring protein-bound GDP, exactly as described in protocol 5-6, except that the sample is 10 µl protein/lipid vesicles in duplicate, and buffer G8 is replaced by reconstitution column buffer. A second set of GDP standards should be prepared in reconstitution column buffer as well.

Receptor-accessible G$_{q\alpha}$ is quantified by measuring receptor-promoted [^{35}S]-GTPγS binding to the vesicles as described below (Section 5.3.3.2). The proportion of G$_{q\alpha}$ that is accessible to receptor varies with the amounts of each protein in the vesicles. Under the conditions described here this is ~50%.

5.3.3 FUNCTIONAL ASSAYS

5.3.3.1 Receptor-Promoted Hydrolysis of PIP$_2$

This method measures hydrolysis of [^{3}H]-PIP$_2$ into [^{3}H]-IP$_3$ upon addition of carbachol (receptor agonist), GTP, Mg^{2+}, Ca^{2+}, and PLC-β1.[18] Its success relies on the incorporation of [^{3}H]-PIP$_2$ into the vesicle bilayer with M$_1$ and G$_q$, as would occur in cells. Detergent-free PLC-β1 is added and it adsorbs to the outer vesicle surface where it is activated by G$_q$ and hydrolyzes its substrate. This is a steady-state assay, as receptor agonist and GTP are present continuously. It is limited primarily by the low concentrations of PIP$_2$ available for hydrolysis, meaning that assay times are rather short.

Protocol 5-13: M$_1$ Muscarinic Receptor-Stimulated [^{3}H]-PIP$_2$ Hydrolysis by PLC-β1

---

1. *Solutions*
   a. Protein/lipid vesicles (1.6 ml, two reconstitutions) containing M$_1$ receptor, G$_q$, and [^{3}H]-PIP$_2$ prepared as described above.
   b. Basic buffer (3 ml): 20 mM HEPES (pH 7.5), 200 mM NaCl, 8.0 mM MgCl$_2$, 9.0 mM EGTA, and 1.2 mM CaCl$_2$. It is convenient to prepare a stock of this buffer and store at 4°C.

   c. PLC dilution buffer (2 ml): 20 mM HEPES (pH 7.5), 100 mM NaCl, 1 mM DTT, 10% glycerol, and 0.1 mg/ml BSA.

   d. GTPγS/Ca^{2+} buffer (1 ml): this buffer is used to determine the total amount of [^{3}H]-PIP$_2$ available in the outer leaflet of the vesicle bilayer for hydrolysis by PLC, termed "accessible PIP$_2$." Mix 500 μl basic buffer, 167 nM GTPγS (16.7 μl of 10 μM stock; see Section 5.4.4), 1.67 nM carbachol (16.8 μl of 100 mM stock), 3.87 mM CaCl$_2$ (38.7 μl of 100 mM stock), and water to give 1 ml final volume. This will result in 1 μM free Ca^{2+}, 1 mM carbachol, and 10 μM GTPγS final in the assay.

   e. GTP/carbachol buffer (2 ml): 1 ml of basic buffer, 16.7 μM GTP (33.4 μl of 1 mM stock; see Section 5.4.4), 1.67 mM carbachol (33.4 μl of 100 mM stock), 0.1 mg/ml BSA (20 μl of 10 mg/ml stock), and water to give 2 ml total volume.

   f. GTP/atropine buffer (2 ml): all the components of GTP/carbachol buffer except carbachol. Add 16.7 μM atropine sulfate (33.4 μl of 1 mM stock) instead.

   g. Diluted PLC-β1 (625 μl): dilute 1 μl of 2 mg/ml stock of PLC-β1 with 99 μl of PLC dilution buffer. Mix 52 μl of this diluted PLC-β1 with 573 μl of PLC dilution buffer. Use 10 μl in the assay to give 1 nM final.

   h. TCA: 10% TCA (w/v) stock solution.

   i. BSA: 3.33 mg/ml BSA stock solution.

2. A protocol is presented for doing a time course of carbachol- and GTP-stimulated PIP$_2$ hydrolysis. It assumes 11 time points are taken at 30 s intervals from 0 to 5 min, along with relevant controls. A total of 50 tubes are used. Sixty microliters of an activating buffer is mixed with 30 μl of vesicles, warmed for several minutes, then 10 μl of diluted PLC-β1 is added to initiate the reaction. Samples are quenched at the indicated time with acid, and processed for scintillation counting.

3. Label polypropylene tubes (12 × 75 mm) on ice. Tubes 1 to 48 are to be used for the time course; tubes 49 and 50 are to measure accessible PIP$_2$.

4. In tubes 1, 2, 5, 6, etc. add 60 μl GTP/carbachol buffer. In tubes 3, 4, 7, 8, etc. add 60 μl GTP/atropine buffer. Add 60 μl of GTPγS/Ca^{2+} buffer to tubes 49 and 50.

5. Add 30 μl of vesicles to all tubes and mix.

6. Prewarm the tubes at 30°C for 3 min. This allows the slow activation of G$_q$ by the receptor to occur prior to addition of PLC-β1.[45]

7. At the end of the 3 min, add 10% TCA to tubes 1 to 4, the zero time points, then add 10 μl diluted PLC-β1 and place on ice. For all other samples, add 10 μl diluted PLC-β1 (tubes 5 to 44) or 10 μl of PLC dilution buffer (tubes 45 to 48), mix quickly by vortexing and return the tube to 30°C for the allotted time. To perform this time course with reproducibility and accuracy, you need uniform temperature treatment for each tube. This is best accomplished by working initially with a few tubes, as in the following example:

a. Place tubes into the water bath at 15 s intervals
b. When the first tube has been incubated for 3 min exactly, quickly add PLC or buffer, mix and return to the bath. Do this within 15 s so that you are ready to process the next tube at 3 min 15 s, etc.
c. If you are doing the 1 min time point, quench the first tube at 4 min, the second tube at 4 min 15 s, etc. With focus and practice, you can manage several time points, nested within each other.

8. Quench PLC activity by adding 200 µl cold 10% TCA and placing the tube on ice. After all samples are quenched, add 300 µl 3.3 mg/ml BSA and process for counting as described in Protocol 5-7.

9. For tubes 49 and 50, no prewarming is necessary. Add 10 µl PLC-β1 to each, incubate at 30°C for 40 min, and quench.

10. Process assay samples for scintillation counting exactly as described in Protocol 5-7.

11. *Additional comments*
    a. The final concentration of components when mixed in these proportions are 70 mM HEPES, 100 mM NaCl, 3 mM MgCl$_2$, 10 nM free Ca^{2+} (0.36 mM CaCl$_2$ and 3 mM EGTA), 1 mM carbachol or 10 µM atropine, 10 µM GTP, and 0.1 mg/ml BSA. The specific activity is on the order of 10 pmol IP$_3$/min/pmol PLC-β1.
    b. The first two tubes at each time point reflect agonist-stimulated PLC activity, while the next two reflect basal PLC activity. Tubes 1 to 4 are the zero time point; tubes 44 to 48 are the assay blanks and should be averaged and deducted from the other samples. Tubes 49 and 50 measure the amount of PIP$_2$ present in the outer vesicle leaflet that can be hydrolyzed by PLC-β1. This number, divided by the total concentration of PIP$_2$, gives the fraction of PLC-accessible PIP$_2$. This is the relevant number for determining what fraction of substrate can be hydrolyzed while maintaining a linear reaction.
    c. Depending on the concentrations of M$_1$, G$_q$, and PIP$_2$ in the vesicles, PLC activity is usually linear only 2 to 3 min under these conditions (Figure 5.1).
    d. This assay can be modified to accommodate different types of experiments, such as dose dependencies or effects of unknown compounds on PLC activity, provided these final concentrations are met. Particularly important is the free Ca^{2+} concentration and changes in temperature or pH which will affect it. If substantial changes are necessary, it is imperative to establish that [^{3}H]-IP$_3$ formation remains linear over a given time interval.

5.3.3.2 Receptor-Promoted Guanine Nucleotide Binding to G Protein

This assay is based on the ability of agonist-bound receptor to promote GDP/GTP exchange on the G protein α subunit.[16,18,56] [^{35}S]-GTPγS, a radiolabeled GTP analog

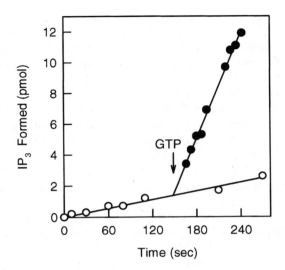

FIGURE 5.1 Kinetics of PLC-β1 activation. Vesicles that contained M_1 receptor, G_q, and [³H]-PIP₂ were incubated initially in buffer that contained 1 nM PLC-β1, 10 nM free Ca^{2+}, and 1 mM carbachol. At the time shown by the arrow, the reaction was split into two tubes, and either ddH₂O (open circles) or 10 μM GTP (solid circles) was added and the assay was allowed to continue. Key concentrations were 0.3 nM M_1 receptor, 1.6 nM G_q, and 0.35 βM accessible PIP₂. Addition of GTP increased the rate of PLC activity ~12 fold, from 0.009 to 0.117 pmol/s. (From Biddlecome, G. H., Berstein, G., and Ross, E. M., *J. Biol. Chem.*, 271, 7999, 1996. With permission.)

which is hydrolyzed extremely slowly by the G protein, is used instead of GTP. Conceptually this assay is similar to an enzymatic assay for receptor activity where receptor catalyzes guanine nucleotide exchange on the G protein. Thus, this is a direct biochemical measurement of receptor function and, because of its relative simplicity it can be used to evaluate test compounds for their ability to activate or block receptor function.

Protocol 5-14: M_1 Muscarinic Receptor-Stimulated [³⁵S]-GTPγS Binding to $G_{qα}$

1. Solutions are prepared and kept on an ice bath until used.
   a. Protein/lipid vesicles (1 ml) containing M_1 receptor and G_q prepared as described above.
   b. Binding solution (5 ml): 20 mM NaHEPES (pH 7.5), 2 mM MgCl₂, 1 mM EGTA (pH 8.0), 100 mM NaCl, 0.2 mg/ml BSA, and 200 nM [³⁵S]-GTPγS (100 cpm/fmol; but see 11d below). Incubate the solution about 10 min on ice, then filter through a prewashed 0.2 μm syringe filter. Dilute as soon as possible into carbachol or atropine solution and start the assay. This step minimizes erroneous background signal, thought to be due to Mg^{2+}-GTPγS aggregate formation. [³⁵S]-GTPγS at

the desired specific activity is prepared by mixing unlabeled GTPγS
(Section 5.4.4; lithium salt, Boehringer Mannheim) with [^{35}S]-GTPγS
(1250 Ci/mmol, 10 μM, New England Nuclear). Counting efficiency
for ^{35}S varies depending on the scintillation counter, but is usually
greater than 70%. Thus, the molar contribution of [^{35}S]-GTPγS in the
mixture is lower than 10%.

  c. Carbachol solution: 1.2 ml binding solution plus 0.6 mM carbachol.
  d. Atropine solution: 1.2 ml binding solution plus 20 μM atropine sulfate.
  e. Quench solution (5 ml): 20 mM Tris-Cl (pH 8.0), 10 mM MgCl$_2$, 1 mM
     GTP, 100 mM NaCl, 0.1 mM DTT, and 0.1% (w/v) C$_{12}$E$_{10}$.
  f. Washing solution (1 l): 20 mM Tris-Cl (pH 8.0), 10 mM MgCl$_2$, and
     100 mM NaCl. Transfer to a bottle furnished with a bottle-top dispenser.
     Keep on an ice bath.

 2. A protocol is presented for doing a time course of carbachol-stimulated
    [^{35}S]-GTPγS binding on G$_{qα}$. It assumes nine time points are taken at 0,
    1, 2, 3, 5, 7, 10, 12, and 15 min, along with relevant controls. A total of
    40 tubes are used. The reaction mixture consists of 25 μl agonist or
    agonist+antagonist solution, and 25 μl protein/lipid vesicles. Reactions are
    terminated by addition of a solution that includes C$_{12}$E$_{10}$ to solubilize the
    lipid/protein vesicles and inactivate the receptor, and a high concentration
    of Mg^{2+} to stabilize the G$_{qα}$-[^{35}S]-GTPγS complex. Unbound [^{35}S]-GTPγS
    is then removed by filtration through nitrocellulose disks which retain the
    G$_{qα}$-[^{35}S]-GTPγS complex.
 3. Label 40 polypropylene tubes (12 × 75 mm) and place them on ice.
 4. Soak 40 nitrocellulose filters (BA85, Schleicher and Schuell) in washing
    solution.
 5. Dispense 25 μl carbachol solution in tubes 1, 2, 5, 6, etc., and 25 μl
    atropine solution in 20 tubes 3, 4, 7, 8, etc.
 6. Prewarm the tubes in a 30°C water bath for 5 min.
 7. Add 100 μl quench solution to tubes 1 to 4, the zero time points, then
    add 25 μl protein/lipid vesicles and place on ice.
 8. Start the reaction by adding 25 μl protein/lipid vesicles (tubes 5 to 36) or
    reconstitution column buffer (tubes 37 to 40), mix quickly by vortexing
    and return to 30°C for the allotted time as described in Protocol 5-13, step 7.
 9. Terminate the reaction by adding 100 μl ice-cold quench solution and
    immediately placing the tubes on an ice bath for at least 10 min.
10. Filter quenched reaction mixtures and process filters as described in
    Protocol 5-6.
11. *Additional comments*
    a. The first two tubes at each time point reflect agonist-stimulated nucle-
       otide binding, while the next two reflect binding due to the intrinsic
       activity of the receptor. In the absence of receptor, G$_{qα}$ will not show
       appreciable nucleotide binding activity at this concentration of [^{35}S]-
       GTPγS. Tubes 1 to 4 are the zero time point; tubes 37 to 40 are the assay
       blanks and should be averaged and deducted from the other samples.

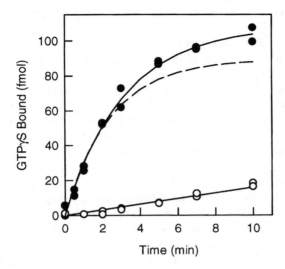

FIGURE 5.2 Carbachol-stimulated binding of GTPγS to M_1 receptor-G_q vesicles. Reactions were initiated by adding vesicles to prewarmed reaction buffers that contained either 1 mM carbachol (solid circles) or 10 μM atropine (open circles) and 100 nM [^{35}S]-GTPγS. The dotted lines are the differences between binding measured in the presence of carbachol and atropine. Reactions shown contained 0.15 nM M_1 receptor and 2.1 nM G_q. The rate constant for GTPγS binding was 0.41 ± 0.03 min^{-1}. (From Biddlecome, G. H., Berstein, G., and Ross, E. M., *J. Biol. Chem.*, 271, 7999, 1996. With permission.)

    b. The $t^{1/2}$ is 1 to 2 min under these conditions (Figure 5.2). Maximal binding represents the total amount of $G_{q\alpha}$ which is accessible to receptor stimulation. This value will be higher than the amount of receptor present in the vesicles, which demonstrates the catalytic nature of receptor action.

    c. This basic protocol described here can easily be modified to measure potency and efficacy of receptor agonists and antagonists, as well as the influence of other factors on receptor-promoted guanine nucleotide exchange on the G protein.

    d. For vesicles that contain [^{3}H]-PIP$_2$ increase the specific activity of [^{35}S]-GTPγS to 200 cpm/fmol and set counting window to exclude ^{3}H.

5.3.3.3 Receptor-Promoted GTP Hydrolysis by G Protein

This assay measures a major biochemical event that occurs after receptor-promoted GDP/GTP exchange on the G protein α subunit.[16,18,56] By virtue of its guanine nucleotide hydrolytic activity the G α subunit cleaves GTP into GDP and inorganic phosphate, the former remaining bound to the G α subunit and the latter being released to the medium. As mentioned in Section 5.1.1, GTP hydrolysis by G α subunits is subject to regulation by other proteins. The assay described here can be used to elucidate mechanistic properties of these regulatory factors.[18,19]

Protocol 5-15: M_1 Muscarinic Receptor-Stimulated $[\gamma^{32}P]$-GTP Hydrolysis by $G_{q\alpha}$

1. Solutions are prepared and kept on an ice bath until used.
   a. Protein/lipid vesicles (1 ml): containing M_1 receptor and G_q prepared as described above.
   b. Substrate solution (5 ml): 20 mM NaHEPES (pH 7.5), 2 mM MgCl$_2$, 1 mM EGTA (pH 8.0), 100 mM NaCl, 0.2 mg/ml BSA, and 1μM $[\gamma^{32}P]$-GTP (100 cpm/fmol). $[\gamma^{32}P]$-GTP at the desired specific activity is prepared by mixing labeled and unlabeled compound. Because of the high specific activity of radiolabeled GTP, its molar contribution to the mixture is negligible. Purity of unlabeled GTP (Boehringer Mannheim) is checked by TLC as described for GTPγS (Section 5.4.4).
   c. Carbachol solution: 1.2 ml substrate solution plus 0.6 mM carbachol.
   d. Atropine solution: 1.2 ml substrate solution plus 20 μM atropine sulfate.
   e. Quench suspension: Suspend activated charcoal (powder, Sigma) in 50 mM NaH$_2$PO$_4$ at 50 g/l and shake well. Centrifuge at 2500 × g 10 min and discard the supernatant. Resuspend the pellet in the same solution at the same concentration. Repeat the centrifugation/resuspension steps two more times. Resuspend final pellet in 50 mM sodium phosphate (pH 3) at 50 g/l and keep at 4°C.

2. A protocol is presented for doing a time course of carbachol-stimulated $[\gamma^{32}P]$-GTP hydrolysis by $G_{q\alpha}$. It assumes nine time points are taken at 0, 1, 2, 3, 5, 7, 10, 12, and 15 min, along with relevant controls. A total of 40 tubes are used. The reaction mixture consists of 25 μl agonist or antagonist solution, and 25 μl protein/lipid vesicles. Reactions are terminated by addition of an acidic suspension of activated charcoal in phosphate buffer followed by centrifugation. ^{32}Pi remains in the supernatant and is quantified by scintillation counting.

3 to 9. Exactly as described in Protocol 5-14, except 750 μl cold quench suspension that has been stirred with a magnetic stirrer is used as a quench solution and omit filters.

10. Spin at 2500 × g in a swinging bucket rotor at 4°C for 15 min.

11. Transfer 400 μl of the supernatant to scintillation vials. Add 10 ml scintillation fluid and count for ^{32}P. Water could be used instead of scintillation fluid to count Cerenkov radiation, but counting efficiency will be substantially reduced.

12. *Additional comments*
   a. See Protocol 5-14, step 11a.
   b. Hydrolysis should be linear with time within this time course. Less than 15% of total $[\gamma^{32}P]$-GTP should be hydrolyzed.
   c. In this particular system the G protein effector, PLC-β1, acts as a GTPase activating protein for $G_{q\alpha}$. This effect can easily be shown using this protocol with the addition of PLC-β1 (Figure 5.3).
   d. For vesicles that contain $[^3H]$-PIP$_2$ increase the specific activity to 200 cpm/fmol, and either set counting window to exclude ^3H or do Cerenkov counting.

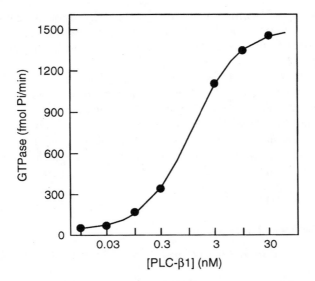

FIGURE 5.3 GTPase activity at increasing concentrations of PLC-β1. M_1 receptor-G_q vesicles were mixed with increasing concentrations of PLC-β1 and assayed for GTP hydrolysis. Assays contained 0.27 nM M_1 receptor, 1.7 nM G_q, 0.44 µM accessible PIP_2, 10 µM GTP, 10 nM Ca^{2+}, and either 1 mM carbachol or 10 µM atropine. Data show increases in activity caused by carbachol. Assay times were adjusted to ensure linear reactions. The EC_{50} for PLC-β1 stimulation of G_q GTPase was 1.1 nM. (From Biddlecome, G. H., Berstein, G., and Ross, E. M., *J. Biol. Chem.*, 271, 7999, 1996. With permission.)

5.4 ADDITIONAL METHODS

5.4.1 PROTEIN DETERMINATION

Protocol 5-16: Amido Black Protein Assay[35]

1. *Solutions*
   a. 1 M Tris + 2% SDS (500 ml): 60.6 g Tris Base, 10.0 g SDS. Dissolve in 400 ml water, pH to 7.5, then bring to 500 ml.
   b. 0.25% amido black dye: mix 0.625 g Naphthol Blue Black (Sigma) with 112.5 MeOH until dissolved. Add 25 ml glacial acetic acid and 112.5 ml water. Store at room temperature.
   c. AB rinse: 2.7 l MeOH, 60 ml acetic acid, and 240 ml water.
   d. Elution solution: 25 ml 1 N NaOH, 0.5 ml 0.1 N EDTA, 526 ml 95% ethanol, and 448.5 ml water.
   e. 90% TCA (w/v).
   f. 6% TCA (w/v).
2. Prepare samples and standards (0.5 to 20 µg protein) in 225 µl volume. BSA can be used as standard (A_{280} of 1 mg/ml BSA = 0.58).
3. Add 30 µl 1 M Tris + 2% SDS.
4. Add 50 µl 90% TCA and vortex; incubate at least 2 min at room temperature.

5. Wet Millipore HA filter with 6% TCA, then apply protein sample to filter under vacuum. A dot blot apparatus is ideal for this purpose.
6. Rinse sample tube with 200 µl 6% TCA and apply to the same spot on the filter.
7. Rinse wells with 500 µl 6% TCA.
8. Carefully remove the filter from apparatus, stain at room temperature for 30 min or more by immersing in 0.25% amido black dye solution.
9. Rinse the filter three times in water, then in AB rinse solution (three times, 1 min each) until background is pure white. Rinse again briefly in water.
11. Punch out spots and transfer to tubes using a hole-puncher.
12. Elute dye with 0.8 ml elution solution, vortexing periodically for 10 min or more.
13. Read A_{630}. Color is stable at least 2 h.

5.4.2 PHOSPHORUS DETERMINATION

Protocol 5-17: Measurement of Phospholipid Concentration by Phosphorus Determination[57]

1. *Solutions*
    a. 70% $HClO_4$ (G.F. Smith Chemical Co.).
    b. 2% and 0.4% ammonium molybdate. Store at room temperature.
    c. 10% ascorbic acid. Store at 4°C; stable up to 1 month.
    d. Phosphate standards (5 to 200 nmol inorganic phosphate, prepared as NaH_2PO_4).
2. Mix sample with 10 µl 2% ammonium molybate and dry under nitrogen (if chloroform solution) or by heating to 80–90°C (if aqueous solution). *Safety note:* if organic solvent is still present, explosion is possible at the next step; therefore, dry the sample thoroughly!
3. Add 300 µl 70% $HClO_4$; heat at 170 to 190°C (in a mineral oil bath) for ~1 h or until colorless.
4. Cool samples; prepare boiling water bath.
5. Add 1.5 ml 0.4% ammonium molybdate and vortex well.
6. Add 225 µl 10% ascorbic acid, vortex well, and immediately place in a boiling water bath.
7. Cool and read A_{830}.
8. Determine phosphorus content of phospholipid stocks by comparison to absorbance values of phosphorus standard curve. Remember that there are 3 mol Pi per mol PIP_2.

5.4.3 GEL FILTRATION COLUMN

Protocol 5-18: Preparation and Calibration of Gel Filtration Column

1. The chromatography column is a siliconized borosilicate glass Omnifit column (6.6 × 250 mm) with 25 µm pore size polyethylene frits, and 1.0 mm I.D. tubing attached at the inlet and outlet.

2. Buffers should be filtered through 0.22 μm filters, degassed, and stored at 4°C.
3. Take precautions to prevent air from entering the inlet tubing of the column during storage and use.
4. Pack column at room temperature with room temperature buffer and resin according to standard methods. Move column to 4°C and, once cold, equilibrate it with reconstitution column buffer.
5. To check for uniformity of packing and to identify the void volume, mix 75 μl of blue dextran solution with 75 μl of lipid/detergent micelles that contain a trace amount of [^{3}H]-PIP$_2$. Apply to the column through the inlet tubing and chromatograph with buffer, collecting 9-drop fractions into microcentrifuge tubes. Visual inspection of blue dextran during gel filtration should reveal a uniform progression of the leading edge of color, although the band will broaden as the various sizes of blue dextran are resolved. The column should be repacked if any major distortions in the leading edge are evident. Once packed properly, a column can be washed and used for several months.
6. The peak of [^{3}H]-PIP$_2$ and blue dextran in the fractions should overlap, but will probably not be identical. The fractions that contain the peak of [^{3}H]-PIP$_2$ indicate which ones should be saved from reconstitutions that lack [^{3}H]-PIP$_2$. In experiments where this phospholipid is present, it is easy to count 1 μl of each fraction to confirm which ones to pool.

5.4.4 CHECKING PURITY OF GUANINE NUCLEOTIDE STOCKS

Purity of unlabeled guanine nucleotides is checked by TLC on polyethyleneimine-cellulose TLC sheets (J.T. Baker) developed with the following solvent: 0.75 M Tris, 0.45 M HCl, and 0.5 M LiCl.[58] After spotting, the chromatograms should be immersed in methanol for 5 min and then allowed to dry at room temperature to remove interfering salts and increase the resolution. A mixture of GMP ($R_f = 0.7$), GDP ($R_f = 0.5$), and GTP ($R_f = 0.2$) is used as a standard. GTPγS migrates at a position similar to that of GTP. Nucleotides appear as ultraviolet (UV)-absorbing spots on the chromatogram. If there is substantial contamination by other guanine nucleotides, the stock can be repurified by ion-exchange chromatography on DEAE-Sephacel or MonoQ resin (Pharmacia) which is eluted with a linear gradient (0 to 0.5 M) LiCl. The concentration of stock solutions is checked by optical density (extinction coefficient at 252 nm is 13,700 M^{-1} cm^{-1} at pH 7.0).

5.4.5 SYNTHESIS OF [γ^{32}P]-GTP

We observed that commercially available preparations of [γ^{32}P]-GTP contained various amounts of contaminating ^{32}P material that decreased the signal-to-noise ratio. Therefore, we synthesized and purified [γ^{32}P]-GTP such that more than 97.7% of its radioactivity was retained by activated charcoal. This procedure is based on the method of Johnson and Walseth.[59] GDP is phosphorylated by ^{32}Pi at the γ position during the enzymatic conversion of L-α-glycerolphosphate to 3-phosphoglycerate, a reaction catalyzed by phosphoglycerate kinase.

Protocol 5-19: Enzymatic Synthesis of [γ ³²P]-GTP

1. *Solutions*
   a. Enzyme mixture stock solution: mix the following enzymes, which are ammonium sulfate suspensions from Boehringer Mannheim, 100 µl glycerol-3-phosphate dehydrogenase (EC 1.1.1.8, 170 U/mg, 2 mg/ml); 1 µl triosephosphate isomerase (EC 5.3.1.1, 5000 U/mg, 2 mg/ml); 20 µl glyceraldehyde-3-phosphate dehydrogenase (EC 1.2.1.12, 80 U/mg, 10 mg/ml); 2 µl 3-phosphoglycerate kinase (EC 2.7.2.3, 450 U/mg, 10 mg/ml); and 20 µl lactate dehydrogenase (EC 1.1.1.27, 550 U/ml, 5 mg/ml, from rabbit muscle). Store this mixture at 4°C.
   b. Tris-DTT solution: 209 µl water, 90 µl 1 M Tris-Cl (pH 9.0), and 21 µl 0.1 M DTT.
   c. Enzyme mixture working solution: centrifuge 10 µl enzyme mixture stock solution for 15 min at 4°C in a microcentrifuge at maximal speed. Remove supernatant carefully and discard it. Resuspend pellet in 130 µl ice-cold Tris-DTT solution.
   d. Substrate mixture: 83 µl water, 50 µl 0.5 M Tris-Cl (pH 9.0), 6 µl 1 M MgCl$_2$, 30 µl 0.1 M DTT, 6 µl 10 mM α-glycerophosphate, 25 µl 10 mM β-NAD, 7 µl 2 mM GDP (Boehringer Mannheim, lithium salt), and 7 µl 4.4 mg/ml sodium pyruvate. Prepare fresh.
   e. ³²Pi: (ICN, Phosphorus-32 High Concentration, H$_3$PO$_4$ in water, HCl-free, 800 mCi/ml, ~90 µM). Dilute 5 mCi with water to 20 µl. Use proper acrylic shielding to provide protection against radioactivity.
2. Carry out the reaction in a microcentrifuge tube. Mix 20 µl ³²Pi, 16 µl substrate mixture, and 2 µl enzyme mixture working solution.
3. Incubate at room temperature for 15 min.
4. Check the completeness of the reaction by the following procedure. Mix 0.1 µl of reaction mixture with 500 µl 50 mM NaH$_2$PO$_4$. Take a 25 µl aliquot to count for total radioactivity. Add a small amount of activated charcoal (powder form, Sigma; a spatula tip will suffice) to the remaining 475 µl, vortex, spin in a microcentrifuge for 15 s at maximal speed, and count 25 µl of the supernatant. Reaction is deemed complete when less than 1% of the radioactivity remains in the supernatant; that is, more than 99% of ³²Pi was incorporated. If the reaction is incomplete continue incubation at room temperature, and check every 5 min until completeness is reached.
5. Heat the reaction mixture at 100°C for 3 min, then transfer to an ice bath for 5 min.
6. Add 40 µl 100% methanol and spin in a microcentrifuge at maximal speed for 10 min at 4°C.
7. Transfer supernatant to a new tube, add 2 µl 0.5 M EDTA and 900 µl 10 mM KPi buffer (pH 7.0).
8. [γ³²P]-GTP is resolved from GDP and ³²Pi by HPLC on a Synchropak SAX column (100 × 2.1 mm; MicroScientific). Test the column performance with a mixture of GMP, GDP, and GTP. A 10-ml linear gradient

of KPi (10 to 100 mM) at a flow rate of 1 ml/min usually results in adequate resolution. Collect 0.5-ml fractions in tubes containing 5 µl 1 M Na-tricine (pH 7.0) on an ice bath. Check ^{32}Pi content of peak fractions at the expected position for GTP by the procedure described in step 4. Less than 0.1% of radioactivity should be present in the supernatant.

9. Dilute active fractions to 2×10^9 cpm/ml with 10 mM Na-tricine, aliquot as needed, and store below $-70°$C. Avoid repeat freezing and thawing because it accelerates nonspecific degradation of the compound.

ACKNOWLEDGMENTS

We learned and/or developed most of the methods described in this chapter during our stay in Elliott Ross' laboratory (University of Texas Southwestern Medical Center, Dallas, TX). We are grateful to Dr. Ross and the members of his group, in particular Karen Chapman, Jimmy Woodson, Suchetana Mukhopadhyay, and the late Tsutomu Higashijima. In addition, we greatly appreciate the help from members of the laboratories of Alfred G. Gilman and Paul Sternweis (University of Texas Southwestern Medical Center, Dallas, TX) in the implementation of many of these techniques, in particular Tohru Kazasa and John Hepler (presently at Emory University, Atlanta, GA).

REFERENCES

1. Gilman, A. G., Nobel Lecture. G proteins and regulation of adenylyl cyclase, *Biosci. Rep.*, 15, 65, 1995.
2. Ross, E. M., G protein GTPase-activating proteins: regulation of speed, amplitude, and signaling selectivity, *Recent Prog. Horm. Res.*, 50, 207, 1995.
3. Bourne, H. R., How receptors talk to trimeric G proteins, *Curr. Opin. Cell Biol.*, 9, 134, 1997.
4. Berman, D. M. and Gilman, A. G., Mammalian RGS proteins: barbarians at the gate, *J. Biol. Chem.*, 273, 1269, 1998.
5. Cerione, R. A., Reconstitution of receptor/GTP-binding protein interactions, *Biochim. Biophys. Acta*, 1071, 473, 1991.
6. Cerione, R. A. and Ross, E. M., Reconstitution of receptors and G proteins in phospholipid vesicles, *Methods Enzymol.*, 195, 329, 1991.
7. Brandt, D. R., Asano, T., Pedersen, S. E., and Ross, E. M., Reconstitution of catecholamine-stimulated guanosine triphosphatase activity, *Biochemistry*, 22, 4357, 1983.
8. Cerione, R. A., Codina, J., Benovic, J. L., Lefkowitz, R. J., Birnbaumer, L., and Caron, M. G., The mammalian β_2-adrenergic receptor: reconstitution of functional interactions between pure receptor and pure stimulatory nucleotide binding protein of the adenylate cyclase system, *Biochemistry*, 23, 4519, 1984.
9. Asano, T., Pedersen, S. E., Scott, C. W., and Ross, E. M., Reconstitution of catecholamine-stimulated binding of guanosine 5'-O-(3-thiotriphosphate) to the stimulatory GTP-binding protein of adenylate cyclase, *Biochemistry*, 23, 5460, 1984.
10. Cerione, R. A., Regan, J. W., Nakata, H., Codina, J., Benovic, J. L. et al., Functional reconstitution of the α_2-adrenergic receptor with guanine nucleotide regulatory proteins in phospholipid vesicles, *J. Biol. Chem.*, 261, 3901, 1986.

11. Kurose, H., Regan, J. W., Caron, M. G., and Lefkowitz, R. J., Functional interactions of recombinant α_2 adrenergic receptor subtypes and G proteins in reconstituted phospholipid vesicles, *Biochemistry*, 30, 3335, 1991.

12. Haga, K., Haga, T., Ichiyama, A., Katada, T., Kurose, H., and Ui, M., Functional reconstitution of purified muscarinic receptors and inhibitory guanine nucleotide regulatory protein, *Nature*, 316, 731, 1985.

13. Haga, K., Haga, T., and Ichiyama, A., Reconstitution of the muscarinic acetylcholine receptor. Guanine nucleotide-sensitive high affinity binding of agonists to purified muscarinic receptors reconstituted with GTP-binding proteins (Gi and Go), *J. Biol. Chem.*, 261, 10133, 1986.

14. Tota, M. R., Kahler, K. R., and Schimerlik, M. I., Reconstitution of the purified porcine atrial muscarinic acetylcholine receptor with purified porcine atrial inhibitory guanine nucleotide binding protein, *Biochemistry*, 26, 8175, 1987.

15. Parker, E. M., Kameyama, K., Higashijima, T., and Ross, E. M., Reconstitutively active G protein-coupled receptors purified from baculovirus-infected insect cells, *J. Biol. Chem.*, 266, 519, 1991.

16. Berstein, G., Blank, J. L., Smrcka, A. V., Higashijima, T., Sternweis, P. C., Exton, J. H., and Ross, E. M., Reconstitution of agonist-stimulated phosphatidylinositol 4,5-bisphosphate hydrolysis using purified m1 muscarinic receptor, $G_{q/11}$, and phospholipase C-β1, *J. Biol. Chem.*, 267, 8081, 1992.

17. Nakamura, F., Kato, M., Kameyama, K., Nukada, T., Haga, T., Kato, H., Takenawa, T., and Kikkawa, U., Characterization of G_q family G proteins $G_{L1\alpha}$ ($G_{14\alpha}$), $G_{L2\alpha}$ ($G_{11\alpha}$), and $G_{q\alpha}$ expressed in the baculovirus-insect cell system, *J. Biol. Chem.*, 270, 6246, 1995.

18. Biddlecome, G. H., Berstein, G., and Ross, E. M., Regulation of phospholipase C-β1 by G_q and m1 muscarinic cholinergic receptor. Steady-state balance of receptor-mediated activation and GTPase-activating protein-promoted deactivation [published erratum appears in *J. Biol. Chem.*, 271(52), 33705, 1996], *J. Biol. Chem.*, 271, 7999, 1996.

19. Berstein, G., Blank, J. L., Jhon, D. Y., Exton, J. H., Rhee, S. G., and Ross, E. M., Phospholipase C-β1 is a GTPase-activating protein for $G_{q/11}$, its physiologic regulator, *Cell*, 70, 411, 1992.

20. Senogles, S. E., Spiegel, A. M., Padrell, E., Iyengar, R., and Caron, M. G., Specificity of receptor-G protein interactions. Discrimination of G_i subtypes by the D2 dopamine receptor in a reconstituted system, *J. Biol. Chem.*, 265, 4507, 1990.

21. Ushikubi, F., Nakamura, K., and Narumiya, S., Functional reconstitution of platelet thromboxane A2 receptors with G_q and G_{12} in phospholipid vesicles, *Mol. Pharmacol.*, 46, 808, 1994.

22. Ohtaki, T., Ogi, K., Masuda, Y., Mitsuoka, K., Fujiyoshi, Y., Kitada, C., Sawada, H., Onda, H., and Fujino, M., Expression, purification, and reconstitution of receptor for pituitary adenylate cyclase-activating polypeptide. Large-scale purification of a functionally active G protein-coupled receptor produced in Sf9 insect cells, *J. Biol. Chem.*, 273, 15464, 1998.

23. Sambrook, J., Fritsch, E. F., and Maniatis, T., *Molecular Cloning: A Laboratory Manual*, Cold Spring Harbor Laboratory, New York, 1989.

24. O'Reilly, D. R., Miller, L. K., and Luckow, V. A., *Baculoviruses Expression Vectors. A Laboratory Manual*, W. H. Freeman and Company, New York, 1992.

25. Peralta, E. G., Ashkenazi, A., Winslow, J. W., Smith, D. H., Ramachandran, J., and Capon, D. J., Distinct primary structures, ligand-binding properties and tissue-specific expression of four human muscarinic acetylcholine receptors, *EMBO J.*, 6, 3923, 1987.

26. Bonner, T. I., Buckley, N. J., Young, A. C., and Brann, M. R., Identification of a family of muscarinic acetylcholine receptor genes [published erratum appears in *Science*, 237(4822), 237, 1987], *Science*, 237, 527, 1987.
27. Strathmann, M. and Simon, M. I., G protein diversity: a distinct class of α subunits is present in vertebrates and invertebrates, *Proc. Natl. Acad. Sci. U.S.A.*, 87, 9113, 1990.
28. Jones, D. T. and Reed, R. R., Molecular cloning of five GTP-binding protein cDNA species from rat olfactory neuroepithelium, *J. Biol. Chem.*, 262, 14241, 1987.
29. Sugimoto, K., Nukada, T., Tanabe, T., Takahashi, H., Noda, M., Minamino, N. et al., Primary structure of the β-subunit of bovine transducin deduced from the cDNA sequence, *FEBS Lett.*, 191, 235, 1985.
30. Gao, B., Gilman, A. G., and Robishaw, J. D., A second form of the β subunit of signal-transducing G proteins, *Proc. Natl. Acad. Sci. U.S.A.*, 84, 6122, 1987.
31. Gautam, N., Baetscher, M., Aebersold, R., and Simon, M. I., A G protein γ subunit shares homology with ras proteins, *Science*, 244, 971, 1989.
32. Suh, P. G., Ryu, S. H., Moon, K. H., Suh, H. W., and Rhee, S. G., Cloning and sequence of multiple forms of phospholipase C, *Cell*, 54, 161, 1988.
33. Hepler, J. R., Biddlecome, G. H., Kleuss, C., Camp, L. A., Hofmann, S. L., Ross, E. M., and Gilman, A. G., Functional importance of the amino terminus of $G_{q\alpha}$, *J. Biol. Chem.*, 271, 496, 1996.
34. Kazasa, T. and Gilman, A. G., Purification of recombinant G proteins from Sf9 cells by hexahistidine tagging of associated subunits. Characterization of α_{12} and inhibition of adenylyl cyclase by α_z, *J. Biol. Chem.*, 270, 1734, 1995.
35. Schaffner, W. and Weissmann, C., A rapid, sensitive, and specific method for the determination of protein in dilute solution, *Anal. Biochem.*, 56, 502, 1973.
36. Ross, E. M. and Schatz, G., Purification of subunit composition of cytochrome c1 from bakers' yeast *Saccharomyces cerevisiae*, *Methods Enzymol.*, 53, 222, 1978.
37. Haga, K. and Haga, T., Purification of the muscarinic acetylcholine receptor from porcine brain, *J. Biol. Chem.*, 260, 7927, 1985.
38. Haga, T., Haga, K., and Hulme, E.C., Solubilization, purification and molecular characterization of muscarinic acetylcholine receptors, in *Receptor-Ligand Interactions*, Hulme, E. C., Ed., IRL Press, Oxford, 1992, 51.
39. Fleming, J. W. and Ross, E. M., Reconstitution of β-adrenergic receptors into phospholipid vesicles: restoration of [^{125}I]iodohydroxybenzylpindolol binding to digitonin-solubilized receptors, *J. Cyclic Nucleotide Res.*, 6, 407, 1980.
40. Mumby, S. M. and Gilman, A. G., Synthetic peptide antisera with determined specificity for G protein α or β subunits, *Methods Enzymol.*, 195, 215, 1991.
41. Kopf, G. S. and Woolkalis, M. J., ADP-ribosylation of G proteins with pertussis toxin, *Methods Enzymol.*, 195, 257, 1991.
42. Carty, D. J., Pertussis toxin-catalyzed ADP-ribosylation of G proteins, *Methods Enzymol.*, 237, 63, 1994.
43. Hepler, J. R., Kozasa, T., Smrcka, A. V., Simon, M. I., Rhee, S. G., Sternweis, P. C., and Gilman, A. G., Purification from Sf9 cells and characterization of recombinant $G_{q\alpha}$ and $G_{11\alpha}$. Activation of purified phospholipase C isozymes by G_α subunits, *J. Biol. Chem.*, 268, 14367, 1993.
44. Hepler, J. R., Kozasa, T., and Gilman, A. G., Purification of recombinant $G_{q\alpha}$, $G_{11\alpha}$, and $G_{16\alpha}$ from Sf9 cells, *Methods Enzymol.*, 237, 191, 1994.
45. Biddlecome, G. H., Regulation of Phospholipase C-β1 by G_q and m1 Muscarinic Cholinergic Receptor, Ph.D. dissertation, University of Texas Southwestern Medical Center, Dallas, TX, 1997.

46. Pang, I. H. and Sternweis, P. C., Purification of unique α subunits of GTP-binding regulatory proteins (G proteins) by affinity chromatography with immobilized βγ subunits, *J. Biol. Chem.*, 265, 18707, 1990.

47. Northup, J. K., Smigel, M. D., and Gilman, A. G., The guanine nucleotide activating site of the regulatory component of adenylate cyclase. Identification by ligand binding, *J. Biol. Chem.*, 257, 11416, 1982.

48. Ferguson, K. M. and Higashijima, T., Preparation of guanine nucleotide-free G proteins, *Methods Enzymol.*, 195, 188, 1991.

49. Ferguson, K. M., Higashijima, T., Smigel, M. D., and Gilman, A. G., The influence of bound GDP on the kinetics of guanine nucleotide binding to G proteins, *J. Biol. Chem.*, 261, 7393, 1986.

50. Blank, J. L., Ross, A. H., and Exton, J. H., Purification and characterization of two G proteins that activate the β1 isozyme of phosphoinositide-specific phospholipase C. Identification as members of the G_q class, *J. Biol. Chem.*, 266, 18206, 1991.

51. Iñiguez-Lluhí, J. A., Simon, M. I., Robishaw, J. D., and Gilman, A. G., G protein βγ subunits synthesized in Sf9 cells. Functional characterization and the significance of prenylation of γ, *J. Biol. Chem.*, 267, 23409, 1992.

52. Lee, K. Y., Ryu, S. H., Suh, P. G., Choi, W. C., and Rhee, S. G., Phospholipase C associated with particulate fractions of bovine brain, *Proc. Natl. Acad. Sci. U.S.A.*, 84, 5540, 1987.

53. Park, D., Jhon, D. Y., Lee, C. W., Ryu, S. H., and Rhee, S. G., Removal of the carboxyl-terminal region of phospholipase C-β1 by calpain abolishes activation by $G_{\alpha q}$, *J. Biol. Chem.*, 268, 3710, 1993.

54. Sternweis, P. C., The purified α subunits of G_o and G_i from bovine brain require βγ for association with phospholipid vesicles, *J. Biol. Chem.*, 261, 631, 1986.

55. Rubenstein, R. C., Linder, M. E., and Ross, E. M., Selectivity of the β-adrenergic receptor among G_s, G_i's, and G_o: assay using recombinant α subunits in reconstituted phospholipid vesicles, *Biochemistry*, 30, 10769, 1991.

56. Brandt, D. R. and Ross, E. M., Catecholamine-stimulated GTPase cycle. Multiple sites of regulation by β-adrenergic receptor and Mg^{2+} studied in reconstituted receptor-G_s vesicles, *J. Biol. Chem.*, 261, 1656, 1986.

57. Ames, B. N., Assay of inorganic phosphate, total phosphate and phosphatases, *Methods Enzymol.*, 8, 115, 1966.

58. Bochner, B. R. and Ames, B. N., Complete analysis of cellular nucleotides by two-dimensional thin layer chromatography, *J. Biol. Chem.*, 257, 9759, 1982.

59. Johnson, R. A. and Walseth, T. F., The enzymatic preparation of [α-^{32}P]ATP, [α-^{32}P]GTP, [^{32}P]cAMP, and [^{32}P]cGMP, and their use in the assay of adenylate and guanylate cyclases and cyclic nucleotide phosphodiesterases, *Adv. Cyclic Nucleotide Res.*, 10, 135, 1979.

6   Mutagenesis Approaches for Studying Receptor–G Protein Interactions and Receptor Assembly

Fu-Yue Zeng, Evi Kostenis, Jan Jakubik, Xiangyang Zhu, Torsten Schöneberg, and Jürgen Wess

CONTENTS

6.1   INTRODUCTION

G protein-coupled receptors (GPCRs) form one of the largest protein families found in nature. Despite the remarkable structural diversity of their activating ligands, all

GPCRs are predicted to share a common three-dimensional fold consisting of seven transmembrane helices (TM I–VII) linked by alternating intracellular (i1–i3) and extracellular loops (o2–o4). The extracellular receptor surface (including the extracellular N-terminal domain, different extracellular loops, and/or the exofacial portions of various TM domains) is known to be critically involved in ligand binding.[1-3] On the other hand, the intracellular receptor surface (including different intracellular loops, the intracellular C-terminal domain, and the cytoplasmic ends of different TM helices) has been shown to be important for G protein recognition and activation.[3-9]

Several different experimental approaches have been used to elucidate the G protein coupling profiles of individual GPCRs (reviewed in Reference 9). These studies have shown that most GPCRs, when activated by agonist ligands, can recognize and activate only a limited set of the many structurally similar G proteins (as defined by their α subunits) expressed within a cell.[9-11] Functional analysis of hybrid receptors constructed between functionally distinct members of a GPCR subfamily has led to a wealth of information regarding the molecular basis of receptor–G protein coupling selectivity.[9] The majority of such studies indicates that the selectivity of G protein recognition is primarily determined by amino acids located within the i2 loop and the N- and C-terminal portions of the i3 domain.[4-6,9] The membrane-proximal portion of the C-terminal "tail" (i4) may also modulate the selectivity of receptor–G protein coupling, as shown for several GPCRs.[9]

In the first part of this chapter, we discuss mutagenesis strategies aimed at identifying receptor–G protein contact sites, as well as residues on the G protein α subunits (Gα) that are critical for determining the selectivity of receptor/G protein interactions. The second part of this chapter deals with "split" GPCRs and their potential usefulness for studying molecular aspects of GPCR assembly. This latter approach takes advantage of the observation that splitting GPCRs in certain extracellular or intracellular loops frequently leads to receptor fragments that are independently stable and can assemble to form functional receptor complexes.

6.2   STRUCTURAL ANALYSIS OF RECEPTOR–G PROTEIN COUPLING SELECTIVITY

6.2.1   IDENTIFICATION OF RECEPTOR–G PROTEIN CONTACT SITES

To understand the structural details involved in receptor–G protein recognition and G protein activation, specific, functionally relevant receptor–G protein contact sites need to be identified. However, despite the recent availability of high-resolution X-ray structures for different G protein heterotrimers,[12,13] the molecular architecture of the receptor–G protein complex still remains only poorly defined.

To address this issue, we have recently developed a "functional complementation approach" that involves the coexpression of hybrid GPCRs with hybrid Gα subunits in cultured mammalian cells.[14] This strategy is based on previous findings showing that it is possible to engineer both hybrid GPCRs as well as hybrid Gα subunits that exhibit profound changes in their coupling properties. For example, studies with many different GPCRs have shown that substitution of the i3 loop of a given receptor subtype into a recipient GPCR (that displays a different G protein

TABLE 6.1
C-Terminally Modified G Protein α Subunits Displaying Altered Receptor Coupling Properties (Examples)

Receptor	Cotransfected G Protein α Subunit	New Activity Gained (Stimulated Effector Enzyme)	Ref.
$G_{i/o}$-coupled			
D2 dopamine, A1 adenosine	qi5,[a] qo5,[b] or qz5[c]	PLCβ	15
α_2-adrenergic	qi5, qo5, or qz5	PLCβ	15
M$_2$ muscarinic	qi5, qo5, or qz5	PLCβ	14
mGluR2, mGluR4	qi5, qo5	PLCβ	16
Chemokine receptors	qi5	PLCβ	46
M$_2$ muscarinic	q(Y$\rightarrow$C);[d] q(N$\rightarrow$G)[e]	PLCβ	17
SSTR3 somatostatin	si5,[f] s14(5),[g] or s16 (5)[h]	Adenylyl cyclase	47
D2 dopamine	13z5[i]	Na$^+$/H$^+$ exchanger	48
α_2-adrenergic, D2 dopamine	13z5, 12z5[j]	Adenylyl cyclase II (coexpressed)	49
G_s-coupled			
V$_2$ vasopressin	qs5[k]	PLCβ	25
$G_{q/11}$-coupled			
V$_{1a}$ vasopressin, BB2 bombesin (GRP)	sq5[l]	Adenylyl cyclase	24, 25
M$_3$ muscarinic	sq5	Adenylyl cyclase	24
V$_{1a}$ vasopressin, M$_3$ muscarinic	s(E$\rightarrow$N);[m] s(Q$\rightarrow$E)[n]	Adenylyl cyclase	24
BB2 bombesin (GRP)	s(E$\rightarrow$N)	Adenylyl cyclase	24

[a] An α_q subunit in which the last five amino acids (aa) were replaced with the corresponding α_i sequence.
[b] An α_q subunit in which the last five aa were replaced with the corresponding α_o sequence.
[c] An α_q subunit in which the last five aa were replaced with the corresponding α_z sequence.
[d] An α_q subunit in which the Tyr residue at position -4 was replaced with the corresponding α_i residue, Cys.
[e] An α_q subunit in which the Asn residue at position -3 was replaced with the corresponding α_i residue, Gly.
[f] An α_s subunit in which the last five aa were replaced with the corresponding α_i sequence.
[g] An α_s subunit in which the last five aa were replaced with the corresponding α_{14} sequence.
[h] An α_s subunit in which the last five aa were replaced with the corresponding α_{s16} sequence.
[i] An α_{13} subunit in which the last five aa were replaced with the corresponding α_z sequence.
[j] An α_{12} subunit in which the last five aa were replaced with the corresponding α_z sequence.
[k] An α_q subunit in which the last five aa were replaced with the corresponding α_s sequence.
[l] An α_s subunit in which the last five aa were replaced with the corresponding α_q sequence.
[m] An α_s subunit in which the Glu residue at position -3 was replaced with the corresponding α_q residue, Asn.
[n] An α_s subunit in which the Gln residue at position -5 was replaced with the corresponding α_q residue, Glu.

Table modified from Reference 9.

coupling preference) can confer on the recipient GPCR the ability to recognize the G proteins normally activated by the donor receptor (reviewed in Reference 9). Similar studies have also been carried out with chimeric Gα subunits (see Table 6.1 for an overview).[9,15] The majority of such studies have focused on the C-terminal portion of the Gα subunits, a region known to be critically involved in receptor–G protein coupling. It was found, for example, that mutant versions of α_q (a subunit that can activate various isoforms of phospholipase Cβ [PLCβ]) in which the last five amino acids of α_q were replaced with the corresponding α_i or α_o sequences allowed stimulation of PLCβ by receptors that otherwise are exclusively coupled to G proteins of the $G_{i/o}$ class which cannot activate PLCβ efficiently.[14-17] In an analogous fashion, G_s- and $G_{q/11}$-coupled receptors gained the ability to stimulate novel signal transduction pathways when coexpressed with appropriately engineered hybrid Gα subunits (see Table 6.1 for a summary).

In the following, the use of the "functional complementation approach" to identify putative receptor–G protein contact sites will be outlined by discussing several examples derived from work done in our laboratory. For most of these studies, the M_2 and M_3 muscarinic receptor subtypes served as model systems. All studies were carried out using COS-7 cells transiently expressing various wild-type (wt) and mutant muscarinic receptors and Gα subunits.

Receptor and G protein mutations were generated using different polymerase chain reaction (PCR)-based mutagenesis strategies. An approach, generally referred to as recombinant or overlap PCR, that can be employed to generate essentially any kind of point mutation, insertion, or deletion, and which can also be adapted to create chimeric constructs, is outlined below (see also Scheme 1).

Protocol 6-1: A Universal PCR Mutagenesis Strategy (Scheme 1)

1. Design the following four oligonucleotides (A-D): primer A (sense) is located upstream of restriction site X. Primer D (antisense) is located downstream of restriction site Y. Preferably, X and Y should represent unique restriction sites in the plasmid to be mutagenized. Primers B and C are complementary oligonucleotides containing the desired mutation.
2. Run two separate PCR reactions with primers A/B and primers C/D using the wild-type plasmid (about 100 ng) as a template (primer concentrations: 1 μM final; reaction volume: 100 μl; use 2.5 U Taq or equivalent DNA polymerase). Use standard buffers and PCR conditions (denature at 94°C for 1 min, anneal at 37°C for 2 min, and extend at 72°C for 3 min, for a total of 25 cycles).
3. Gel-purify the two primary PCR products by standard techniques.
4. Run a secondary PCR reaction with the two "outside" primers A/D using a mixture of the two primary PCR products (about 50 to 100 ng each) as a template (use PCR conditions as indicated above).
5. Extract the secondary PCR product with phenol/chloroform and precipitate with ethanol. Resuspend DNA in TE and digest with restriction enzymes X and Y.
6. Replace X-Y fragment in the wild-type plasmid with the X/Y-digested PCR product using standard molecular biology techniques.
7. Sequence the entire X-Y region in the recombinant construct to verify the presence of the desired mutation and the absence of unwanted mutations introduced during the PCR amplification.

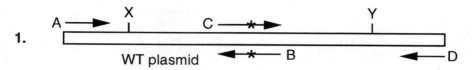

A-D: PCR primers;   X, Y: Unique restriction sites   *Mutation

1.

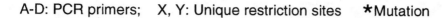

2. PCR with primers A/B

PCR with primers C/D

PCR product 1

PCR product 2

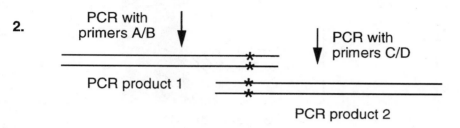

3. Isolate PCR products 1 and 2, mix, denature, and renature

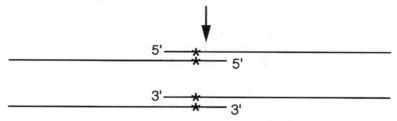

4. Extend 3' ends/PCR with primers A/D

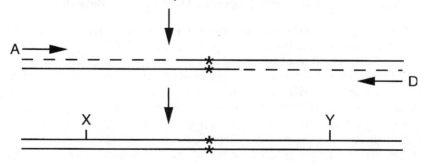

5. Cut with X and Y; replace X-Y segment in WT plasmid with mutated PCR fragment

SCHEME 1 A universally applicable overlap PCP mutagenesis approach (see text for details).

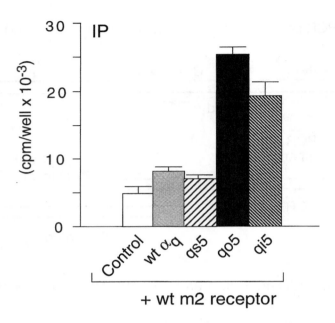

FIGURE 6.1 Activation of mutant α_q subunits upon stimulation of the wt M_2 muscarinic receptor. COS-7 cells were cotransfected with expression plasmids coding for the wt M_2 receptor and wt α_q or mutant α_q subunits in which the C-terminal five amino acids of α_q (EYNLV) were replaced with the corresponding sequences derived from α_o (qo5; GCGLY), $\alpha_{i1,2}$ (qi5; DCGLF), or α_s (qs5; QYELL). Cells were incubated with the muscarinic agonist carbachol (1 mM), and increases in intracellular IP_1 levels were determined as described.[14] Control IP_1 levels were determined in cells coexpressing the wt M_2 muscarinic receptor and wt α_q in the absence of carbachol (data taken from Reference 14).

6.2.1.1 Studies With the $G_{i/o}$-Coupled M_2 Muscarinic Receptor

When expressed in COS-7 or other mammalian cells, activation of the M_3 muscarinic receptor which is selectively linked to G proteins of the $G_{q/11}$ class leads to the stimulation of PLCβ resulting in the breakdown of phosphatidylinositol (PI) lipids.[18] The M_2 muscarinic receptor, on the other hand, which preferentially couples to G proteins of the $G_{i/o}$ family, stimulates PI hydrolysis only poorly, even when overexpressed with wt α_q (Figure 6.1).[18] However, coexpression of the wt M_2 receptor with mutant α_q subunits (qo5, qi5, or qz5) in which the last five amino acids of wt α_q were replaced with the corresponding sequences derived from various members of the $\alpha_{i/o}$ protein family (α_{i2}, α_o, or α_z) led to a pronounced stimulation of the PI cascade (Figure 6.1)[14] (see protocols 6-2 and 6-3 for experimental details). The specificity of this interaction could be demonstrated by the relative lack of PLC activity observed after coexpression of the M_2 receptor with a mutant α_q subunit (qs5) that contained five amino acids of α_s sequence at its C terminus (Figure 6.1).[14]

These data suggested that the M_2 receptor contains a sequence element that can be specifically recognized by the C terminus of $G\alpha_{i/o}$ subunits. It seemed reasonable to assume that this sequence element is located in the i2 or i3 loop of the receptor,

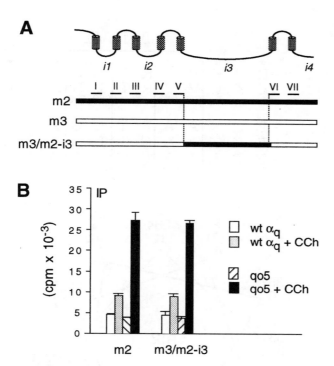

FIGURE 6.2   Interaction of an M_2/M_3 hybrid muscarinic receptor with a C-terminally modified mutant α_q subunit (qo5). (A) Structure of the M_3/M_2-i3 hybrid receptor composed of rat M_3 and human M_2 muscarinic receptor sequences ($M3_{252-493} \rightarrow M2_{208-390}$).[19,20] The positions of TM I-VII are indicated. (B) COS-7 cells coexpressing the indicated receptors and G proteins (qo5 is a mutant α_q subunit in which the C-terminal five amino acids were derived from α_o) were stimulated with carbachol (CCh; 1 mM), and the resulting increases in intracellular IP_1 levels were determined as described.[14] Data were taken from Reference 14.

since these regions are known to be intimately involved in dictating the coupling preference of individual muscarinic receptor subtypes.[18]

Previous mutagenesis studies had shown that a mutant M_3 receptor in which the entire i3 loop (including two amino acids at the N terminus of TM VI) was replaced with the corresponding M_2 receptor sequence (resulting in M_3/M_2-i3; Figure 6.2) almost completely lost the ability to stimulate the PI cascade, but gained the ability to mediate the inhibition of adenylyl cyclase (mediated by $G_{i/o}$ proteins).[19,20] Taken together, these findings strongly suggested that the C terminus of $\alpha_{i/o}$ subunits can interact with residues located in the i3 loop of the M_2 receptor and that coexpression of the M_3/M_2-i3 hybrid receptor with qo5 or qi5 should reestablish receptor-mediated activation of PI hydrolysis (as a measure of receptor-mediated G protein activation). Figure 6.2 shows that this was in fact the case.[14]

Analysis of additional mutant M_3 receptors[14] containing smaller segments of M_2 receptor sequences within their i3 loops showed that only those mutant receptors were able to interact with qi5 or qo5 that contained a four-amino-acid motif derived from the M_2 muscarinic receptor (Val385, Thr386, Ile389, and Leu390; hereafter

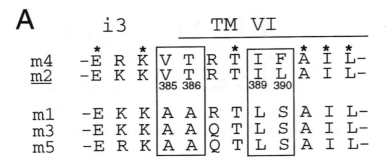

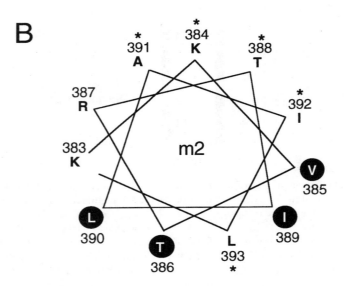

FIGURE 6.3 (A) Comparison of M_1–M_5 muscarinic receptor sequences at the i3 loop/TM VI junction. Numbers refer to amino acid positions in the human M_2 muscarinic receptor.[50] Asterisks indicate positions where all muscarinic receptor subtypes have identical residues. (B) Helical wheel representation of the M_2 receptor sequence at the i3 loop/TM VI junction, as viewed from the cytoplasm. The residues predicted to form the functionally important VTIL motif[14,17] are highlighted. (From Kostenis, E., Conklin, B. R., and Wess, J., *Biochemistry*, 36, 1487, 1997. With permission.)

referred to as the VTIL motif). These four residues are located at the i3 loop/TM VI junction (Figure 6.3), a receptor region thought to be α-helically arranged.[3,21] The VTIL residues are therefore predicted to form a contiguous surface (Figure 6.3) which can recognize the C terminus of $\alpha_{i/o}$ subunits, most likely via hydrophobic interactions. This contact may induce a conformational change in the C terminus of the $\alpha_{i/o}$ subunits which ultimately leads to GDP release and G protein activation.[22]

If the VTIL sequence is indeed critical for recognition of the C terminus of $\alpha_{i/o}$ subunits by the wt M_2 receptor, one would expect that mutational modification of this motif should disrupt the ability of the M_2 receptor to interact with qi5 or qo5.

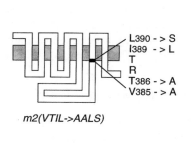

L390 - > S
I389 - > L
T
R
T386 - > A
V385 - > A

m2(VTIL->AALS)

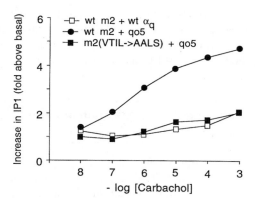

FIGURE 6.4 Lack of interaction of the M_2(VTIL→AALS) mutant M_2 muscarinic receptor with the C-terminally modified mutant α_q subunit, qo5. In M_2(VTIL→AALS), Val385, Thr386, Ile389, and Leu390 were replaced with the corresponding M_3 receptor residues (AALS; see Figure 6.3). In qo5, the C-terminal five amino acids of wt α_q were replaced with the corresponding α_o sequence. COS-7 cells coexpressing the indicated receptor and G protein constructs were incubated with different concentrations of carbachol, and increases in intracellular IP_1 levels were determined as described.[14] Data were taken from Reference 14.

To test this hypothesis, we generated a mutant M_2 receptor (M_2(VTIL→AALS)) in which these four residues were replaced by the corresponding M_3 receptor residues (AALS; Figure 6.4). Consistent with the gain-of-function mutagenesis studies summarized in the previous paragraphs, coexpression studies showed that the M_2(VTIL→AALS) mutant receptor lost the ability to efficiently couple to qo5 (Figure 6.4) or wt G_i (studied in assays measuring receptor-mediated inhibition of adenylyl cyclase).[14]

To study the functional importance of the VTIL residues in greater detail, we created additional mutant M_2 receptors in which these four residues were replaced, either individually or in combination, with the corresponding residues present in the $G_{q/11}$-coupled muscarinic receptors (M_1, M_3, and M_5) (Figure 6.3).[17] Subsequently, the ability of the resulting mutant M_2 receptors to interact with the mutant α_q subunit, qo5, was investigated in cotransfected COS-7 cells. We found that three of the four examined M_2 receptor single point mutants (M_2(V→A), M_2(T→A), and M_2(I→L)) were clearly less efficient than the wt M_2 receptor in mediating qo5-dependent PLC activation (loss in efficacy and/or carbachol potency). However, these three mutant receptors retained the ability to productively couple to G_{15} (upon coexpression with α_{15}),[17] a G protein known to be activated by most GPCRs, including the M_2 muscarinic receptor.[23] This observation suggested that the inability of the three mutant receptors to efficiently interact with qo5 is not caused by a generalized misfolding of the intracellular receptor surface. On the other hand, the M_2(L→S) point mutant was able to couple to qo5 in a fashion virtually identical to that found with the wt M_2 receptor.[17] Taken together, these results strongly support the concept that Val385, Thr386, and Ile389 (but not Leu390) play key roles in the recognition of the C terminus of $\alpha_{i/o}$ subunits by the M_2 muscarinic receptor. This view is also supported by studies analyzing the ability of a series of M_2 receptor double point mutants

(VT→AA, VI→AL, IL→LS, and TL→AS) to interact with the qo5 subunit.[17] The most straightforward explanation for these findings is that activation of the M_2 muscarinic receptor leads to conformational changes that allow selective binding of the VTI(L) motif to the C terminus of $\alpha_{i/o}$ subunits, eventually triggering G protein activation.

Protocol 6-2: Cotransfection of COS-7 Cells with GPCRs and G Protein α Subunits

1. Grow COS-7 cells to be transfected in Dulbecco's modified Eagle's medium (DMEM) supplemented with 10% fetal bovine serum (FBS) at 37°C in a 5% CO_2 incubator to about 70% confluence.
2. One day before the transfection, split cells into 100-mm dishes at a density of about 1×10^6 cells/dish (a 70% confluent 175 cm^2 tissue culture flask yields about 13 100-mm dishes).
3. Transfect cells about 18 to 24 h after seeding into 100-mm dishes.
4. For transfections, prepare the following DNA mix (amounts are given per 100-mm dish):
   4 µg receptor DNA
   1 µg G protein α subunit DNA
   850 µl phosphate-buffered saline (PBS) with Ca^{2+} and Mg^{2+}
   55 µl DEAE-dextran (10 mg/ml in PBS)
   Mix well and let mixture sit at room temperature for 30 to 60 min.
5. Wash cells once with PBS containing Ca^{2+} and Mg^{2+} and slowly add the DNA-dextran mixture along the side of the dish.
6. Tilt the dish to spread the mixture evenly and incubate at 37°C (5% CO_2) for 2 h. Tilt the plate carefully every 20 min.
7. Add 7 ml of DMEM containing 10% FBS and chloroquine (80 to 100 µM final concentration) and incubate for 3 to 5 h (37°C, 5% CO_2).
8. Suck off the chloroquine-containing medium and add new DMEM containing 10% FBS.
9. Culture the cells for the appropriate amount of time, depending on the type of experiment to be performed. For functional assays (e.g., PI or cAMP assays), use cells about 18 to 24 h after transfection.

Protocol 6-3: Inositol Phosphate Assay Using Transfected COS-7 Cells

1. About 18 to 24 h after transfection (see Protocol 6-2), split cells into 6-well dishes (approximately 0.4×10^6 cells per well) and label with 3 µCi/ml [^{3}H]-myoinositol (20 Ci/mmol).
2. Label cells for 24 to 36 h (37°C, 5% CO_2).
3. After the labeling period, wash cells once with Hanks' balanced salt solution (HBSS) containing 20 mM HEPES (HBSS + HEPES) and add 1 ml of HBSS + HEPES containing 10 mM LiCl to each well.
4. Shake cells gently on an orbital shaker for 20 min at room temperature.

5. Add the appropriate amount of agonist and/or antagonist ligand and incubate for 1 h at 37°C.
6. Suck off the ligand-containing medium and add 750 µl of ice-cold 20 mM formic acid.
7. Incubate for 40 min at 4°C.
8. Neutralize with 250 µl of 60 mM ammonium hydroxide.
9. Collect cell extracts and isolate the inositol monophosphate (IP$_1$) fraction via anion exchange chromatography as originally described by Berridge et al.[54]

6.2.1.2   Studies With the G$_{q/11}$-Coupled M$_3$ Muscarinic Receptor

To examine whether the results obtained with the G$_{i/o}$-coupled M$_2$ muscarinic receptor (see Section 6.2.1.1) are of general relevance, we recently extended the "functional complementation approach" to other classes of GPCRs and Gα subunits. First, we wanted to identify the site(s) on G$_{q/11}$-coupled receptors that can functionally interact with the C terminus of α$_{q/11}$ subunits.[24] To address this question, we initially demonstrated that G$_{q/11}$-coupled receptors, including the M$_3$ muscarinic receptor, while unable to efficiently interact with wt α$_s$, can productively couple to a mutant version of α$_s$ (sq5) in which the last five amino acids of α$_s$ (QYELL) were replaced with the corresponding α$_q$ sequence (EYNLV) (Figure 6.5)[24] (see also Reference 25). Essentially similar results were obtained with two other G$_{q/11}$-coupled receptors, the V1a vasopressin and the gastrin-releasing peptide (GRP) receptors,[24,25] consistent with the concept that the C terminus of Gα subunits is generally important for the selectivity of receptor recognition (Table 6.1).

As expected, the G$_{i/o}$-coupled M$_2$ muscarinic receptor was unable to activate wt α$_s$ or sq5 to a significant extent.[24] We therefore speculated that substitution into the M$_2$ receptor of the M$_3$ receptor domain(s) capable of recognizing the C terminus of α$_q$ may enable the resulting hybrid receptor(s) to interact with the sq5 subunit. Since neither the wt M$_2$ nor the wt M$_3$ receptor can productively couple to wt α$_s$, this strategy appeared to be particularly attractive (measure of productive coupling: increase in adenylyl cyclase activity). A mutant M$_2$ receptor (M$_2$(VTIL→AALS)) in which the functionally critical VTIL motif at the i3 loop/TM VI junction was replaced with the corresponding M$_3$ receptor sequence (AALS) was unable to productively interact with wt α$_s$, but gained the ability to couple to sq5 (Reference 24). This observation indicated that the C terminus of the i3 loop is generally important for recognition of the C terminus of Gα subunits. To our surprise, however, two additional mutant M$_2$ receptors, which contained M$_3$ receptor sequences in the i2 loop and at the N terminus of the i3 domain, were identified that also gained efficient coupling to sq5, similar to the M$_2$(VTIL→AALS) mutant receptor.[24] Previous studies have shown that the i2 loop as well as the N- and C-terminal segments of the i3 domain contain the key residues determining the G$_{q/11}$ coupling selectivity of the M$_3$ receptor subtype.[18] The "Baldwin model"[26,27] suggests that these residues cluster together in a "triangle" defined by the positions of TM III, VI, and VII, forming a well-defined G protein binding surface. One possibility, therefore, is that the C-terminal portion of α$_q$ is simultaneously contacted by amino acids located

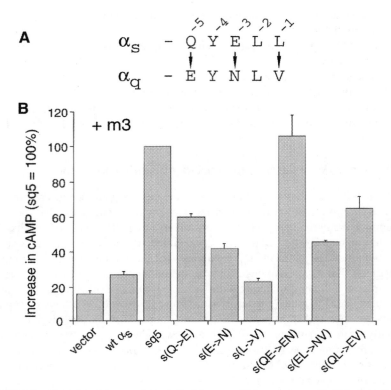

FIGURE 6.5 Functional interaction of the M_3 muscarinic receptor with C-terminally modified mutant α_s subunits. (A) Mutant α_s subunits were created by replacing the marked α_s residues (arrows) with the corresponding α_q residues, either individually or in combination. (B) COS-7 cells were cotransfected with wt M_3 muscarinic receptor DNA and the indicated G protein constructs (or pCDNAI vector DNA as a control). In sq5, the last five amino acids of wt α_s were replaced with the corresponding α_q sequence. Carbachol (0.5 mM)-induced increases in intracellular cAMP levels were determined as described.[24] Data were taken from Reference 24.

on different intracellular receptor regions. This view is also consistent with a recent mutagenesis study examining the ability of a C-terminal α_t peptide to bind to wt and mutant versions of rhodopsin.[28]

6.2.2 IMPORTANCE OF THE N TERMINUS OF Gα_q FOR RECEPTOR COUPLING SELECTIVITY

It is likely that other regions of Gα, besides the C-terminal domain, also contribute to receptor binding and the selectivity of receptor–G protein interactions.[29,30] Specifically, several lines of evidence suggest that the N-terminal portion of Gα may be in contact with the receptor protein.[31–33] To test the hypothesis that the N-terminal segment of Gα subunits is also involved in determining the selectivity of receptor–G protein interactions, we again employed a coexpression strategy involving the use

A

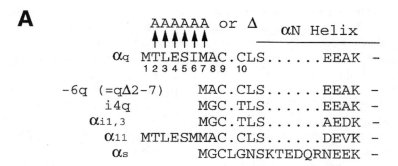

B

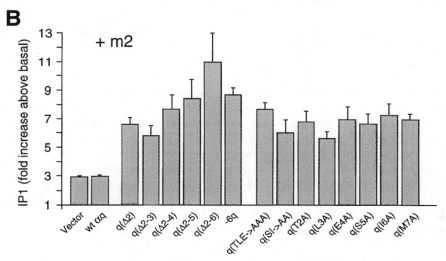

FIGURE 6.6 (A) Residues 2 to 7 of α_q were progressively deleted or replaced, either individually or in combination, with alanine residues. In i4q, the first ten amino acids of α_q were replaced with the first four amino acids of α_i. (B) Functional interaction of the wt M_2 muscarinic receptor with mutant α_q subunits. COS-7 cells were cotransfected with wt M_2 receptor DNA and the indicated G protein constructs (or pCDNAI vector DNA as a control). Carbachol (0.5 mM)-induced increases in intracellular IP_1 levels were determined as described.[35] Data were taken from Reference 35.

of hybrid Gα subunits.[34] Initially, we created a mutant α_q subunit in which the first ten amino acids of α_q, including a six-amino-acid extension that is characteristic for $\alpha_{q/11}$ subunits, were replaced with the N-terminal four amino acids (MGCT) present in α_i/α_o subunits (referred to as i4q; Figure 6.6). In addition, we also generated a mutant α_q subunit (–6q) that lacked the six-amino-acid extension present in α_q and α_{11} (Figure 6.6). The ability of the $G_{i/o}$-linked M_2 muscarinic receptor to gain coupling to these mutant G proteins was then studied in cotransfected COS-7 cells by measuring receptor-mediated stimulation of PI hydrolysis. These studies led to the surprising observation[34] that the wt M_2 receptor gained the ability to couple to both mutant G proteins, i4q as well as –6q, suggesting that the ability of the M_2 receptor to interact with the hybrid i4q subunit is due primarily to the lack of the six-amino-acid extension rather than the presence of four amino acids of α_i sequence. More detailed studies[34] showed that several other $G_{i/o}$- as well as G_s-coupled receptors were also capable of interacting with the –6q subunit (but not with wt α_q), indicating that the N-terminal extension characteristic for $\alpha_{q/11}$ subunits is required for restraining the receptor coupling selectivity of these G proteins.

Based on these results, our next goal was to identify specific amino acids within the N-terminal segment of α_q that are of particular importance for constraining the receptor coupling selectivity of this subunit. Toward this end, this region was subjected to systematic deletion and alanine scanning mutagenesis.[35] Two series of N-terminally modified mutant α_q subunits were prepared. In one series, the N terminus of α_q was progressively shortened by deletion mutagenesis, and in the other one, amino acids 2 to 7 were replaced, either individually or in combination, with alanine residues (Figure 6.6A). Coexpression studies led to the unexpected finding that the wt M_2 muscarinic receptor was able to productively couple to all 14 mutant α_q subunits (but not to wt α_q) (Figure 6.6B). Analogous findings were obtained with the $G_{i/o}$-linked D_2 dopamine and the G_s-coupled β_2-adrenergic receptors.[35]

Additional experiments indicated that the functional promiscuity displayed by all mutant α_q constructs investigated was not due to overexpression (as compared with wt α_q) or internal initiation of translation at a downstream ATG codon (codon seven).[35] Since α_q and α_{11}, like most other Gα subunits, are known to be palmitoylated at N-terminal cysteine residues (corresponding to Cys9 and Cys10 in Figure 6.6A), we also wanted to examine whether the promiscuous mutant α_q subunits differed from wt α_q in their palmitoylation patterns. For these studies, wt α_q and selected mutant α_q subunits were expressed in COS-7 cells and metabolically labeled with [^{3}H]-palmitic acid, followed by immunoprecipitation, sodium dodecyl sulfate-polyacrylamide gel electrophoresis (SDS-PAGE), and fluorography.[35] We found that all promiscuous mutant α_q subunits studied incorporated significantly smaller amounts of [^{3}H]-palmitate (the palmitoylation signal was reduced by about 40 to 75%) than α_q (Reference 35).

In summary, these data are consistent with a model in which the six-amino-acid extension characteristic for $\alpha_{q/11}$ subunits forms a tightly folded protein subdomain which may serve a gate function by selectively preventing access of $G_{i/o}$- and G_s-coupled receptors. Moreover, since all promiscuous mutant α_q subunits studied exhibited reduced [^{3}H]-palmitate incorporation, the possibility exists that palmitoylation of Gα subunits may contribute to regulating the selectivity of receptor–G protein interactions.

6.2.3 C-TERMINAL Gα RESIDUES CRITICAL FOR RECEPTOR–G PROTEIN COUPLING SELECTIVITY

As summarized in Table 6.1, a considerable body of evidence indicates that the C-terminal five amino acids of Gα subunits play a key role in dictating the specificity of receptor–G protein coupling. Based on these studies, we have employed a gain-of-function mutagenesis strategy to identify individual amino acids within the C-terminal tail of different Gα subunits that are critical for the

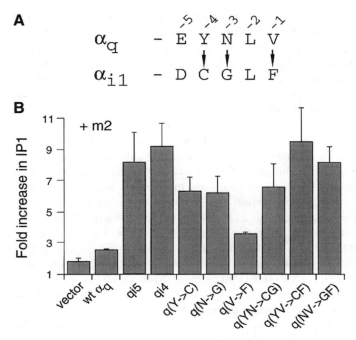

FIGURE 6.7 Functional interaction of the M_2 muscarinic receptor with C-terminally modified mutant α_q subunits. (A) Mutant α_q subunits were created by replacing the marked α_q residues (arrows) with the corresponding α_{i1} residues, either individually or in combination. (B) COS-7 cells were cotransfected with wt M_2 muscarinic DNA and the indicated G protein constructs (or pCDNAI vector DNA as a control). In qi5/qi4, the last five/four amino acids of wt α_q were replaced with the corresponding α_{i1} sequence. Carbachol-induced increases in intracellular IP_1 levels were determined as described.[17] Maximum responses are shown (data taken from Reference 17).

selectivity of receptor recognition. Specific examples are given in the following two sections.

6.2.3.1 Functionally Critical Residues at the C Terminus of $\alpha_{i/o}$ Subunits

Initially, we wanted to evaluate the functional importance of individual amino acids within the C-terminal segment of $\alpha_{i/o}$ subunits. Toward this goal, specific residues at the C terminus of wt α_q were replaced, either alone or in combination, with the corresponding $\alpha_{i1,2}$ residues (Figure 6.7). We then examined, using cotransfected COS-7 cells, whether the $G_{i/o}$-coupled M_2 muscarinic receptor would gain the ability to interact with the resulting mutant Gα subunits by measuring receptor-mediated stimulation of PI hydrolysis.

As already discussed in Section 6.2.1.1, the wt M_2 receptor did not efficiently recognize wt α_q but could productively interact with qi5, a mutant α_q subunit in which the last five amino acids of wt α_q (EYNLV) were replaced with the corresponding $\alpha_{i1,2}$ sequence (DCGLF) (Figures 6.1 and 6.7).[14,17] The mutant α_q subunit, qi4, in which only the last four amino acids of wt α_q were replaced with the corresponding $\alpha_{i1,2}$ sequence, could be activated by the M_2 receptor in a fashion very similar to qi5 (maximum responses are shown in Figure 6.7).[17] Since the C-terminal four amino acids of α_q differ in only three residues from the corresponding $\alpha_{i1,2}$ sequence (the leucine residue at position –2 is conserved among all mammalian α subunits), three α_q single point mutants were prepared (q(Y→C), q(N→G), and q(V→F); Figure 6.7). As shown in Figure 6.7, all three mutant α_q subunits gained the ability to be activated by the wt M_2 receptor, although with different efficiencies. Whereas q(Y→C) and q(N→G) were able to mediate a rather robust (sixfold) increase in PLC activity (as compared to an eight- to ninefold increase observed with qi4 or qi5), coexpression of q(V→F) with the wt M_2 receptor resulted in a relatively modest increase in PLC activity (three- to fourfold).[17] To examine whether the functional effects caused by these single amino acid substitutions were additive, we generated three α_q double point mutants, q(YN→CG), q(YV→CF), and q(NV→GF). Figure 6.7 shows that the wt M_2 receptor was able to interact with q(YV→CF) and q(NV→GF) more efficiently than with either of the corresponding α_q single point mutants (q(Y→C), q(N→G), or q(V→F)). The q(YN→CG) mutant subunit (upon coexpression with the wt M_2 receptor and incubation with carbachol) could stimulate PLC activity to a maximum extent similar to that found with q(Y→C) and q(N→G) (Figure 6.7) but with significantly increased (three- to fourfold) carbachol potency.[17] Taken together, these studies indicate that the amino acids at positions –4 (Cys), –3 (Gly), and –1 (Phe/Tyr) play key roles in determining proper receptor recognition by $\alpha_{i/o}$ subunits.

6.2.3.2 Functionally Critical Residues at the C Terminus of $\alpha_{q/11}$ Subunits

By using an approach analogous to that described in the previous section, we also studied the functional importance of specific amino acids present at the C terminus of $\alpha_{q/11}$ subunits.[24] In this case, amino acids at the C terminus of wt α_s were replaced, either alone or in combination, with the corresponding α_q residues (Figure 6.5). The ability of the M_3 muscarinic or other $G_{q/11}$-coupled receptors (which do not couple well to wt G_s) to gain coupling to these mutant α_s subunits was then examined in cotransfected COS-7 cells. As a measure of productive receptor–G protein coupling, agonist-induced increases in intracellular cAMP levels were determined.[24]

As discussed in Section 6.2.1.2, the M_3 muscarinic receptor did not efficiently interact with wt α_s but was able to productively couple to the mutant α_s subunit, sq5, in which the last five amino acids of α_s were replaced with the corresponding α_q sequence (Figure 6.5). Figure 6.5 also shows that the C-terminal five amino acids of α_s differ in only three residues from the corresponding α_q sequence. We therefore prepared three α_s single point mutants, s(Q→E), s(E→N), and s(L→V), and all three possible α_s double point mutants, s(QE→EN), s(EL→NV), and s(QL→EV) (Figure 6.5). Western analysis showed that wt α_s and all mutant α_s subunits were expressed at similar levels.[24]

$$
\begin{array}{llccccccc}
 & & -7 & -6 & -5 & -4 & -3 & -2 & -1 \\
\alpha_s & - & \boxed{L} & R & Q & Y & E & \boxed{L} & L \\
\alpha_{olf} & - & L & K & Q & Y & E & L & L \\
\\
\alpha_{q,11} & - & L & K & \blacksquare E & Y & \blacksquare N & L & V \\
\alpha_{14} & - & L & R & E & F & N & L & V \\
\alpha_{15,16} & - & L & D & E & I & N & L & L \\
\\
\alpha_{i1,2} & - & L & K & D & \blacksquare C & G & L & \blacksquare F \\
\alpha_{i3} & - & L & K & E & C & G & L & Y \\
\alpha_{o1,2} & - & L & R & G & C & G & L & Y \\
\alpha_{t1,2} & - & L & K & D & C & G & L & F \\
\alpha_z & - & L & K & Y & I & G & L & C \\
\alpha_{gust} & - & L & K & D & C & G & L & F \\
\\
\alpha_{12} & - & L & K & D & I & M & L & Q \\
\alpha_{13} & - & L & K & Q & L & M & L & Q \\
\end{array}
$$

FIGURE 6.8 Alignment of the C-terminal sequences of different classes of mammalian G protein α subunits. The residues highlighted in black have been shown to be critically involved in determining the selectivity of receptor–G protein interactions (see text for details).[14,15,17,24,25] The boxed leucine residues at positions –2 and –7 are conserved among all mammalian α subunits and appear to be of general importance for efficient receptor–G protein coupling.[51–53] (From Wess, J., *Pharmacol. Ther.*, 80, 231, 1998. With permission.)

Strikingly, functional studies with cotransfected COS-7 cells showed that the wt M_3 receptor displayed significantly improved coupling (as compared to wt α_s) to two of the three studied α_s single point mutants, s(Q→E) and s(E→N) (Figure 6.5). On the other hand, coexpression of the wt M_3 receptor with s(L→V) did not result in a cAMP response that was significantly different from that seen with cells cotransfected with wt α_s. Coexpression of the wt M_3 receptor with the s(QE→EN) double point mutant led to a cAMP response that was similar in magnitude to that observed with sq5 (Figure 6.5), indicating that the functional effects of the Q→E and E→N point mutations were additive. In contrast, the s(EL→ NV) and s(QL→EV) double point mutants showed functional responses similar in magnitude to those found with the s(E→N) and s(Q→E) single point mutants, respectively. Similar results were obtained with two other $G_{q/11}$-coupled receptors, the V_{1a} vasopressin and the GRP receptors.[24]

In summary, these findings indicate that two C-terminal $\alpha_{q/11}$ residues, an Asn at position –3 and a Glu at position –5, play key roles in determining the receptor selectivity of $\alpha_{q/11}$ subunits. Consistent with this notion, these two residues are found, with no exception, in all members of the $\alpha_{q/11}$ protein family (α_q, α_{11}, α_{14}, α_{15}, and α_{16}; Figure 6.8).

Figure 6.8 also shows that the precise positions of the C-terminal Gα residues that are functionally important for determining the selectivity of receptor–G protein

TABLE 6.2
Summary of Studies Using "split GPCRs"

GPCR	Site of "split"[a]	Ref.
β_2-adrenergic	i3	37
M_2 muscarinic	i2, i3, o3 (2nd outer loop)	38, Wess (unpublished data)
M_3 muscarinic	i2, i3, o3 (2nd outer loop)	40
Rhodopsin	i2, i3, o3 (2nd outer loop)	39, 41
V_2 vasopressin	i3	42
GnRH	i3	43
Neurokinin-1	i3	44

[a] Only those "split sites" are listed where coexpression of the resulting receptor fragments resulted in ligand binding and/or functional activity.

interactions vary between different functional classes of Gα subunits. However, as outlined in this and the previous section, the −3 residue is of fundamental importance for proper receptor recognition in both $\alpha_{i/o}$ and $\alpha_{q/11}$ subunits. The residues present at this position are perfectly conserved within individual Gα subfamilies (Figure 6.8) and can correctly predict the receptor coupling profile of a given Gα subunit. It is therefore likely that the corresponding residues in α_s and $\alpha_{12/13}$ subunits (Glu and Met, respectively) also play key roles in dictating receptor–G protein coupling selectivity. This hypothesis can be tested experimentally by applying coexpression strategies similar to those described here.

6.3   USE OF "SPLIT RECEPTORS" AS TOOLS TO STUDY GPCR STRUCTURE

Several studies have shown that GPCRs, like other polytopic transmembrane proteins,[36] can be assembled from two or more independently stable receptor fragments.[37-44] For these studies, GPCRs were "split" in various intracellular and extracellular loops by using recombinant DNA techniques. The resulting fragment pairs were then coexpressed in cultured mammalian cells or *Xenopus* oocytes, followed by their pharmacological, biochemical, and functional characterization. Such studies showed that several polypeptide pairs were capable of assembling into functional GPCR complexes (see Table 6.2 for a summary of such studies). These observations strongly support the notion that the folding of GPCRs (like that of other polytopic transmembrane proteins[36]) occurs in two consecutive steps: in step I, individual folding domains are established across the lipid bilayer, which, in step II, are then assembled, most likely by helix–helix interactions, to form a functional receptor protein.

Over the past few years, we have carried out studies with split M_3 muscarinic[38,40] and V_2 vasopressin receptors[42] to learn more about GPCR structure and assembly. For example, the rat M_3 muscarinic receptor was "split" in all three intracellular (i1–i3) and all three extracellular loops (o2–o4) using PCR-based mutagenesis techniques (Figure 6.9A).[40] Fragment expression was studied via enzyme-linked

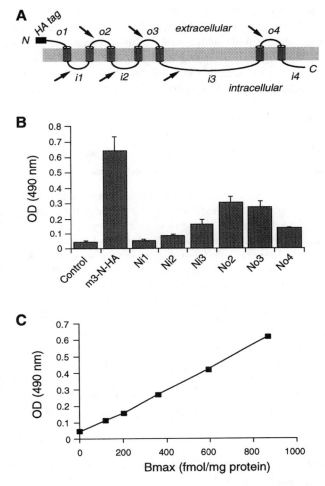

FIGURE 6.9   (A) The rat M_3 muscarinic receptor was "split" (arrows) in all three intracellular (i1–i3) and all three extracellular loops (o2–o4), resulting in six polypeptide pairs (Ni1+Ci1, No2+Co2, etc.). Note that an HA epitope tag was present at the N terminus of the wt receptor (resulting in M_3-N-HA) and all N-terminal receptor fragments.[40] (B) Cell surface expression of N-terminal M_3 receptor fragments studied via ELISA. COS-7 cells were transfected with plasmid DNAs (4 μg per 100-mm dish) coding for the wt M_3 receptor (untagged; control), M_3-N-HA, and N-terminal M_3 receptor fragments (Ni1–No4; the Ni1 fragment is truncated within the i1 loop, etc.). ELISA measurements were carried out using nonpermeabilized COS-7 cells as detailed in Protocol 6-4. (C) Linear relationship between M_3 muscarinic receptor (M_3-N-HA) density (B_{max}) and OD readings determined via ELISA. COS-7 cells were transfected in 100-mm dishes with increasing amounts of M_3-N-HA receptor DNA (0.125 to 4 μg, supplemented with vector DNA to keep the amount of transfected plasmid DNA constant at 4 μg). ELISA measurements were performed using nonpermeabilized COS-7 cells as described in Protocol 6-4. B_{max} values were determined in [3H]-NMS saturation binding studies. ELISA experiments and radioligand binding studies were carried out with the same batch of cells (data taken from Reference 40).

immunosorbent assay (ELISA) and confocal immunofluorescence microscopy, using the monoclonal antibody 12CA5 (which recognizes the HA epitope tag that was added to the N terminus of the full-length receptor and all N-terminal M_3 receptor fragments; Figure 6.9A), and the rabbit polyclonal antibody, anti-C-M_3, which interacts with the C-terminal sequence of the M_3 receptor protein. Immunofluorescence microscopic studies showed that most of the individual receptor fragments could be stably expressed in cultured COS-7 cells, apparently in the proper transmembrane topology.[40]

6.3.1 AN ELISA STRATEGY TO DETERMINE CELL SURFACE LOCALIZATION OF RECEPTOR FRAGMENTS

Expectedly, none of the individual M_3 receptor fragments was able to bind muscarinic radioligands when expressed alone.[40] We therefore developed an indirect cellular ELISA in order to quantitate the amount of N-terminal receptor fragments (Ni1–No4) present on the cell surface (see protocol 6-4 for experimental details).[40] In brief, COS-7 cells (nonpermeabilized) expressing the wt M_3 receptor or the various N-terminal M_3 receptor fragments were first incubated with the 12CA5 antibody, followed by the addition of a peroxidase-conjugated secondary antibody and the photometric determination of peroxidase activity. For control purposes, a wt M_3 receptor containing an HA epitope at its (intracellular) C terminus (M_3-C-HA) was also included in these experiments. Whereas cells expressing the N-terminally tagged wt M_3 receptor (M_3-N-HA) gave a robust ELISA signal (Figure 6.9B), cells transfected with M_3-C-HA led to OD readings similarly low as those found with cells expressing the non-tagged version of the wt M_3 receptor (data not shown). This observation clearly demonstrated the intactness of the plasma membrane barrier under the experimental conditions used for conducting the ELISA described in Protocol 6-4.

Protocol 6-4: ELISA to Measure Cell Surface Expression of GPCRs/GPCR Fragments (Example)

1. Transfect COS-7 cells with 4 μg of plasmid DNA coding for M_3-trunk (an N-terminal M_3 receptor fragment containing an HA epitope tag at its N terminus; Figures 6.9 and 6.10).
2. One day after transfections, split cells into 96-well plates (about 2.5×10^4 cells/well) and culture in a 5% CO_2 incubator at 37°C for 48 h.
3. Fix cells with 4% formaldehyde (Sigma) in PBS for 30 min at room temperature and wash cells twice with PBS (180 μl/well).
4. Add 180 μl 10% FBS in DMEM to each well and incubate at 37°C for 1 h.
5. Add 100 μl of 12CA5 monoclonal antibody (Boehringer Mannheim) (20 μg/ml in DMEM with 10% FBS) to each well and incubate at 37°C for 2 h (the 12CA5 antibody detects the HA epitope tag).
6. Wash plates four times with PBS (180 μl/well).

7. Add 100 µl of goat anti-mouse IgG antibody conjugated with horseradish peroxidase (Amersham Life Science) (1:1000 dilution in DMEM with 10% FBS) and incubate at 37°C for 1 h.
8. Wash plates four times with PBS (180 µl/well).
9. Add 100 µl of "color development solution" (2.5 mM H_2O_2 and 2.5 mM o-phenylenediamine (Sigma) in 0.1 M phosphate-citrate buffer, pH 5.0) and incubate at room temperature for 15 to 30 min.
10. Add 50 µl of 1 M H_2SO_4 containing 0.05 M Na_2SO_3 to terminate color development.
11. Measure optical densities at 490 and 630 nm (background) using the BioKinetics reader (EL 312, Bio Tek Instruments, Inc.).

In ELISA studies (Protocol 6-4), COS-7 cells transfected with most N-terminal M_3 receptor fragments (Ni2, Ni3, No2, No3, and No4) yielded OD readings that were significantly higher than background values (Figure 6.9B),[40] indicating that these polypeptides are present on the cell surface. To correlate ELISA signals with actual amounts of cell surface receptor/fragment protein, the M_3-N-HA receptor was expressed at different densities (B_{max}, determined in [^{3}H]-NMS saturation binding studies) by stepwise reduction of the amount of transfected plasmid DNA (Figure 6.9C). These experiments revealed a linear relationship between the amount of transfected receptor DNA and the number of detectable receptor binding sites (data not shown). In parallel, ELISA experiments were carried out with nonpermeabilized, M_3-N-HA-expressing COS-7 cells derived from the same batch of cells used for the B_{max} measurements. Figure 6.9C shows that the OD values observed in the ELISA were directly proportional to receptor densities (B_{max}) determined in the radioligand binding studies.[40] Based on this observation, it can be estimated that cell surface expression of most N-terminal M_3 receptor fragments is reduced by about 2- to 10-fold, as compared with the wt receptor (M_3-N-HA).

The ELISA strategy described in this section is also more generally useful to examine whether mutant GPCRs which are unable to bind radioligands are properly transported to the cell surface (see, for example, Reference 45). Such information is crucial for the proper interpretation of mutagenesis data since mutant GPCRs are frequently retained in an intracellular compartment.

6.3.2 ANALYSIS OF COEXPRESSED RECEPTOR FRAGMENTS IN RADIOLIGAND BINDING AND FUNCTIONAL ASSAYS

As already stated in the previous section, individual M_3 receptor fragments (when expressed in COS-7 cells) are functionally inactive (lack of radioligand binding and agonist-dependent G protein activation). However, we previously reported[40] that coexpression of three M_3 receptor polypeptide pairs (Ni2+Ci2, No3+Co3, and Ni3+Ci3) led to the appearance of a significant number of specific high-affinity [^{3}H]-NMS binding sites (Figure 6.10). As shown in Figure 6.10, these polypeptides were created by splitting the M_3 receptor in the i2, i3, or o3 regions. To study whether the three polypeptide complexes (Ni2+Ci2, No3+Co3, and Ni3+Ci3) were still

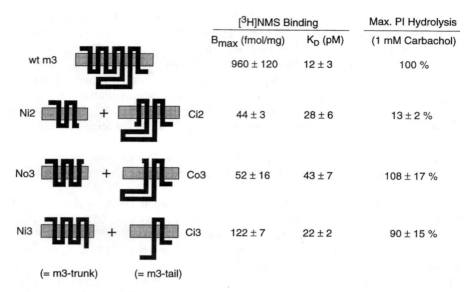

| | [3H]NMS Binding | | Max. PI Hydrolysis |
	B_{max} (fmol/mg)	K_D (pM)	(1 mM Carbachol)
wt m3	960 ± 120	12 ± 3	100 %
Ni2 + Ci2	44 ± 3	28 ± 6	13 ± 2 %
No3 + Co3	52 ± 16	43 ± 7	108 ± 17 %
Ni3 + Ci3	122 ± 7	22 ± 2	90 ± 15 %

(= m3-trunk)   (= m3-tail)

FIGURE 6.10 Radioligand binding and functional properties of polypeptide complexes formed upon coexpression of N- and C-terminal M_3 muscarinic receptor fragments. The precise amino acid composition of the various receptor fragments is given in Reference 40. COS-7 cells cotransfected with the indicated polypeptide pairs were studied for their ability to bind the radioligand, [3H]-NMS, and to mediate carbachol-induced (1 mM) increases in IP_1 levels.[40] For PI assays, the wt M_3 receptor was expressed at B_{max} levels comparable to those found with the coexpressed polypeptides (data taken from Reference 40).

capable of activating G proteins, their ability to mediate carbachol-induced stimulation of PI hydrolysis was examined. These studies showed that the No3+Co3 and Ni3+Ci3 complexes gave maximum PI responses similar to those observed with the wt M_3 receptor (Figure 6.10).[40] In contrast, splitting the wt M_3 receptor within the i2 loop resulted in a polypeptide complex that showed only residual functional activity,[40] indicating that the structural integrity of the i2 loop is critical for proper receptor–G protein coupling, at least within the muscarinic receptor family (see, however, References 39 and 41).

6.3.3 A SANDWICH ELISA STRATEGY TO STUDY GPCR ASSEMBLY

From a mechanistic point of view, it seems reasonable to assume that the appearance of functional receptors observed in the coexpression experiments summarized in Figure 6.10 is due to a direct association of the two corresponding receptor fragments. To demonstrate that this is in fact the case, we developed the following sandwich ELISA (Figure 6.11; see Protocol 6-5 for experimental details): in brief, microtiter plates are coated with the anti-C-M_3 antibody which is directed against the C terminus of the M_3 receptor protein. To the coated plates, lysates prepared from COS-7 cells cotransfected with an N-terminal M_3 receptor fragment and the corresponding C-terminal M_3 receptor polypeptide are added. The C-terminal

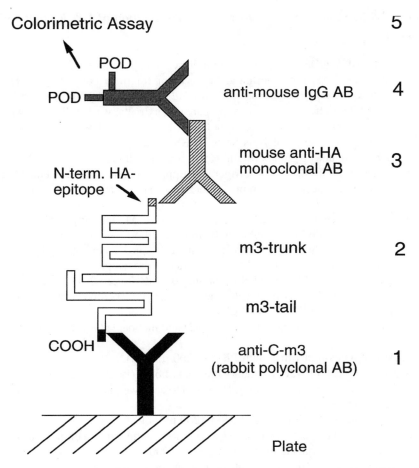

Colorimetric Assay 5

POD

POD — anti-mouse IgG AB 4

mouse anti-HA monoclonal AB 3

N-term. HA-epitope

m3-trunk 2

m3-tail

COOH anti-C-m3 (rabbit polyclonal AB) 1

Plate

FIGURE 6.11 Sandwich ELISA strategy useful for studying the association of N- and C-terminal M_3 muscarinic receptor fragments. (1) 96-well plates were coated with a rabbit polyclonal antibody (anti-C-M_3) directed against the C-terminal segment of the rat M_3 muscarinic receptor (highlighted in black). (2) Membrane lysates prepared from COS-7 cells cotransfected with M_3-trunk ($M3_{1-272}$) and M_3-tail ($M3_{388-589}$) (see Figure 6.10) were added to the coated plates, followed by extensive washing. (3) The M_3-trunk fragment bound to the M_3-tail polypeptide was detected with a monoclonal antibody directed against the N-terminal HA epitope tag (hatched). (4) Anti-mouse IgG antibody, linked to horseradish peroxidase (POD), was added, followed by extensive washing. (5) POD activity was determined by a simple color reaction following the addition of substrate (H_2O_2) and chromogen (o-phenylenediamine) (measure absorbance at 490 nm). For experimental details, see Protocol 6-5.

polypeptide is bound to the plate via the coating antibody, and the associated N-terminal receptor fragment can then be detected with the 12CA5 monoclonal antibody directed against the HA epitope tag attached to its N terminus. Finally, anti-mouse IgG antibody linked to horseradish peroxidase is added, and enzymatic activity is determined by a simple color reaction.

Protocol 6-5: Sandwich ELISA to Measure Association of GPCR Fragments (Example)

1. Cotransfect COS-7 cells in 100-mm dishes with plasmid DNAs coding for M_3-trunk DNA (an N-terminal M_3 receptor fragment containing an HA epitope tag at its N terminus) and M_3-tail (4 µg each) (see Protocol 6-2).
2. About 72 h after transfections, prepare cell extracts: wash cells once with PBS (10 ml) and collect cells in 1.5-ml Eppendorf tubes. Incubate cells with 150 to 300 µl of lysis buffer (PBS containing 1% digitonin, 0.5% deoxycholate, 1 µg/ml pepstatin A, 1 µg/ml aprotinin, and 0.1 mM PMSF) at 4°C for 1 h. Spin in a refrigerated Eppendorf 5417R microcentrifuge at maximal speed for 30 min. Collect supernatants for immediate use or storage at −80°C.
3. "Antibody coating" of ELISA plates: add anti-C-M_3 antibody (100 µl/well, 5 µg/ml in PBS) (a rabbit polyclonal antibody directed against the C terminus of the M_3 receptor protein) to 96-well plates (Nunc-Immuno Plate, MaxiSorp™) and incubate overnight at room temperature. Wash plates four times with PBS-T (PBS containing 0.075% Tween 20) (200 µl/well).
4. Add 200 µl of 5% BSA in PBS and incubate at 37°C for 1 h. Wash plates four times with PBS-T (200 µl/well).
6. Add 100 µl of cell extract (from step 2) and incubate at room temperature for 2 h.
7. Wash plates four times with PBS-T (200 µl/well).
8. Add 100 µl of 12CA5 monoclonal antibody (Boehringer Mannheim) (1 µg/ml in PBS-T containing 1% BSA) to each well and incubate at room temperature for 2 h.
9. Wash plates six times with PBS-T (200 µl/well).
10. Add 100 µl of goat anti-mouse IgG antibody conjugated with horseradish peroxidase (Amersham Life Science) (1:3000 dilution in PBS-T containing 1% BSA) and incubate at room temperature for 2 h.
11. Wash plates four times with PBS-T (200 µl/well).
12. Add 100 µl of "color development solution" (2.5 mM H_2O_2 and 2.5 mM o-phenylenediamine (Sigma) in 0.1 M phosphate-citrate buffer, pH 5.0) and incubate at room temperature for 15 to 30 min.
13. Add 50 µl of 1 M H_2SO_4 containing 0.05 M Na_2SO_3 to terminate color development.
14. Measure absorbance at 490 and 630 nm (background) using the BioKinetics reader (EL 312, Bio Tek Instruments, Inc.).

Figure 6.12 depicts the results of a typical sandwich ELISA. Cell lysates prepared from cells transfected with the M_3-trunk polypeptide (which corresponds to fragment Ni3 shown in Figure 6.10) alone gave only a very weak OD signal that was not significantly different from background determined with vector-transfected cells (data not shown). On the other hand, lysates prepared from cells cotransfected with M_3-trunk and M_3-tail (which corresponds to fragment Ci3 shown in Figure 6.10) resulted in a marked increase in OD readings (Figure 6.12A), indicative of fragment

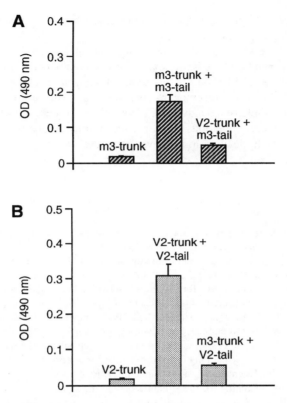

FIGURE 6.12   Sandwich ELISA demonstrating the receptor specificity of fragment association. (A) In one set of experiments, M_3-tail was coexpressed in COS-7 cells with M_3-trunk or V_2-trunk ($V2_{1-241}$; a human V_2 vasopressin receptor fragment that is structurally homologous to M_3-trunk). Fragment association was studied using the sandwich ELISA depicted in Figure 6.11. For these studies, plates were coated with a rabbit polyclonal antibody (anti-C-M_3) directed against the C-terminal portion of the M_3 receptor. (B) V_2-tail ($V2_{242-371}$; a human V_2 vasopressin receptor fragment that is structurally homologous to M_3-tail) was coexpressed with V_2-trunk or M_3-trunk. In this set of experiments, plates were coated with a rabbit polyclonal antibody (anti-C-V_2) directed against the C terminus of the V_2 receptor.

association. When the M_3-tail polypeptide was cotransfected with the V_2-trunk fragment (a fragment derived from the human V_2 vasopressin receptor that is structurally homologous to the M_3-trunk polypeptide[42]), the OD signal was strongly reduced (Figure 6.12A), indicating that efficient association of receptor fragments only occurs when they are derived from the same receptor. Very similar results were obtained when we studied the association of trunk and tail fragments derived from the V_2 vasopressin receptor (Figure 6.12B). Again, a pronounced increase in OD readings was seen with cell lysates prepared from cells cotransfected with V_2-trunk and V_2-tail (a fragment derived from the human V_2 vasopressin receptor that is structurally homologous to the M_3-tail polypeptide[42]). On the other hand, when the V_2-tail fragment was coexpressed with the M_3-trunk polypeptide, this signal was drastically reduced (Figure 6.12B), indicating that V_2-trunk recognizes the V_2-tail fragment with high specificity.

At present, little is known about the molecular interactions between individual TM helices that maintain GPCRs in a functionally competent conformation. However, the sandwich ELISA described here, which can measure the association between N- and C-terminal receptor fragments, should prove useful to shed light on this issue. For example, the sandwich ELISA offers a quick and sensitive method to evaluate the effect of mutations on proper receptor assembly. It is likely that such studies will eventually lead to an improved structural model of the transmembrane core of GPCRs.

REFERENCES

1. Coughlin, S. R., Expanding horizons for receptors coupled to G proteins: diversity and disease, *Curr. Opin. Cell Biol.*, 6, 191, 1994.
2. Schwartz, T. W., Locating ligand-binding sites in 7TM receptors by protein engineering, *Curr. Opin. Biotechnol.*, 5, 434, 1994.
3. Strader, C. D., Fong, T. M., Tota, M. R., Underwood, D., and Dixon, R. A. F., Structure and function of G protein-coupled receptors, *Annu. Rev. Biochem.*, 63, 101, 1994.
4. Dohlman, H. G., Thorner, J., Caron, M. G., and Lefkowitz, R. J., Model systems for the study of seven-transmembrane-segment receptors, *Annu. Rev. Biochem.*, 60, 653, 1991.
5. Savarese, T. M., and Fraser, C. M., In vitro mutagenesis and the search for structure-function relationships among G protein-coupled receptors, *Biochem. J.*, 283, 1, 1992.
6. Hedin, K. E., Duerson, K., and Clapham, D. E., Specificity of receptor-G protein interactions: searching for the structure behind the signal, *Cell. Sign.*, 5, 505, 1993.
7. Bourne, H. R., How receptors talk to trimeric G proteins, *Curr. Opin. Cell Biol.*, 9, 134, 1997.
8. Wess, J., G protein-coupled receptors: molecular mechanisms involved in receptor activation and selectivity of G protein recognition, *FASEB J.*, 11, 346, 1997.
9. Wess, J., Molecular basis of receptor-G protein-coupling selectivity, *Pharmacol. Ther.*, 80, 231, 1998.
10. Gudermann, T., Kalkbrenner, F., and Schultz, G., Diversity and selectivity of receptor-G protein interaction, *Annu. Rev. Pharmacol. Toxicol.*, 36, 429, 1996.
11. Gudermann, T., Schöneberg, T., and Schultz, G., Functional and structural complexity of signal transduction via G protein-coupled receptors, *Annu. Rev. Neurosci.*, 20, 395, 1997.
12. Wall, M. A., Coleman, D. E., Lee, E., Iniguez-Lluhi, J. A., Posner, B. A., Gilman, A. G., and Sprang, S.R., The structure of the G protein heterotrimer $G_{i\alpha1}\beta_1\gamma_2$, *Cell*, 83, 1047, 1995.
13. Lambright, D. G., Sondek, J., Bohm, A., Skiba, N. P., Hamm, H. E., and Sigler, P. B., The 2.0 Å crystal structure of a heterotrimeric G protein, *Nature*, 379, 311, 1996.
14. Liu, J., Conklin, B. R., Blin, N., Yun, J., and Wess, J., Identification of a receptor/G protein contact site critical for signalling specificity and G protein activation, *Proc. Natl. Acad. Sci. U.S.A.*, 92, 11642, 1995.
15. Conklin, B. R., Farfel, Z., Lustig, K. D., Julius, D., and Bourne, H. R., Substitution of three amino acids switches receptor specificity of $G_q\alpha$ to that of $G_i\alpha$, *Nature*, 363, 274, 1993.

16. Gomeza, J., Mary, S., Brabet, I., Parmentier, M.-L., Restituito, S., Bockaert, J., and Pin, J.-P., Coupling of metabotropic glutamate receptors 2 and 4 to $G_{\alpha15}$, $G_{\alpha16}$, and chimeric $G_{\alpha q/i}$ proteins: characterization of new antagonists, *Mol. Pharmacol.*, 50, 923, 1996.

17. Kostenis, E., Conklin, B. R., and Wess, J., Molecular basis of receptor/G protein coupling selectivity studied by coexpression of wild type and mutant m2 muscarinic receptors with mutant $G\alpha_q$ subunits, *Biochemistry*, 36, 1487, 1997.

18. Wess, J., Molecular biology of muscarinic acetylcholine receptors, *Crit. Rev. Neurobiol.*, 10, 69, 1996.

19. Wess, J., Brann, M. R., and Bonner, T. I., Identification of a small intracellular region of the muscarinic m3 receptor as a determinant of selective coupling to PI turnover, *FEBS Lett.*, 258, 133, 1989.

20. Wess, J., Bonner, T. I., Dörje, F., and Brann, M. R., Delineation of muscarinic receptor selectivity of coupling to guanine-nucleotide binding proteins and second messengers, *Mol. Pharmacol.*, 38, 517, 1990.

21. Altenbach, C., Yang, K., Farrens, D. L., Farahbakhsh, Z. T., Khorana, H. G., and Hubbell, W. L., Structural features and light-dependent changes in the cytoplasmic interhelical E-F loop region of rhodopsin: a site-directed spin labeling study, *Biochemistry*, 35, 12470, 1996.

22. Dratz, E. A., Furstenau, J. E., Lambert, C. G., Thireault, D. L., Rarick, H., Schepers, T., Pakhlevaniants, S., and Hamm, H. E., NMR structure of a receptor-bound G protein peptide, *Nature*, 363, 276, 1993.

23. Offermanns, S. and Simon, M. I., $G\alpha_{15}$ and $G\alpha_{16}$ couple a wide variety of receptors to phospholipase C, *J. Biol. Chem.*, 270, 15175, 1995.

24. Kostenis, E., Gomeza, J., Lerche, C., and Wess, J., Genetic analysis of receptor-$G\alpha_q$ coupling selectivity, *J. Biol. Chem.*, 272, 23675, 1997.

25. Conklin, B. R., Herzmark, P., Ishida, S., Voyno-Yasenetskaya, T. A., Sun, Y., Farfel, Z., and Bourne, H. R., Carboxyl-terminal mutations of $G_q\alpha$ and $G_s\alpha$ that alter the fidelity of receptor activation, *Mol. Pharmacol.*, 50, 885, 1996.

26. Baldwin, J. M., The probable arrangement of the helices in G protein-coupled receptors, *EMBO J.*, 12, 1693, 1993.

27. Baldwin, J. M., Schertler, G. F. X., and Unger, V. M., An alpha-carbon template for the transmembrane helices in the rhodopsin family of G protein-coupled receptors, *J. Mol. Biol.*, 272, 144, 1997.

28. Acharya, S., Saad, Y., and Karnik, S. S., Transducin-α C-terminal peptide binding site consists of C-D and E-F loops of rhodopsin, *J. Biol. Chem.*, 272, 6519, 1997.

29. Conklin, B. R. and Bourne, H. R., Structural elements of $G\alpha$ subunits that interact with $G\beta\gamma$, receptors, and effectors, *Cell*, 73, 631, 1993.

30. Rens-Domiano, S. and Hamm, H. E., Structural and functional relationships of heterotrimeric G proteins, *FASEB J.*, 9, 1059, 1995.

31. Hamm, H. E., Deretik, D., Arendt, A., Hargrave, P. A., Koenig, B., and Hofmann, K. P., Site of G protein binding to rhodopsin mapped with synthetic peptides from the α subunit, *Science*, 241, 832, 1988.

32. Higashijima, T. and Ross, E. M., Mapping of the mastoparan-binding site on G proteins: cross-linking of [^{125}I-Tyr^3,Cys^{11}]mastoparan to G_o, *J. Biol. Chem.*, 266, 12655, 1991.

33. Taylor, J. M., Jacob-Mosier, G. G., Lawton, R. G., Remmers, A. E., and Neubig, R. R., Binding of an α_2 adrenergic receptor third intracellular loop peptide to $G\beta$ and the amino terminus of $G\alpha$, *J. Biol. Chem.*, 269, 27618, 1994.

34. Kostenis, E., Degtyarev, M. Y., Conklin, B. R., and Wess, J., The N-terminal extension of $G\alpha_q$ is critical for constraining the selectivity of receptor coupling, *J. Biol. Chem.*, 272, 19107, 1997.

35. Kostenis, E., Zeng, F.-Y., and Wess, J., Functional characterization of a series of mutant G protein α_q subunits displaying promiscuous receptor coupling properties, *J. Biol. Chem.*, 273, 17886, 1998.

36. Popot, J.-L. and Engelman, D. M., Membrane protein folding and oligomerization: the two-stage model, *Biochemistry*, 29, 4031, 1990.

37. Kobilka, B. K., Kobilka, T. S., Daniel, K., Regan, J. W., Caron, M. G., and Lefkowitz, R. J., Chimeric α_2-, β_2-adrenergic receptors: delineation of domains involved in effector coupling and ligand binding specificity, *Science*, 240, 1310, 1988.

38. Maggio, R., Vogel, Z., and Wess, J., Reconstitution of functional muscarinic receptors by coexpression of amino and carboxyl terminal receptor fragments, *FEBS Lett.*, 319, 195, 1993.

39. Ridge, K. D., Lee, S. S. J., and Yao, L. L., In vivo assembly of rhodopsin from expressed polypeptide fragments, *Proc. Natl. Acad. Sci. U.S.A.*, 92, 3204, 1995.

40. Schöneberg, T., Liu, J., and Wess, J., Plasma membrane localization and functional rescue of truncated forms of a G protein-coupled receptor, *J. Biol. Chem.*, 270, 18000, 1995.

41. Yu, H., Kono, M., McKee, T. D., and Oprian, D. D., A general method for mapping tertiary contacts between amino acid residues in membrane-embedded proteins, *Biochemistry*, 34, 14963, 1995.

42. Schöneberg, T., Yun, J., Wenkert, D., and Wess, J., Functional rescue of mutant V2 vasopressin receptors causing nephrogenic diabetes insipidus by a coexpressed receptor polypeptide, *EMBO J.*, 15, 1283, 1996.

43. Grosse, R., Schöneberg, T., Schultz, G., and Gudermann, T., Inhibition of gonadotropin-releasing hormone receptor signaling by expression of a splice variant of the human receptor, *Mol. Endocrinol.*, 11, 1305, 1997.

44. Nielsen, S. M., Elling, C. E., and Schwartz, T. W., Split-receptors in the tachykinin neurokinin-1 system—mutational analysis of intracellular loop 3, *Eur. J. Biochem.*, 251, 217, 1998.

45. Liu, J., Schöneberg, T., van Rhee, M., and Wes, J., Mutational analysis of the relative orientation of transmembrane helices I and VII in G protein-coupled receptors, *J. Biol. Chem.*, 270, 19532, 1995.

46. Arai, H. and Charo, I. F., Differential regulation of G protein-mediated signaling by chemokine receptors, *J. Biol. Chem.*, 271, 21814, 1996.

47. Komatsuzaki, K., Murayama, Y., Giambarella, U., Ogata, E., Seino, S., and Nishimoto, I., A novel system that reports the G protein linked to a given receptor: a study of type 3 somatostatin receptor, *FEBS Lett.*, 406, 165, 1997.

48. Voyno-Yasenetskaya, T., Conklin, B. R., Gilbert, R. L., Hooley, R., Bourne, H. R., and Barber, D. L., $G\alpha 13$ stimulates Na-H exchange, *J. Biol. Chem.*, 269, 4721, 1994.

49. Tsu, R. C., Ho, M. K. C., Yung, L. Y., Joshi, S., and Wong, Y. H., Role of amino- and carboxyl-terminal regions of $G_{\alpha z}$ in the recognition of G_i-coupled receptors, *Mol. Pharmacol.*, 52, 38, 1997.

50. Bonner, T. I., Buckley, N. J., Young, A. C., and Brann, M. R., Identification of a family of muscarinic acetylcholine receptor genes, *Science*, 237, 527, 1987.

51. Garcia, P. D., Onrust, R., Bell, S. M., Sakmar, T. P., and Bourne, H. R., Transducin-α C-terminal mutations prevent activation by rhodopsin: a new assay using recombinant proteins expressed in cultured cells, *EMBO J.*, 14, 4460, 1995.

52. Osawa, S. and Weiss, E. R., The effect of carboxyl-terminal mutagenesis of $G_t\alpha$ on rhodopsin and guanine nucleotide binding, *J. Biol. Chem.*, 270, 31052, 1995.

53. Martin, E. L., Rens-Domiano, S., Schatz, P. J., and Hamm, H., Potent peptide analogues of a G protein receptor-binding region obtained with a combinatorial library, *J. Biol. Chem.*, 271, 361, 1996.

54. Berridge, M. J., Dawson, M. C., Downes, C. P., Heslop, J. P., and Irvine, R. F., Changes in the levels of inositol phosphates after agonist-dependent hydrolysis of membrane phosphoinositides, *Biochem. J.*, 212, 473, 1983.

7 G Protein-Coupled Receptor Kinases and Desensitization of Receptors

Hitoshi Kurose

CONTENTS

7.1 OVERVIEW OF RECEPTOR DESENSITIZATION

The responses induced by G protein-coupled receptors wane during prolonged stimulation by an agonist, which is referred to as desensitization. Desensitization is a phenomenon that prevents cells from making unfavorable responses. There are two types of desensitization: homologous and heterologous desensitization. Homologous desensitization is defined as the reduced response of a receptor induced only by the agonist that has stimulated the receptor (by the agonist that has stimulated the receptor). In contrast with homologous desensitization, heterologous desensitization is defined as the reduced response of a receptor, when cells are stimulated by agonists of different signaling pathways.[1] Desensitization consists of three processes (Figure 7.1). The first is receptor uncoupling from G proteins through phosphorylation of the receptors, which occurs in seconds to minutes. The second is internalization or sequestration that removes the receptors from the cell surface, which occurs in minutes. The third process is downregulation, i.e., the receptors disappear from the cell surface when agonist stimulation continues for hours.[1] Phosphorylation is mediated by second messenger-activated kinases and G protein-coupled receptor kinases (GRKs). Although second messenger-activated kinases such as cAMP-dependent protein kinase (PKA) and protein kinase C (PKC) phosphorylate sequences irrespective of the receptor conformation, GRKs phosphorylate the receptors in the active conformation. The phosphorylation of the receptors by GRKs is not enough to quench the coupling with G proteins. The binding of β-arrestins to the phosphorylated receptors efficiently prevents coupling with G proteins. It is believed that heterologous desensitization is mediated by second messenger-activated kinases, and homologous desensitization by GRKs and β-arrestins.[1]

So far, the cDNAs of six members of the GRK family have been cloned from several species, which exhibited common structural features (Table 7.1). They belong to the serine/threonine kinase family, and can be subdivided into three parts.[2,3] The central part consisting of ~270 amino acids, is the catalytic domain, and the ~180 amino acid amino-terminal part is assumed to be the receptor binding domain. The carboxyl-terminal part has various lengths (80 to 240 amino acids) and contains the regulatory domains. Rhodopsin kinase (GRK1) is post-translationally modified through farnesylation and carboxyl methylation at the carboxyl terminus, which is necessary for reversible translocation from the cytosol to the membrane in a light-dependent manner.[4] GRK2 is also called βARK1 due to its discovery as the kinase that can phosphorylate β-adrenergic receptors in an agonist-dependent manner, and has a pleckstrin homology (PH) domain and a βγ subunit binding domain at the carboxyl terminus.[3] The βγ subunit binding domain does not completely overlap the PH domain. The PH domain can bind to phosphatidylinositol-4,5-bisphosphate, which is also important for the translocation of GRK2 to the membrane. This βγ subunit dependency also supports homologous desensitization, because the βγ subunit is only released from G proteins that interact with activated receptors. GRK5 has polybasic regions in its amino and carboxyl terminals that interact with acidic phospholipids. Phospholipid-dependent autophosphorylation of GRK5 increased its phosphorylating activity.[5] GRK4 and GRK6 are palmitoylated, which is assumed to promote the translocation of these kinases to the membrane.[6,7] GRK4 has a splice site at both its amino and carboxyl terminuses, and thus exists as four spliced variants.[6] It is unknown whether the four spliced variants exhibit different substrate selectivities or activation mechanisms.

There is cross talk between GRKs and PKC.[3] PKC phosphorylates GRK2 and promotes the translocation of GRK2 to the membrane, resulting in enhanced phosphorylation activity of the kinase.[8] PKC also phosphorylates GRK5 and decreases its phosphorylation activity due to both decreased affinity for receptors and decreased phosphorylation activity.[9]

The consensus sequence(s) for GRK phosphorylation is largely unknown. GRK1 and GRK2 prefer amino acids serine and threonine in a neutral or acidic environment. Phosphorylation sites by GRK1 or GRK2 were identified as Ser334, Ser338, and Ser343 in the carboxyl terminus of rhodopsin or Ser396, Ser401, Ser407, and Thr384 in the carboxyl terminus of the $β_2$-adrenergic receptor, respectively.[10,11] In contrast with GRK1 and GRK2, GRK5 and GRK6 prefer serine and threonine in a basic environment.

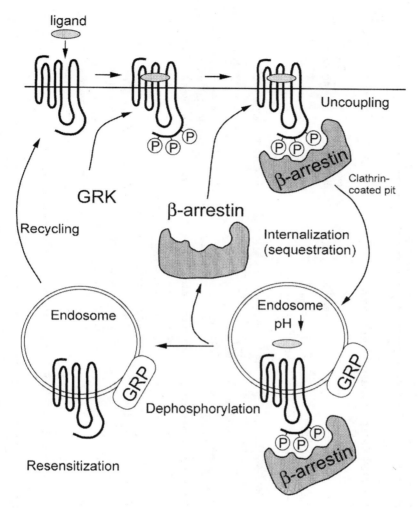

FIGURE 7.1 Early phase of desensitization. When a ligand binds to the receptor, it induces phosphorylation by GRK and subsequent binding of β-arrestin. The receptor–β-arrestin complex loses the ability to couple with G protein, and is internalized via clathrin-coated vesicles. The acidic conditions in the endosome induce conformational change of the receptor and facilitate dephosphorylation by G protein-coupled receptor phosphatase (GRP). The dephosphorylated receptor returns to the cell surface, and responds to a new stimulus.

So far, four arrestins have been cloned, and arrestin, β-arrestin 1, and β-arrestin 2 (also referred to as arrestin 3) have three, two, and two splice variants, respectively[12] (Table 7.2). In contrast with the restricted expression of arrestin and cone arrestin in rods and cones, both β-arrestin 1 and β-arrestin 2 exist ubiquitously. It is thought that β-arrestins bind to receptors phosphorylated by GRK but not second messenger-activated kinases such as PKA and PKC.[12] The subsequent binding of β-arrestins to the phosphorylated receptors then inhibits the coupling of the receptors with

TABLE 7.1
Molecular Properties of the G Protein-Coupled Receptor Kinase (GRK) Family

Family Name	Amino Acid	Common Name	Gene Structure	In Vitro Substrate	Regulators	Chromosome Localization	Tissue Distribution
GRK1	561	Rhodopsin kinase	7 exon	Rhodopsin, β_1^*, β_2^*, acidic peptide, and protein	$\beta\gamma$ ($\downarrow$), recoverin ($\downarrow$), autophosphorylation, farnesylation	13q34	Retina
GRK2	689	βARK1	21 exon	β_2, M_2^*, M_3^*, β_1, α_1, α_2^*, thrombin, substance P, N-formyl peptide, tubulin	$\beta\gamma$ ($\uparrow$), PH domain, PKC ($\uparrow$), calmodulin ($\beta\gamma$-dependent activation $\downarrow$)	11q13	Ubiquitous
GRK3	688	βARK2	Unknown	β_2, M_2, olfactory	PH domain	22q11	Ubiquitous (expression is ~10% of GRK2)
GRK4	578 (α) 546 (β) 532 (γ) 500 (δ)	IT11 (= GRK4β)	16 exon	β_2, chorionic gonadotropin, rhodopsin (α but not β, γ, δ)	Four splice variants (α to δ), palmitoylation, calmodulin ($\alpha\downarrow$), calmodulin does not bind to β, γ, δ isoforms	4q16.3	Testis, brain
GRK5	590	—	Unknown	Rhodopsin, β_2, thrombin, thyrotropin	Palmitoylation calmodulin ($\downarrow$), calmodulin ($\downarrow$), phospholipid, actin ($\downarrow$)	10q24-qter	Heart, lung, retina (associated with membrane)
GRK6	576	—	Unknown	Rhodopsin, β_2	Palmitoylation	5q35	Brain, pancreas, lung, heart, skeletal muscle

$^*\beta_1$, β_1-adrenergic receptor; β_2, β_2-adrenergic receptor; M_2, M_2 muscarinic acetylcholine receptor; M_3, M_3 Muscarinic acetylcholine receptor; α_1, α_1-adrenergic receptor; α_2, α_2-adrenergic receptor.

G proteins. Thus, it is believed that GRKs and β-arrestins play major roles in uncoupling at an early stage of desensitization. Recently, it was reported that GRK and β-arrestin have another function, i.e., they facilitate agonist-induced internalization. Overexpression of GRK restored the phosphorylation and internalization of a mutant of the β$_2$-adrenergic receptor (Y326A-β$_2$AR) that is not internalized at all upon agonist stimulation.[13] In contrast with GRK, overexpression of β-arrestin restored the internalization but not the phosphorylation of Y326A-β$_2$AR.[14] β-arrestin-mediated internalization is assumed to occur through clathrin-coated pits, because a dominant negative dynamin mutant (K44A-dynamin) can inhibit the internalization and β-arrestins can interact with clathrin at the carboxyl terminus.[15,16] Since the inhibition of internalization by high osmolarity, concanavalin A, or disruption of clathrin-coated vesicles prevented resensitization of receptors, internalization is a necessary process for phosphorylated receptors to restore the responsiveness.[17] Thus, the phosphorylation of the receptors by GRK plays two roles, i.e. desensitization and resensitization. The acidic conditions in the endosomes help to enhance dephosphorylation of the receptors by inducing a conformational change in them and facilitating attack by a phosphatase.[18]

Different tissues or cells express different combinations of GRKs and β-arrestins at different levels. This means that the extent and kinetics of receptor desensitization may differ from one cell to another. Thus, it is necessary to analyze the desensitization in native or isolated cells to understand the physiological regulation. Since there are more than 1000 G protein-coupled receptors, and less than 10 GRKs and arrestin isoforms in the genome, it is obvious that GRK phosphorylates and β-arrestins bind to many receptors. There are several methods for analyzing the involvement of GRK in desensitization at the cellular level. It has been reported that overexpression of dominant negative GRK2 impairs desensitization of the A$_2$ adenosine receptor and the β$_2$-adrenergic receptor, and also inhibits phosphorylation of the M$_2$ muscarinic acetylcholine receptor in intact cells.[19–21] Another approach is the expression of antisense oligonucleotides complementary to the sequences of GRK, PKA, and PKC to eliminate the endogenous kinases.[22] It is possible to examine the contribution of each kinase to the desensitization by analyzing the desensitization of cells. A third approach is co-transfection of the receptor with a GRK and/or a β-arrestin, which allows us to examine the roles of these proteins in the phosphorylation of the receptors and signaling evoked by receptor stimulation.[23,24] However, the results of overexpression of GRKs or β-arrestins indicate the potential interaction of receptors and GRK, which does not necessarily mean that the receptor is phosphorylated and desensitized by a specific GRK and β-arrestin combination in a cell. Furthermore, we cannot estimate the relative contribution of each GRK or β-arrestin to the desensitization in a particular tissue or cell. A novel approach is the intracellular application of polyclonal or monoclonal antibodies selective for one GRK or β-arrestin. The intracellular introduction of such selective antibodies revealed the involvement of GRK3 in the desensitization of the olfactory receptor and GRK3 in the desensitization of α$_2$-adrenergic receptor-mediated Ca^{2+} channel inhibition in chicken embryonic neurons.[25,26]

There are no low-molecular-weight compounds that can enter cells and selectively inhibit one GRK. Although heparin is often used as an inhibitor of GRKs *in vitro*, it cannot cross the membrane due to its positive charge. The carboxyl terminus

TABLE 7.2
Molecular Properties of the Arrestin Family

Name (Proposed nomenclature)	Amino Acid	Specificity	Tissue Distribution	Chromosome Localization	Regulation	Splice Variants
Arrestin (arrestin-1)	404 396 370	rhodopsin >> β_2,[a] M_2[b]	Rod cell	2q37	Unknown	3
β-arrestin 1 (arrestin-2)	418 410	β_2, M_2 > rhodopsin	Ubiquitous	11q13	Clathrin binding, phosphorylation	2
β-arrestin 2 (arrestin-3)	420 409	β_2, M_2 > rhodopsin	Ubiquitous, olfactory	17q13	Clathrin binding	2
Cone arrestin (arrestin-4)	388	Unknown	Cone cell	Xcen-q21	Unknown	Unknown

[a]β_2-adrenergic receptor; [b]M_2 muscarinic acetylcholine receptor.

of GRK2 interacts with the $\beta\gamma$ subunit, which is important for the translocation and activation of GRK2. Overexpression of the carboxyl terminus of GRK2 (βARK1-ct) is expected to impair the $\beta\gamma$ subunit-dependent processes in cells, including activation of GRK2 and GRK3 (GRK2- and GRK3-mediated phosphorylation of receptors).[27] Since the $\beta\gamma$ subunit activates several effector molecules, such as phospholipase C-β, phosphatidylinositol 3-kinase, adenylyl cyclase type II/IV, and MAP kinase, the inhibition of desensitization by βARK1-ct does not necessarily mean the involvement of GRK2 and GRK3 in desensitization.

I would like to focus on methods for analyzing the interaction of GRK, β-arrestin and receptor in desensitization at the cellular level.

7.2   RECEPTOR EXPRESSION IN MAMMALIAN CELLS

7.2.1   EXPRESSION IN CULTURED CELL LINES

7.2.1.1   Stable Transfection

It is important to establish stable cell lines for examining the function and regulation of receptors. The receptor cDNA is subcloned into expression vectors, and then transfected into cells. When an expression vector does not contain a drug-resistant gene, the receptor should be co-transfected with a drug-resistant gene. The G418 (Geneticin)-resistant gene is most commonly used for such a purpose, due to its high efficiency in the production of colonies. Selection drugs and corresponding genes other than G-418 are available for blasticidin S (Kaken Pharmaceutical Co., Japan), zeocin (Invitrogen), and hygromycin (Invitrogen). Every cell exhibits different sensitivities to the drugs used for selection. The concentration of a selection drug which kills all of the cells that do not express enough of the drug-resistant gene must be determined before the transfection. Several clones should be analyzed after the selection, because there is the possibility of picking up a clone which exhibits a different phenotype from the rest of the clones. When two genes are expressed in the same cell, it is simple to use two different drug-resistant genes. Otherwise, two genes with the same selection marker are co-transfected and the cells are screened with two indexes.

Compared with transient transfection, stable cells have several advantages. With transient transfection, a small percentage of the cells expresses the recombinant proteins, and the expression levels are usually high. This may not represent physiological interactions between the expressed proteins. Although it is time-consuming to establish stable cell lines, stable cells can be used for experiments to examine the long-term effects of agonists or other reagents.

Chinese hamster ovary (CHO) cells are often used to stably express recombinant proteins, because they grow rapidly and there is a correlation between growth rate and efficiency of producing resistant colonies. Agonist stimulation of the α_2-adrenergic receptor expressed in CHO cells revealed the biphasic responses of adenylyl cyclase. At a low agonist concentration, the α_2-adrenergic receptor inhibited the cyclase activity, and at a higher concentration, it stimulated the cyclase.[28] When CHO cells expressing α_2-adrenergic receptors were pretreated with an agonist, epinephrine, the α_2-adrenergic receptor-induced inhibition of adenylyl cyclase was desensitized. The dose–response curve of epinephrine shifted to the right.[23] As the level of expression is critical for observing desensitization of α_2-adrenergic receptor in CHO cells, it is important to establish cell lines that express the receptor at high levels. In contrast with inhibitory receptors, desensitization of stimulatory receptors such as the β_2-adrenergic receptor can be observed with a relatively low level of the receptors.

Protocol 7-1: Stable Transfection of CHO Cells

1. The day before transfection, plate CHO cells at a density of 3×10^5 cells in two 25-cm^2 flasks (for one transfection, prepare two flasks).
2. On the day of transfection, make two solutions for transfection. Solution I comprises 280 mM NaCl, 50 mM HEPES (pH 7.13), and 1.5 mM Na$_2$HPO$_4$; solution II comprises 25 mM CaCl$_2$, 30 µg DNA mixture of cDNA, and 3 µg drug-resistant marker gene (if the expression construct of cDNA does not have a drug-resistant gene). Add 500 µl of solution II to 500 µl of solution I dropwise while stirring solution I. Leave the mixture undisturbed for 30 min at room temperature.
3. Add 500 µl of the mixture to each flask containing 5 ml of F-12 medium with 10% fetal bovine serum (FBS).
4. Incubate at 37°C overnight.
5. Aspirate the medium off and wash the flasks with F-12 medium without FBS. Add 3 ml of 15% glycerol in 140 mM NaCl, 25mM HEPES (pH 7.13), and 0.75 mM Na$_2$HPO$_4$, and then incubate for 2 min at 37°C.
6. Remove the medium, wash the flasks twice with 5 ml of F-12 medium without FBS, then add 5 ml of F-12 medium with FBS.
7. The next day, change the medium.
8. Two days after transfection, wash the cells twice with 5 ml of phosphate-buffered saline (PBS), and then incubate in 0.5 ml of 0.25% trypsin and 1 mM ethylene diamine tetraacetic acid (EDTA) in PBS for 1 min at 37°C. Add 5 ml of F-12 medium with 10% FBS and a selection drug (use 1 mg/ml of G-418 for CHO cell) (selection medium) to each flask, combine the two flasks (total volume, ~12 ml), and dispense 4 ml and 1 ml of the cell suspension into two 100-mm dishes, and 0.5 ml into four 100-mm dishes. Add the selection medium to give a total volume of 8 to 10 ml.
9. Change the medium to new selection medium every 3 or 4 days.
10. When the colonies are visible with the eyes, they can be picked up by scratching and sucking with a Pipetman tip (20 to 200 µl). Typically 50 to 100 µl is enough to suck up the cells from a colony. We routinely pick up 48 colonies. Add 2 ml of the selection medium to each well.
11. When the cells have grown and become confluent, they are transferred to the wells of two separate 6-well plates for assay and further expansion.
12. Determine the expression level of the receptor with membrane prepared from one 6-well plate. Keep one plate for further expansion and stock. Typically ~30% of colonies express receptors. Once the stable cell line

is established, the concentration of G-418 can be reduced to 0.5 mg/ml to maintain the selection.

7.2.1.2 Transient Transfection

Although transient transfection has several limitations, it provides a quick assay for analyzing receptor mutants, and the interaction between receptors, kinases, and β-arrestins. The transient transfection assay takes only 3 to 4 d, as compared with at least 1 month to establish stable cell lines. The following protocol involves the transfection of HEK 293 cells to measure internalization of the β_2-adrenergic receptor.

Protocol 7-2: Transient Transfection of HEK 293 Cells

1. Grow HEK 293 cells in Dulbecco's modified Eagle's medium (DMEM) containing 10% FBS (DMEM + FBS) to confluency in a 75-cm^2 flask.
2. Wash the cells with PBS and then incubate with PBS containing 0.5 mM EDTA at 37°C for 1 min. The cells will become detached from the plates. After diluting the cells in the DMEM + FBS, centrifuge at 500 rpm for 3 min. High-speed centrifugation will have a detrimental effect on HEK 293 cells. Resuspend the cell pellet in the DMEM + FBS, determine the cell number, and plate onto 6-well plates at a density of 2×10^5 cells/well in a final volume of 2 to 3 ml. One 75-cm^2 flask is enough for two 6-well paltes. Regular tissue culture dishes or plates work well to maintain HEK 293 cells, but the 6-well plates for transfection should be collagen-coated (for instance, Falcon, catalog no. 3846). The cells are split with 0.5 mM EDTA in PBS but not with trypsin-EDTA solution. When HEK 293 cells are continually cultured for more than 2 months, good expression and responses such as agonist-induced internalization are lost. Thus, it is important to stock a lot at an early stage.
3. Next day, the cells are transfected with lipofectamine (Gibco). Make two solutions for transfection. Solution I comprises cDNA (total amount, 1 μg) in 100 μl of Opti-MEM medium (Gibco), and solution II comprises 5 μl of lipofectamine in 100 μl of Opti-MEM. The total amount of cDNA is adjusted with the vector plasmid. Mix them together and leave the mixtures undisturbed for 15 to 45 min. While waiting, the DMEM + FBS is removed from the cultured flask and the cell is washed once with 2 ml of FBS-free DMEM, and then the medium is replaced with 0.8 ml of Opti-MEM.
4. Add the lipofectamine cDNA mixture to the cells, and incubate for 3 h at 37°C, then add 1 ml of DMEM containing 20% FBS.
5. Two days after the transfection, the cells are ready for the experiments.

7.2.2 EXPRESSION IN PRIMARY CULTURES BY RECOMBINANT ADENOVIRUS

As mentioned earlier, it is necessary to examine the regulation of a receptor in the native environment. However, it is difficult to transfect primary cultures of cells isolated from tissues with methods such as calcium phosphate precipitation and lipid-mediated transfection, as compared with immortalized cell lines. Virus-mediated expression has several advantages as compared with other methods. Recombinant proteins can be expressed at high levels in almost all kinds of cells, including primary cultures of cells such as myocytes and neurons. The effect of a recombinant protein on physiological responses can then be determined. Retrovirus is one of the choices for a virus-mediated expression system.[29] However, retrovirus can infect only dividing cells, and is not suitable for primary cultures of cells.[30] In contrast with retrovirus, adenovirus can infect and express the proteins in cells that are not dividing. It has been thought that it is time-consuming and labor intensive to produce recombinant adenoviruses at high titers because of the inefficiency of production of the viruses. However, a couple of groups recently reported simpler and more efficient methods for producing the recombinant adenoviruses.[31,32]

One of the methods utilizes a cosmid vector for homologous recombination with the adenovirus genome in HEK 293 cells.[31] Adenovirus vector pAxCAwt is 45 kb in size, and lacks the E1a, E1b, and E3 genes. The E1a gene product is essential for the adenovirus promoter to work in the cells. HEK 293 cells constitutively express the E1a and E1b proteins, and thus adenovirus can only grow in HEK 293 cells. The recombinant virus does not have the E1a gene in its genome, and thus it can enter cells but cannot produce viruses in cells other than HEK 293 cells.

Figure 7.2 is an outline of the method for producing recombinant adenoviruses, for which a kit is available commercially (TAKARA, Japan) (Figure 7.3). We modified their protocol because some parts did not work well. There are three steps for producing recombinant adenoviruses. The first step is insertion of cDNA into a cosmid vector. The second step is transfection of the cosmid vector with adenovirus DNA, and screening for the recombinant viruses. The third step is amplification and determination of the virus titer.

Protocol 7-3: Ligation of cDNA With a Cosmid Vector

1. Remove the entire coding region of cDNA from the vector, blunt both ends, and then purify the fragment (~0.2 µg). When the 5′-untranslated region is long and/or does not have a Kozak sequence, the 5′-untranslated region around the first methionine should be changed by polymerase chain reaction (PCR) to optimize the expression. The typical sequence of the 5′-untranslated region is a restriction site followed by sequence CCAC-CATG (the underlined sequence is the first methionine).

2. Digest 1 µg of cosmid vector pAxCAwt with 40 to 50 U SwaI in a total volume of 50 µl for 1 h at 25°C.

3. Stop the reaction by phenol/chloroform extraction and then the supernatant is purified with a 1 ml spun-column. See Appendix E.37 to E.38 of Reference 31a, for spun-column chromatography.

4. Ligate 1 µg of SwaI-digested pAxCAwt vector with 0.2 µg of cDNA fragment in a total volume of 20 µl by incubation with 1 µl of T4 DNA ligase for more than 16 h at 16°C.

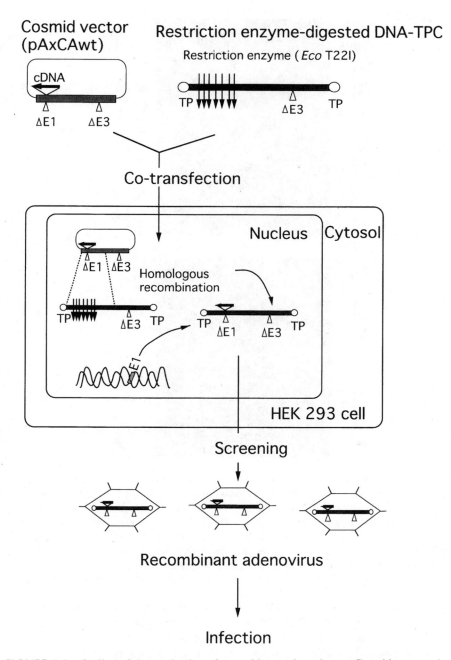

FIGURE 7.2   Outline of the production of recombinant adenoviruses. Cosmid vector pAx-CAwt containing an insert was co-transfected with the *EcoT22I*-digested adenovirus DNA-terminal protein complex into HEK 293 cells. The recombinant viruses were screened by PCR and amplified, then their titers were determined for infection of the cells.

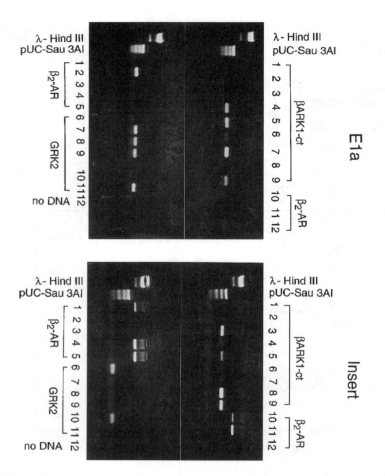

FIGURE 7.3 Screening of recombinant adenoviruses by PCR. (A) The inserts for βARK1-ct, β₂-adrenergic receptor and GRK2 were amplified by PCR, and the PCR products were subjected to 1.5% agarose gel electrophoresis. The sizes of the inserts are ~0.46 kb (βARK1-ct), ~1.9 kb (β₂-adrenergic receptor), and ~2.1 kb (GRK2), respectively. The sequence encoding E1a was also amplified by PCR with two primers as described in the text. The PCR products were subjected to 1.0% agarose gel electrophoresis. The size of the E1a sequence is ~0.45 kb. The insert-positive but E1a-negative clones are 3 and 8 for βARK1-ct, and 10, 11, 1, 4, and 5 for the β₂-adrenergic receptor. There were no positive clones for GRK2.

5. Add 0.5 µl of 2 M NaCl, 4 µl of 10× high-salt buffer (TAKARA), and 21 µl of H₂O, and then heat at 70°C for 10 min to inactivate the T4 DNA ligase.
6. Add 20 to 30 U SwaI, and then incubate for 1 h at 25°C.
7. Repeat the phenol/chloroform extraction and then purify by spun-column chromatography. The eluate is ~45 µl.
8. Repeat the SwaI digestion by adding 5 µl of the high-salt buffer and 30 to 40 U SwaI, followed by incubation for 3 h at 25°C. At this point, the total volume is ~50 µl.

9. Take 2 μl from 50 μl reaction mix and package it with GIGAPACK III XL-4 (Stratagene) according to the manufacturer's instructions.
10. Incubate for 90 to 120 min at room temperature.
11. Add 200 μl of SM buffer (NaCl, 5.8g; $MgSO_4 \cdot 7H_2O$, 2 g; 1 M Tris (pH 7.5), 50 ml; and 2% gelatin solution, 5 ml per 1000 ml), and then take 100 μl of packaged cosmid for infection of the treated *Escherichia coli* DH5α. The treated DH5α for infection was prepared as follows. Grow DH5α in 1 ml of LB medium containing 0.2% maltose overnight. The maltose solution should not be autoclaved, and should be prepared separately by filtering 20% stock solution, and then add to LB medium. Centrifuge and resuspend the *E. coli* in 0.5 ml of 10 mM $MgCl_2$.
12. After mixing the cosmid with the treated DH5α, incubate for 15 min at room temperature.
13. Add 1 ml of LB medium and incubate at 37°C for 30 min, and then plate 1 ml, 100 μl, and 10 μl of *E. coli* suspension onto three separate LB plates containing 50 μg/ml ampicillin.
14. Incubate at 37°C overnight, pick up the colonies, and incubate in 2 ml of LB medium containing ampicillin. It is important to pick up entire colonies together with the agar.
15. Prepare the cosmid by the alkali lysis method according to the protocol described in Reference 31a, page 1.25 to 1.28 (finally suspend the pellet in 50 μl of 0.5 mM EDTA). Check the insertion and orientation by PCR with 10 μl of cosmid. The entire coding sequence of cDNA inserted into pAxCAwt can be amplified with two primers (sense, 5′-GCT-CCT-GGG-CAA-CGT-GCT-GGT-3′; antisense, 5′-GGC-AGC-CTG-CAC-CTG-AGG-AG-3′, which are located near the *Swa*I site). The orientation can be determined with one of the above primers and an internal primer of cDNA.
16. Package the cosmid containing an appropriate insert in the proper orientation. The cosmids cannot be kept in a form of *E. coli*, because they will lose the internal repeat regions during storage. Grow *E. coli* in 100 to 150 ml of LB medium containing ampicillin, and purify the cosmid with QIAGEN maxi plasmid kits.

Protocol 7-4: Co-transfection of Cosmid Vector and Adenovirus DNA into HEK 293 Cells

1. HEK 293 cells are grown in DMEM containing 10% FBS, but can be grown in DMEM containing 5% FBS for the preparation of and infection with viruses. DMEM in the following protocols means DMEM containing 5% FBS unless otherwise stated.
2. Prepare HEK 293 cells in 60- and 100-mm dishes for transfection. The cells are nearly 100% confluent.
3. Mix 8 μg of cosmid and 5 μl (0.2 to 1 μg) of *Eco*T22I-digested DNA-TPC (terminal protein complex) that are provided by the supplier, and transfect the cells in a 60-mm dish by the calcium phosphate precipitation method (Promega), and incubate at 37°C overnight.

4. The next morning, replace the medium. When many cells are floating, they should be collected and replated.

5. In the evening, remove the cells and plate onto three collagen-coated 96-well plates at three different dilutions (original, 10-fold, and 100-fold dilution). To adjust the cell number for 10-fold and 100-fold dilution, use one-third of the HEK 293 cells grown in a 100-mm dish. The final volume of DMEM per well is 100 µl.

6. Add 50 µl of DMEM with 10% FBS to each well at 4 and 8 d after transfection.

7. At around 15 d after transfection, determine the infectivity and collect the cells together with the medium from positive wells, and release the viruses from the cells by passing the cell suspension through a 26G needle 15 times (crude virus lysate) instead of freezing-thawing cycles. When the freezing-thawing cycle is repeated many times, it will kill all of the viruses. Centrifuge at 5000 rpm for 10 min, and keep a portion of the supernatant as the *first seed virus* at –80°C.

8. Plate HEK 293 cells on a 24-well plate and infect the cells in two wells with 10 µl of the first seed virus. For infection, remove the medium, add 10 µl of virus with 100 µl of DMEM, and incubate for 1 h at 37°C with gentle rocking every 20 min to avoid drying the cells. Then add 0.5 ml of DMEM and continue the culture at 37°C.

9. After 3 d, almost all of the cells should have died. Prepare the virus separately from both wells according to the above-mentioned protocol and keep one as the *second seed virus*. Take 25 µl of the crude virus lysate prepared from another well, add 12.5 µl 4 *M* guanidine thiocyanate, and then leave for 10 min at room temperature.

10. Add 62.5 µl of H_2O, and take 2.5 µl to check the E1a sequence by PCR with two primers (sense, 5′-GAG-ACA-TAT-CTG-CCA-CGG-AGG-3′; antisense, 5′-TTG-GCA-TAG-AAA-CCG-GAC-CCA-AGG-3′). At the same time, check the insert by PCR (Figure 7.3).

11. Infect HEK 293 cells with the insert-positive but E1a-negative second seed viruses. Plate the HEK 293 cells in a 25-cm² flask, remove the medium, and then add 15 µl of the second seed virus. Incubate for 1 h at 37°C with gentle rocking every 20 min to avoid drying the cells, and then add 5 ml of DMEM. After 3 d, prepare the crude virus lysate according to the above-mentioned protocol. Dispense the virus into small aliquots and keep at –80°C as the *third seed virus*.

12. Prepare the fourth seed viruse with HEK 293 cells in a 75-cm² flask and 50 µl of the third seed virus. The volume of DMEM is 2 ml at infection and 15 ml at culture. Keep the virus as the *fourth seed virus* and determine the titer.

Protocol 7-5: Determination of the Adenovirus Titer

1. Dispense 50 µl of DMEM into each well of a collagen-coated 96-well plate.

2. Dilute the fourth seed virus to 10^{-1}–10^{-4} with DMEM.
3. Add 25 μl of 10^{-4}-diluted virus to the first row (8 wells).
4. Transfer 25 μl of the first row to the second row by use of a multichannel pipet. Repeat the dilution to the eleventh row. It is very important to use a new tip for each well, to avoid contamination of the virus. The resultant dilution is $3^{-1} \times 10^{-4}$ to $3^{-11} \times 10^{-4}$. The last row (twelfth row) is kept for noninfected control cells.
5. Prepare HEK 293 cells from a 100-mm dish and resuspend them in 6 ml of DMEM medium. Add 50 μl of the cell suspension to each well from the most diluted row.
6. Add 50 μl of DMEM at 4 to 5 d and 7 to 8 d after infection.
7. At 12 d after infection, determine the infectivity by examining the morphology and integrity of the cells.
8. Determine the $TCID_{50}$ (50% tissue culture infectious dose) value with Karber's equation. The $TCID_{50}$ value is essentially the same as the value determined by means of the conventional plaque forming method. $TCID_{50} = 10^x$; $x = \log$ (dilution of first row; $3^{-1} \times 10^{-4}$ in this case) $- \{(\Sigma a/b) - 0.5\} \times \log$ (dilution factor; 3 in this case), where a = number of wells in a row in which more than 50% of the cells died, and b = number of total wells in a row, 8 in this case. When the volume of the virus at infection is 50 μl, the titer is calculated with the following equation: titer (pfu/ml) = 1 ml ÷ 50 μl ÷ $TCID_{50}$.
9. The virus is now ready for infection.

Protocol 7-6: Expression of Recombinant Protein by Adenovirus

The susceptibility of cells to adenovirus infection varies with the cell type. The amount of virus (MOI = pfu/ml) that is enough for expressing the protein in almost all of the cells should be determined. Generally, cells derived from rat or mouse require ten times higher MOI for full infection as compared with cells of human origin. The following protocol is a method for expressing the β_2-adrenergic receptor in myocytes isolated from rabbit heart. However, the method of infection for other cells is essentially the same as this method.

1. Prepare ventricular myocytes from rabbit heart by the Langendorf method,[33] and plate onto laminin-coated dishes at a density of 1×10^6 cells/100-mm dish or at 10^5 cells/well onto a 6-well plate. The plates and dishes should be coated with 1.5 μg/cm² laminin in PBS. After 2 h, the plates and dishes are washed three times with PBS. Two to 3 h after plating, the cells become attached to the bottom of the dishes. Remove the medium and add the minimum volume of the fourth seed virus diluted in culture medium. The volume at infection is 1 ml and 0.3 ml for a 100-mm dish and a 6-well plate, respectively. Leave the dishes or plates at 37°C in CO_2 incubator.
2. Gently rock the plates or dishes three times every 20 min to avoid drying the cells.

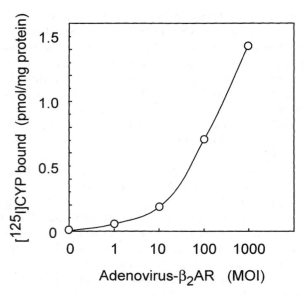

FIGURE 7.4   Expression of recombinant β_2-adrenergic receptors in rabbit myocytes. Myocytes were isolated from rabbit heart, and infected with recombinant adenovirus-β_2AR at the indicated MOI. After 2 d, the β-adrenergic receptor-binding activities of the myocytes were determined with 250 pM [^{125}I]-iodocyanopindolol.

3. One hour after infection, add DMEM and continue the culture for 2 d. One to 2 h is enough for the virus to infect the cells.
4. Prepare the crude membrane fraction and determine the binding activity (Figure 7.4). Label the cells with [^{3}H]-adenine (2 to 3 µCi/ml) from the day before the experiment to determine the intracellular cAMP accumulation. Serum such as calf serum may contain antibodies against adenovirus. Use serum-free medium for the infection.

7.3   DESENSITIZATION OF ADENYLYL CYCLASE ACTIVITY

Desensitization of adenylyl cyclase is easily measured by *in vitro* enzyme assay. However, desensitization of adenylyl cyclase in intact cells can be measured by cellular cAMP accumulation. The following protocol is a method for detecting desensitization by adenylyl cyclase activity.

Protocol 7-7: Measurement of Receptor Desensitization by Adenylyl Cyclase Assay

1. Before starting the experiments, we prepare Dowex and alumina columns to isolate the generated cAMP.[34] Prepare two 100 polyprep-columns (Bio-Rad) on a sheet of plexiglass which has poles at the four corners. The

size of the plexiglass sheet is adjusted to that of a rack of scintillation vials. Now we can easily handle the columns and collect the eluted cAMP directly in the scintillation vials. The bottom of all the columns should be replaced with glass wool to increase the flow rate. Dispense 1 ml of Dowex AG50W-X8 (100 to 200 mesh, H$^+$-form, Bio-Rad) or 1 ml of alumina powder (Alumina N-Super I, ICN or equivalent) into 100 columns. Regenerate the Dowex columns by washing once with 10 ml of 1 N HCl followed by twice with 10 ml of H$_2$O, and alumina columns by washing twice with 10 ml of 0.1 M imidazole (pH 7.65). The elution profile is also determined before the assay. It is determined by applying the cyclase mix including the stop solution and radioactive ATP without the membrane to the Dowex and alumina columns, and collecting the eluates every 1 ml.

2. Plate CHO cells expressing the receptor in two 100-mm dishes, and grow the cells to ~80% confluency.

3. Replace the medium with F-12 medium without FBS and G-418, and incubate 1 to 2 h at 37°C.

4. Add the agonist to the cells. Isoproterenol at a concentration of 1 to 10 μM fully activates the β$_2$-adrenergic receptor, and 100 μM epinephrine or norepinephrine is used for full activation of the α$_2$-adrenergic receptor. Hydrophilic agonists are better than hydrophobic synthetic agonists due to the easy washing out.

5. Place the plate on ice to stop the stimulation, and wash the cells five times with 10 ml ice-cold PBS. Then add 2 to 3 ml of lysis buffer (5 mM Tris, pH 7.4, and 5 mM EDTA) to the plates. Collect the cells with a cell lifter and centrifuge at 18,000 rpm for 10 min at 4°C to obtain a crude cell membrane fraction. Resuspend the membrane in 1 ml of buffer comprising 40 mM HEPES, pH 7.4, 0.8 mM EDTA, and 1.6 mM MgCl$_2$ with a Teflon pestle, and then use for the assay. These mild lysis conditions work well for measurement of the inhibition and stimulation of adenylyl cyclase.

6. Determine the adenylyl cyclase activity as follows. Dispense the adenylyl cyclase assay mixture containing the following components at the indicated concentrations, 0.12 mM ATP, 50 μM GTP, 2.8 mM phosphoenolpyruvate, 0.1 mM cAMP, 1 U/tube myokinase, 0.2 U/tube pyruvate kinase, 0.8 mM EDTA, 1.6 mM MgCl$_2$, 40 mM HEPES, pH 7.4, 1 to 2 μCi/tube [α-^{32}P]-ATP, and an appropriate concentration of agonist. The concentration of MgCl$_2$ should be increased to more than 10 mM to measure the stimulation of adenylyl cyclase.

7. After adding the membrane (20 to 30 μg) to each tube, mix and then transfer the tubes to a water bath at 37°C to start the reaction.

8. Place the tubes on ice after an appropriate time (typically 30 min). Then add 1 ml of the stop solution (0.5 mM ATP, 0.5 mM cAMP, and ~100,000 cpm [^{3}H]-cAMP) to each tube.

9. Apply the whole contents of the tubes to Dowex columns, and wash with 2 ml of H$_2$O. Place the rack of Dowex columns on the rack of alumina columns. Then elute the bound cAMP with 3 ml of H$_2$O. The eluates from

individual Dowex columns now directly pour into the corresponding alumina columns. Wash the alumina columns with 1 ml of 0.1 M imidazole (pH 7.65), and elute cAMP into vials with 3 ml of imidazole (pH 7.65). Elution profile from the columns may change when we make new columns.

10. Add 11 ml of the scintillation cocktail (Clearsol I from Wako, Japan) to each tube and count. We can use radiolabeled [^{3}H]-ATP (2.5 μCi/tube) instead of [α-^{32}P]-ATP, in which case we should use a stop solution containing [^{14}C]-cAMP (~10,000 dpm) instead of [^{3}H]-cAMP to calculate the recovery of the radiolabeled cAMP from each column.

11. After finishing the experiments, regenerate the Dowex and alumina columns.

7.4 RECEPTOR PHOSPHORYLATION

As phosphorylation is involved in the uncoupling of the receptors with G proteins and internalization of the receptors, determination of the phosphorylation state in cells is essential for understanding the regulation of the receptors. To analyze the phosphorylation of the receptors, anti-receptor antibodies recognizing some portion of the receptors are necessary to immunoprecipitate the phosphorylated receptors. Otherwise, the receptors should be tagged with specific sequences recognized by commercially available monoclonal antibodies. Epitope tags such as myc, flag, and HA (hemagglutinin) are frequently used for this purpose. It should be noted that the recognition of an antibody is dependent on the sequence around the inserted tag, for instance glycosylation may interfere with the binding of antibody. The phosphorylation conditions have been described for the α_2-adrenergic receptor[23] (Figure 7.5). However, the protocol works well for other receptors.

Protocol 7-8: Determination of Receptor Phosphorylation State

1. Wash cells once with phosphate-free DMEM (Gibco), and then incubate them for 1 to 2 h at 37°C with radiolabeled [^{32}P]-orthophosphoric acid (~9000 Ci/mmol). For a stable cell line, use 1 to 5 mCi for a 100-mm dish. For transient transfection, such as Cos-7 cells, 100 μCi to 1 mCi is used per 100-mm dish.

2. Stimulate the cells with 100 μM epinephrine for 20 min at 37°C.

3. Place the plate on ice to stop the stimulation, and then wash the cells five times with 10 ml of ice-cold PBS.

4. Add 1 ml of lysis buffer (10 mM Tris, pH 7.4, 5 mM EDTA, 10 mM sodium pyrophosphate, 10 mM sodium fluoride, 10 μg/ml benzamidine, 10 μg/ml trypsin inhibitor, and 5 μg/ml leupeptin) to each plate, and collect the cells with a cell lifter. Homogenize the cells by passing them several times through a needle (smaller than 24G) or brief sonication, and then centrifuge at 18,000 rpm for 10 min (or 15,000 rpm for 15 min) at 4°C. It is important to use tubes with rubber-sealed cups and to briefly spin the tubes before opening them.

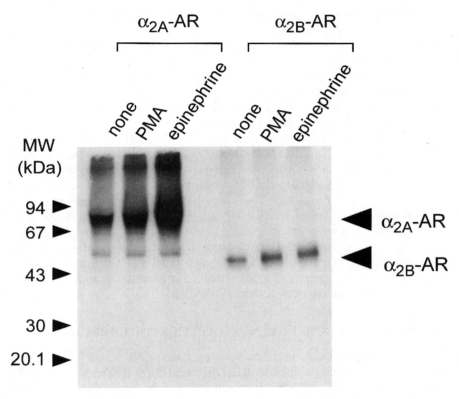

FIGURE 7.5 Phosphorylation of the α_2-adrenergic receptors in COS-7 cells. COS-7 cells expressing α_{2A}- or α_{2B}-adrenergic receptors were labeled with [^{32}P]-PO$_4$ (1 mCi/dish), and then stimulated with 100 μM epinephrine or 100 nM PMA for 20 min at 37°C. The phosphorylated α_2-adrenergic receptors were solubilized and immunoprecipitated with anti-receptor antisera. Both the α_{2A}- and α_{2B}-adrenergic receptors were phosphorylated through epinephrine (probably mediated by GRK) and PMA (probably mediated by PKC) stimulation. It should be noted that the background depends on the serum used for the immunoprecipitation.

5. Resuspend the pellet in 1 ml of the lysis buffer with the aid of a syringe with a needle or sonication, and then centrifuge it. Repeat this step two or three times.
6. Solubilize the membrane proteins with 1 ml of solubilization buffer (1% Triton X-100, 0.05% sodium dodecyl sulfate (SDS), 5 mM EDTA, 1 mM EGTA, 10 mM sodium fluoride, 10 mM sodium pyrophosphate, 50 mM Tris, pH 7.5, 10 μg/ml benzamidine, 10 μg/ml trypsin inhibitor, and 5 μg/ml leupeptin), and then continuously mix for 1 h at 4°C with an inversion wheel.
7. Centrifuge at 100,000 rpm for 15 min in a Beckman TL100 Ultracentrifuge or at 15,000 rpm for 30 min at 4°C.
8. Add 100 μl of protein A-Sepharose, and then mix continuously for 1 h at 4°C with an inversion wheel.

9. Briefly centrifuge to pellet the Protein A-Sepharose, and then transfer the supernatant to a new tube.

10. Incubate with 4 µl of anti-receptor antiserum and then incubate for 1 h overnight at 4°C.

11. Add 100 µl protein A-Sepharose, and then mix continuously with an inversion wheel for 1 h at 4°C.

12. Wash the Protein A-Sepharose four times with 1 ml of solubilization buffer, and then once with 1 ml of PBS.

13. Add 100 µl of SDS-sample buffer (50 mM Tris, pH 6.8, 10% glycerol, 1% SDS, 1% 2-mercaptoethanol, and 0.02% bromophenol blue) to the receptor–Protein A–Sepharose complexes, and then incubate for 1 h at 37°C to elute the receptors. It is important to elute the receptors at a relatively low temperature. The receptor–Protein A–Sepharose complexes should not be boiled because of aggregation of the receptors. As the expression level of the receptor co-transfected with a GRK is usually lower than that of the receptor alone, the receptor binding activities should be measured on separate plates for each transfection to normalize the amount of the receptors for SDS-PAGE. The samples are now ready for SDS-PAGE and autoradiography (Figure 7.5).

7.5   INTERACTION BETWEEN RECEPTORS AND ARRESTIN

To measure the physical coupling between receptors, GRKs, and β-arrestins, we should use a cross-linking reagent to detect the labile interaction between them. The following protocol is a method for detecting the interaction between β_1- or β_2-adrenergic receptors and β-arrestin 1.

Protocol 7-9: Cross-Linking of β-adrenergic Receptors and β-arrestin

1. Plate HEK 293 cells on 100-mm collagen-coated dishes, and then transfect with cDNAs of HA epitope-tagged β_1- or β_2-adrenergic receptors and β-arrestin 1 in a total amount of 8 µg, with ratios of the receptor to β-arrestin of 1:1 to 1:3, by means of the above-mentioned lipofectamine method. Incubate the cells in 6.4 ml of Opti-MEM at transfection, and then add 10 ml of DMEM containing 20% FBS after 3 h incubation with 20 µl lipofectamine and cDNAs diluted with 1.6 ml Opti-MEM. Receptor cDNA should be tagged or you should have an antibody that recognizes the amino-terminal portion of the receptor. The HA tag is recognized by monoclonal antibody 12CA5.

2. After 2 d, the medium is changed to 10 ml of FBS-free DMEM, followed by incubation at 37°C for 2 h.

3. Add isoproterenol or propranolol to a final concentration of 10 µM, and then incubate for 10 min at 37°C.

4. Remove the medium, and then add 3 ml of cross-linking buffer (10 mM HEPES, pH 7.4, 2.5 mM dithiobis(succinimidylpropionate) (DSP, Pierce),

and 10% (v/v) dimethylsulfoxide (DMSO) in PBS). The stock DSP solution is made in DMSO at a concentration of 25 mM. Incubate for 30 min at room temperature with occasional shaking to avoid drying the cells.

5. Place the cells on ice, and then replace the medium with 1 ml of lysis buffer (1% Triton X-100, 10% glycerol, 50 mM HEPES, pH 7.5, 10 mM NaCl, and 1 mM EDTA, 1 mM EGTA).

6. Collect the cells in a microcentrifuge tube, sonicate briefly to disperse them, and then centrifuge them for 10 min at 14,000 rpm at 4°C.

7. 10% of the supernatant is used for Western blotting to confirm that β-arrestin 1 is expressed in an equal amount. Add 100 μl of Protein G-Sepharose to the rest of the supernatant, and then incubate for 1 h at 4°C with continuous mixing with an inversion wheel.

8. Briefly centrifuge, transfer the supernatant to a new tube, and add 3 μg of anti-HA antibodies. Then incubate for 1 h at 4°C.

9. Add 60 μl of Protein G-Sepharose, and then incubate for 1 h at 4°C with continuous mixing with an inversion wheel.

10. Wash the Protein G-Sepharose three times with 1 ml of lysis buffer, add 60 μl of SDS-sample buffer, and then boil for 3 min. The samples are now ready for Western blotting.

7.6   RECEPTOR INTERNALIZATION

Internalization of G protein-coupled receptors can be measured in HEK 293 cells, because of their high expression of β-arrestin. HEK 293 cells are derived from human embryonic kidneys and consistently express adenovirus E1 proteins. The features of HEK 293 cells are high efficiency of transfection and human origin. A series of experiments involving HEK 293 cells demonstrated that internalization of the β_2-adrenergic receptor requires GRK-mediated phosphorylation and the subsequent binding of β-arrestin.

There are several methods for quantitatively measuring internalization. When a radiolabeled hydrophilic antagonist is commercially available, it will be very simple and easy to detect internalization by measuring the cell surface receptors remaining after agonist stimulation with this reagent at low temperature (typically at 4°C). It is important to perform the binding experiments at low temparature, because the internalization process is blocked and the reaction will not proceed further at low temperature (below 16°C). When radiolabeled hydrophobic and unlabeled hydrophilic antagonists are available, instead of a radiolabeled hydrophilic antagonist, the receptors located on the cell surface and intracellular vesicles can be measured at the same time by determining the binding activities in the presence of an unlabeled hydrophilic antagonist (or agonist) that displaces the ligand from only cell surface receptors or the presence of an unlabeled hydrophobic antagonist that displaces the ligand from both the cell surface and intracellular receptors. When appropriate radiolabeled ligands are not available, the receptors are tagged at the amino terminus. Internalization can then be measured by flow cytometry or enzyme-linked immunoassay by use of antibodies recognizing the tags.[35,36] As the expression level and/or kinds of components of the internalization machinery in one cell type are different from those in another, it is possible for the receptors to behave differentially between the cell types.

Green fluorescent protein (GFP) is used widely for monitoring the subcellular localization of receptors and β-arrestin.[37] It allows real-time measurement of movement. GFP should be fused to the carboxyl terminus but not to the amino terminus of G protein-coupled receptors. We constructed two fusion proteins of GFP with the β_2-adrenergic receptors; one fused to the amino terminus (GFP-β_2AR), and the other to the carboxyl terminus (β_2AR-GFP) of the β_2-adrenergic receptor (β_2-AR). When both fusion proteins were expressed separately in HEK 293 cells, only β_2AR-GFP was expressed and localized in the membrane. GFP-β_2AR was not expressed at all. It has been reported that glycosylation of the amino terminus is important for the trafficking of the β_2-adrenergic receptor to the membrane.[38] These data also suggest the importance of the amino terminus of the receptor for proper expression and trafficking. Some sequences among several GFP constructs are adjusted to increase the expression in mammalian cells. However, increased GFP luminescence intensity means that the number of luminescable GFP molecules increased, but does not necessarily mean that the intensity of the luminescence of the GFP molecule itself is increased. At present, several GFP mutants are available that exhibit different excitation and emission spectra. It may be possible to examine protein–protein interactions such as receptor–receptor dimer formation, receptor–GRK interaction, and receptor–β-arrestin interaction by the use of these GFP molecules.[39]

Protocol 7-10: Measurement of Receptor Internalization Using a Hydrophilic Ligand

1. Replace the medium with 1 ml of FBS-free DMEM, and leave for 1 h at 37°C.
2. Stimulate the cells with isoproterenol at appropriate concentrations and then incubate at 37°C in a CO_2 incubator.
3. After stimulation, place the plates on ice, and wash three times gently with 1 ml of ice-cold PBS.
4. Incubate the cells with 6 nM [^{3}H]-CGP-12177 (hydrophilic β-adrenergic antagonist) in the presence or absence of 10 μM propranolol in 1 ml of KRH buffer (136 mM NaCl, 4.7 mM KCl, 1.25 mM $MgSO_4$, 1.25 mM $CaCl_2$, 20 mM HEPES, pH 7.4, and 2 mg/ml BSA) for at least 3 h at 4°C.
5. Wash the cells three times gently with 1 ml of ice-cold PBS, and then dissolve the cells with 1 ml of 1% SDS.
6. Transfer to scintillation vials, add the scintillation cocktail, and count.

Protocol 7-11: Detection of β-adrenergic Receptor GFP Fusion Protein by Confocal Microscopy

1. Plate HEK 293 cells directly to cover a glass-bottom culture 35-mm dish (Mat Tek Corp., Ashland, MA) at a density of 1×10^5 cells/well, and then

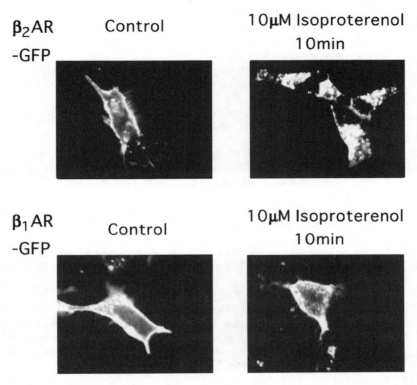

FIGURE 7.6   Redistribution of β_2- but not β_1-adrenergic receptors on isoproterenol stimulation detected by confocal microscopy. The β_1- and β_2-adrenergic receptor-GFP fusion proteins were expressed in HEK 293 cells, and stimulated with 10 μM isoproterenol for 10 min. The β_2- but not the β_1-adrenergic receptors were internalized through the agonist stimulation. This different sensitivity to agonist-induced internalization was confirmed by the receptor binding assay.

transfect with the GFP-tagged β_2- or β_1-adrenergic receptor cDNAs by means of the above-mentioned method.

2. The day after transfection, change the medium. A large number of cells may have became detached from the dishes, but continue the culture.

3. Two days after transfection, the cells are ready for the experiments.

4. The cells are stimulated with 10 μM isoproterenol for 10 min. Then the location of the β-adrenergic receptor-GFP fusion protein can be measured by confocal microscopy. The use of a glass-bottom culture dish eliminates the fixation step with paraformaldehyde, and allows us to detect the movement of the receptors in real time (Figure 7.6).

REFERENCES

1. Böhm, S. K., Grady, E. F., and Bunnett, N. W., Regulatory mechanisms that modulate signalling by G protein-coupled receptors, *Biochem. J.*, 322, 1, 1997.

2. Palczewski, K., GTP-binding-protein-coupled receptor kinases: two mechanistic models, *Eur. J. Biochem.*, 248, 261, 1997.

3. Pitcher, J. A., Freedman, N. J., and Lefkowitz, R. J., G protein-coupled receptor kinases, *Annu. Rev. Biochem.*, 67, 653, 1998.

4. Inglese, J., Koch, W. J., Caron, M. G., and Lefkowitz, R. J., Isoprenylation in regulation of signal transduction by G protein-coupled receptor kinases, *Nature*, 359, 147, 1992.

5. Kunapuli, P., Gurevich, V. V., and Benovic, J. L., Phospholipid-stimulated autophosphorylation activates the G protein-coupled receptor kinase GRK5, *J. Biol. Chem.*, 269, 10209, 1994.

6. Premont, R. T., Macrae, A. D., Stoffel, R. H., Chung, N., Pitcher, J. A., Ambrose, C., Inglese, J., MacDonald, M. E., and Lefkowitz, R. J., Characterization of the G protein-coupled receptor kinase GRK4: identification of four splice variants, *J. Biol. Chem.*, 271, 6403, 1996.

7. Stoffel, R. H., Randall, R. R., Premont, R. T., Lefkowitz, R. J., and Inglese, J., Palmitoylation of G protein-coupled receptor kinase, GRK6: lipid modification diversity in the GRK family, *J. Biol. Chem.*, 269, 27791, 1994.

8. Chuang, T. T., Levine, H., III, and De Blasi, A., Phosphorylation and activation of β-adrenergic receptor kinase by protein kinase C, *J. Biol. Chem.*, 270, 18660, 1995.

9. Pronin, A. N. and Benovic, J. L., Regulation of the G protein-coupled receptor kinase GRK5 by protein kinase C, *J. Biol. Chem.*, 272, 3806, 1997.

10. Ohguro, H., Van Hooser, J. P., Milam, A. H., and Palczewski, K., Rhodopsin phosphorylation and dephosphorylation *in vivo*, *J. Biol. Chem.*, 270, 14259, 1995.

11. Fredericks, Z. L., Pitcher, J. A., and Lefkowitz, R. J., Identification of the G protein-coupled receptor kinase phosphorylation sites in the human β_2-adrenergic receptor, *J. Biol. Chem.*, 271, 13796, 1996.

12. Krupnick, J. G. and Benovic, J. L., The role of receptor kinases and arrestins in G protein-coupled receptor regulation, *Annu. Rev. Pharmacol. Toxicol.*, 38, 289, 1998.

13. Ferguson, S. S. G., Ménard, L., Barak, L. S., Koch, W. J., Colapietro, A.-M., and Caron, M. G., Role of phosphorylation in agonist-promoted β_2-adrenergic receptor sequestration: rescue of a sequestration-defective mutant receptor by βARK1, *J. Biol. Chem.*, 270, 24782, 1995.

14. Ferguson, S. S. G., Downey, W. E., III, Colapietro, A.-M., Barak, L. S., Ménard, L., and Caron, M. G., Role of β-arrestin in mediating agonist-promoted G protein-coupled receptor internalization, *Science*, 271, 363, 1996.

15. Zhang, J., Ferguson, S. S. G., Barak, L. S., Ménard, L., and Caron, M. G., Dynamin and β-arrestin reveal distinct mechanisms for G protein-coupled receptor internalization, *J. Biol. Chem.*, 271, 18302, 1996.

16. Goodman, O. B., Jr., Krupnick, J. G., Santini, F., Gurevich, V. V., Penn, R. B., Gagnon, A. W., Keen, J. H., and Benovic, J. L., β-Arrestin acts as a clathrin adaptor in endocytosis of the β_2-adrenergic receptor, *Nature*, 383, 447, 1996.

17. Yu, S. S., Lefkowitz, R. J., and Hausdorff, W. P., β-adrenergic receptor sequestration: a potential mechanism of receptor resensitization, *J. Biol. Chem.*, 268, 337, 1993.

18. Krueger, K. M., Daaka, Y., Pitcher, J. A., and Lefkowitz, R. J., The role of sequestration in G protein-coupled receptor resensitization: regulation of β_2-adrenergic receptor dephosphorylation by vesicular acidification, *J. Biol. Chem.*, 272, 5, 1997.

19. Kong, G., Penn, R., and Benovic, J. L., A β-adrenergic receptor kinase dominant negative mutant attenuates desensitization of the β_2-adrenergic receptor, *J. Biol. Chem.*, 269, 13084, 1994.

20. Mundell, S. J., Benovic, J. L., and Kelly, E., A dominant negative mutant of the G protein-coupled receptor kinase 2 selectively attenuates adenosine A_2 receptor desensitization, *Mol. Pharmacol.*, 51, 991, 1997.

21. Tsuga, H., Kameyama, K., Haga, T., Kurose, H., and Nagao, T., Sequestration of muscarinic acetylcholine receptor m2 subtypes: facilitation by G protein-coupled receptor kinase (GRK2) and attenuation by a dominant-negative mutant of GRK2, *J. Biol. Chem.*, 269, 32522, 1994.

22. Shih, M. and Malbon, C. C., Oligonucleotides antisense to mRNA encoding protein kinase A, protein kinase C, and β-adrenergic receptor kinase reveal distinctive cell-type-specific roles in agonist-induced desensitization, *Proc. Natl. Acad. Sci. U.S.A.*, 91, 12193, 1994.

23. Kurose, H. and Lefkowitz, R. J., Differential desensitization and phosphorylation of three cloned and transfected α_2-adrenergic receptor subtypes, *J. Biol. Chem.*, 269, 10093, 1994.

24. Freedman, N. J., Ament, A. S., Oppermann, M., Stoffel, R. H., Exum, S. T., and Lefkowitz, R. J., Phosphorylation and desensitization of human endothelin A and B receptors: evidence for G protein-coupled receptor kinase specificity, *J. Biol. Chem.*, 272, 17734, 1997.

25. Oppermann, M., Diversé-Pierluissi, M., Drazner, M. H., Dyer, S. L., Freedman, N. J., Peppel, K. C., and Lefkowitz, R. J., Monoclonal antibodies reveal receptor specificity among G protein-coupled receptor kinases, *Proc. Natl. Acad. Sci. U.S.A.*, 93, 7649, 1996.

26. Dawson, T. M., Arriza, J. L., Jaworsky, D. E., Borisy, F. F., Attramadal, H., Lefkowitz, R. J., and Ronnett, G. V., β-adrenergic receptor kinase-2 and β-arrestin-2 as mediators of odorant-induced desensitization, *Science*, 259, 825, 1993.

27. Koch, W. J., Hawes, B. E., Inglese, J., Luttrell, L. M., and Lefkowitz, R. J., Cellular expression of the carboxyl terminus of a G protein-coupled receptor kinase attenuates $G_{\beta\gamma}$-mediated signaling, *J. Biol. Chem.*, 269, 6193, 1994.

28. Eason, M. G., Kurose, H., Holt, B. D., Raymond, J. R., and Liggett, S. B., Simultaneous coupling of α_2-adrenergic receptors to two G proteins with opposing effects: subtype-selective coupling of α_2C10, α_2C4, and α_2C2 adrenergic receptors to G_i and G_s, *J. Biol. Chem.*, 267, 15795, 1992.

29. Günzburg, W. H. and Salmons, B., Development of retroviral vectors as safe, targeted gene delivery system, *J. Mol. Med.*, 74, 171, 1996.

30. Wilson, J. M., Adenoviruses as gene-delivery vehicles, *New Engl. J. Med.*, 334, 1185, 1996.

31. Miyake, S., Makimura, M., Kanegae, Y., Harada, S., Sato, Y., Takamori, K., Tokuda, C., and Saito, I., Efficient generation of recombinant adenoviruses using adenovirus DNA-terminal protein complex and a cosmid bearing the full-length virus genome, *Proc. Natl. Acad. Sci. U.S.A.*, 93, 1320, 1996.

31a. Sambrook, J., Fritsh, E. F., and Maniatis, T., Eds., *Molecular Cloning: A laboratory manual*, 2nd ed., Cold Spring Harbor Laboratory Press, 1989.

32. He, T.-C., Zhou, S., Da Costa, L. T., Yu, J., Kinzler, K. W., and Vogelstein, B., A simplified system for generating recombinant adenoviruses, *Proc. Natl. Acad. Sci. U.S.A.*, 95, 2509, 1998.

33. Neely, J. R. and Rovetto, M. J., Techniques for perfusing isolated rat hearts, *Methods Enzymol.*, 39, 43, 1975.

34. Johnson, R. A. and Salomon, Y., Assay of adenylyl cyclase catalytic activity, *Methods Enzymol.*, 195, 3, 1991.

35. Barak, L. S., Tiberi, M., Freedman, N. J., Kwatra, M. M., Lefkowitz, R. J., and Caron, M. G., A highly conserved tyrosine residue in G protein-coupled receptors is required for agonist-mediated β₂-adrenergic receptor sequestration, *J. Biol. Chem.*, 269, 2790, 1994.
36. Schöneberg, T., Sandig, V., Wess, J., Gudermann, T., and Schultz, G., Reconstitution of mutant V2 vasopressin receptors by adenovirus-mediated gene transfer: molecular basis and clinical implication, *J. Clin. Invest.*, 100, 1547, 1997.
37. Barak, L. S., Ferguson, S. S. G., Zhang, J., and Caron, M. G., A β-arrestin/green fluorescent protein biosenser for detecting G protein-coupled receptor activation, *J. Biol. Chem.*, 272, 27497, 1997.
38. Rands, E., Candelore, M. R., Cheung, A. H., Hill, W. S., Strader, C. D., and Dixon, R. A. F., Mutational analysis of β-adrenergic receptor glycosylation, *J. Biol. Chem.*, 265, 10759, 1990.
39. Tsien, R. Y., The green fluorescent protein, *Annu. Rev. Biochem.*, 67, 509, 1998.

8 Internalization and Downregulation of G Protein-Coupled Receptors

Jelveh Lameh

CONTENTS

0-8493-3384-9/00/$0.00+$.50
© 2000 by CRC Press LLC

8.1   INTRODUCTION

As the number of receptors cloned and expressed in cell lines has increased, studies on the regulation of these receptors have also increased. Among the processes that are being studied in more detail are the trafficking of G protein-coupled receptors (GPCRs). The first of these processes is that of receptor internalization. Receptor internalization is the process by which cell surface receptor is translocated into intracellular compartments after exposure to the agonist. The tools that are used to measure this process are primarily those that allow for the distinction of two populations of the receptors, the ones at the cell surface and those inside the cell. The second trafficking process is that of downregulation. Receptor downregulation refers to a decrease in total receptor number. This process generally occurs by degradation of the receptors in the lysozomal compartments following prolonged agonist treatment.

The methods that are currently used for studying receptor internalization primarily depend on the specific receptor under study and the tools available for that specific receptor. For example, if hydrophilic ligands are available for the receptor of interest, measuring the binding of this ligand to the receptor can be used successfully for quantitating the cell surface receptors. Since the hydrophilic ligand will not cross the membrane, the sites that it will bind to in the intact cells will be only those at the cell surface. On the other hand, if a radioactive agonist is available for the receptor of interest, translocation of the agonist into the cell, along with the receptor, can be used as a measure of the extent of receptor internalization. More recently, advances in cell biology and the advent of fluorescent confocal microscopy have allowed the actual visualization of the internalized receptors in the intracellular vesicles.

Studying receptor downregulation has been more limited. For many years the two terms of receptor internalization and downregulation were used interchangeably and the studies did not clearly distinguish between the two processes. It was believed that once internalized, the receptors would only undergo downregulation. Thus, measuring either process would be a measure of receptor degradation. As detailed studies of receptor internalization have emerged, it has become clear that not all receptors that are translocated into the cytoplasmic compartments (internalized) will undergo downregulation. The internalized receptors can be resensitized and recycled to the cell surface[1,2] or be processed to the lysozome for degradation.[1] There is also some evidence that the internalized receptors can continue to signal.[3,4] At this time, the primary method for measuring receptor downregulation is carrying out binding assays with lipophilic ligand in intact cells or hydrophilic ligands in cell homogenates.

In this chapter the most common techniques used for assessing receptor internalization and downregulation are discussed in detail with reference to several studies applying these techniques. Care must be taken when applying these protocols to other receptors since most protocols may need to be customized for the specific receptors and cell lines used.

8.2 INTERNALIZATION

Receptor internalization is the process by which the receptor is removed from the cell surface and translocated into an intracellular compartment. The half time of this receptor translocation is around 7.5 to 10 min for most GPCRs.[5,6] This translocation of the receptor allows for distinguishing the two receptor populations by several different techniques, as detailed below.

8.2.1 BINDING STUDIES

Measuring the receptors remaining at the cell surface after agonist exposure has long been the method of choice for quantitating receptor internalization. The specific technique used will depend on the receptor under study and the tools available for that receptor, as discussed below. The protocols described below assume that a basic knowledge of the optimal binding conditions for the receptor under study is available.

8.2.1.1 Use of Hydrophilic Ligands

In the cases where radioactively labeled hydrophilic ligands are available for the receptor under study, the method detailed below can be used for quantitating the extent of receptor internalization in intact cells.

The main factors to consider for these assays are summarized as follows.

1. Saturating concentrations of the radioactive ligand must be used for all binding assays. Thus, first, saturating binding assays need to be carried out with the radioactive ligand. Once the concentrations of the labeled ligand leading to saturation of all the receptor sites is determined, that tracer concentration has to be used for the internalization assay.
2. The binding assay has to be carried out at temperatures below 10°C to prevent recycling of the internalized receptors to the cell surface.
3. Depending on the cell line used, after the binding assay the cell monolayer can be washed without significant loss of the cells. However, when cell lines such as human embryonic kidney cells (HEK 293) are used, the cell monolayer will come off at the end of the binding assay. In this case, the cells will have to be harvested and washed on a filter or by centrifugation.
4. For the binding assays of intact cells, 6-, 12-, or 24-well tissue culture plates can be used. The choice of the plate depends on the cell line used. If the cells attach firmly and express enough binding sites such that the binding from a single well of a 24-well plate is sufficient for data analysis, the smaller wells can be used. Otherwise, 12-well or even 6-well plates may have to be utilized.

Protocol 8-1: Assay of Receptor Internalization Using Hydrophilic Ligands[7-10]

1. Seed exponentially growing cells expressing the receptor on 12-well tissue culture plates.
2. Allow cells to reach confluency.
3. Remove old medium and replace with 0.5 to 1 ml serum-free medium containing the desired concentration(s) of the agonist. Triplicate or quadruple samples are recommended.
4. Incubate the cells with the drug for desired times at 37°C. For most GPCRs maximum internalization is reached after 15 to 30 min of agonist treatment.
5. Transfer the plates onto ice and allow the cells to cool for about 5 min.
6. Aspirate the medium containing the drug and wash the cell monolayer three to four times with ice-cold phosphate-buffered saline (PBS), pH 7.4, or serum-free medium.
7. Add binding buffer containing saturating concentrations of the radioactive ligand.
8. Use three wells for the measurement of nonspecific binding and three wells for measurement of total binding.
9. Incubate the cells at 0 to 10°C for the optimal time, depending on the specific binding requirements of the receptor and ligand used.
10. At the end of incubation, visually inspect the cells. If the cells are firmly attached, transfer the plates onto ice. Aspirate the radioactive tracer into specific radioactive waste container. Gently wash the cell monolayer three to four times with ice-cold binding buffer or any other buffer appropriate for the receptor under study.
11. Harvest the cells from the wells using buffer, transfer into scintillation vials, add scintillation fluid, and determine radioactivity. The cells can be lysed using 0.5 N NaOH or 1% sodium dodecyl sulfate (SDS) prior to counting.
12. If the cell monolayer does not remain firmly attached to the plate at the end of incubation, or it cannot withstand multiple washes, the cells can be washed using either one of the following two methods:
    a. Harvest the cells and transfer into properly labeled microcentrifuge tubes. Centrifuge the cells at low speed (less than $400 \times g$) to prevent cell lysis at 4°C. Remove supernatant and resuspend the cells in ice-cold buffer and centrifuge again. Repeat until the cells are washed three to four times. Transfer cells to scintillation vials and determine radioactivity.
    b. Harvest the cells from one well at a time and transfer onto a filter on a cell harvester with the vacuum off. Rinse the wells once with ice-cold buffer and add to the cells on the filter. Start the vacuum, aspirate the liquid, and rapidly wash the filter containing the cells three more times with ice-cold buffer. Go to the next sample. Keep the plates on ice while filtering the samples. If a Brandel-type harvester is available, cells from multiple wells can be dislodged at the same time and filtered and washed all at once. Ice-cold buffer has to be used at all times.

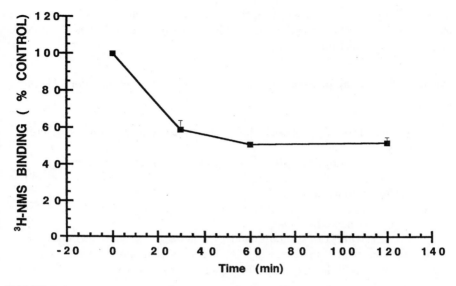

FIGURE 8.1   Time course of receptor internalization using hydrophilic ligand: cells stably expressing human muscarinic receptor subtype 3 were treated in the presence or absence of agonist for different times. After washing of the cells and binding with hydrophilic ligand, cells were harvested and radioactivity determined. See text for details of the assay.

Figure 8.1 shows internalization of muscarinic receptors as assessed by loss of the binding activity with a hydrophilic ligand, $[^3H]$-N-methylscopolamine, from the cell surface.

In the cases where hydrophilic ligand is not available in radiolabeled form, the nonradioactive form can be used.[11] Receptor binding is carried out with lipophilic radioactive ligand in the presence or absence of large excess of unlabeled hydrophilic ligand. Receptor internalization is defined as the percentage of binding that is not displacable with the nonradioactive hydrophilic ligand.

Another method for assessing receptor internalization with lipophilic ligand in intact cells is to carry out the reaction on ice where receptor internalization does not occur.[12] The problem with this technique is that a small amount of the radioactive ligand can still cross the membrane by simple diffusion and this will make accurate determination of the extent of receptor internalization difficult. Experimental procedures are as follows: (a) treat cells with the agonist at 37°C; (b) cool cells on ice and wash off the agonist; (c) carry out binding reaction on ice; (d) wash cells and measure radioactivity as described above.

8.2.1.2   Fluorescent Ligand Uptake, Live Cells

If a fluorescent agonist is available for the receptor under study, the translocation of the agonist during receptor internalization can be followed. This will be much like the methods following the endocytosis and recycling of proteins such as transferrin together with their receptor. For localization of these receptors and their ligands, a fluorescent microscopy technique needs to be employed. This method is mainly qualitative. For quantitation of this technique, see the section below for quantitative confocal microscopy.

Protocol 8-2: Assay of Receptor Internalization Using Fluorescent Ligand Uptake[13–16]

1. Grow cells on slides or coverslips as discussed below for confocal microscopy.
2. Wash cells with PBS, pH 7.4, and cool cells on ice for 10 min.
3. Label cells with the fluorescently labeled agonist at 4°C to achieve equilibrium binding, at least 1 h.
4. Wash the cells with ice-cold buffer and incubate at 37°C for desired time to induce internalization.
5. Fix cells with 2 to 4% paraformaldehyde in PBS, pH 7.4, for 20 min at 4°C.
6. Incubate with PBS containing 1% normal serum and 0.1% saponin for 5 min. Repeat this incubation process two more times to permeabilize the cells.
7. Wash the cells, mount the slides or coverslips, and visualize under appropriate settings of the microscope.

Fluorescent agonists can be used for following the receptor in live cells. For this protocol, the cells are treated as follows:[16] (a) cells are treated with fluorescent agonist and washed as described above; (b) cells are then incubated at 37°C on an inverted microscope equipped with a microincubator (PDMN-2, Medical Systems Corp., Greenvale, NY); (c) the endocytosis of the receptor and its ligand are visualized under the fluorescent microscope.

8.2.1.3 Radioactive Ligands

For GPCRs that have peptides as agonists, studying the process of internalization can be carried out in intact cells by the use of radioactively labeled agonists. In this method the agonist is allowed to bind the receptor. Following incubation the ligand that is bound to the cell surface receptors are stripped off using acidic medium, leaving only the ligands that are internalized into the cell and are inaccessible to acid. The radioactivity remaining associated with the cells is then determined, giving a measure of the internalized receptors.

Protocol 8-3: Assay of Receptor Internalization Using Radioactive Agonists[16–21]

1. Grow cells in 24-well plates until confluent.
2. Incubate with the radioactive tracer at a concentration of 0.2 to 2 nM for appropriate time at 37°C to induce internalization.
3. For nonspecific binding, use a 1000-fold excess of nonradioactive tracer.

4. At the end of incubation, transfer the cells to ice and rapidly wash twice with ice-cold PBS, pH 7.4.
5. Determine total binding by solubilizing the cells in 0.2 N NaOH and 1% SDS.
6. For assessing internalized receptors, strip the extracellular associated binding by washing once with 1 ml of acid solution (50 mM acetic acid and 150 mM NaCl, pH 2.8) for 12 min.
7. Collect acid wash to determine cell surface binding.
8. Quantitate internalized radioactivity after solubilizing the cells in SDS/NaOH solution.

Incubation with acid solution can be in any of the following conditions: (a) 0.2 M acetic acid, 500 mM NaCl, pH 2.5 or 2.0 for 5 min at 4°C;[18,20,22] (b) PBS-BSA (bovine serum albumin) solution adjusted to pH 2.0 with HCl, 3 min at 4°C;[23] (c) 50 mM glycine, 150 mM NaCl, pH 3.0 for 10 min at 4°C.[15,24]

8.2.2 Subcellular Fractionation

This technique takes advantage of the fact that different cellular organelles have different sizes and buoyant densities. Once lysed, different cellular components will be in suspension in a solution. When this solution is subjected to centrifugal force, the heavier particles will move faster and farther than the lighter particles. If a large size difference exists between two components, the two can be separated by differential centrifugation. The larger and heavier particles, such as the nuclei, can be sedimented by low-speed centrifugation. In most cases, however, the organelles to be separated are more homogeneous in size and the above method cannot be employed. In these cases, the particles can be separated on the basis of their buoyant density. A continuous or discontinuous gradient can be used in the range covering the buoyant densities of the particle of interest. The principle behind this type of gradient is that a particle will sediment, or float up, through the gradient until it reaches a point where the density of the medium equals that of the particle. In discontinuous gradients, the particle will reach an interface at which the density of the gradient exceeds or equals that of the particle. The denser particles continue to sediment. For example on a sucrose density gradient, the heavier membrane fraction of the cells will equilibrate with heavier, or denser gradient fraction.

The major problem associated with this technique is preparation of an ideal homogenate such that all organelles are released in the suspension as individual elements. Organelles can remain associated with the cytoskeletal elements surrounding the nucleus and sediment with the nuclear fraction.

Certain factors have to be kept in mind when using subcellular fractionation techniques. First, confluent cells are more difficult to homogenize successfully. Cells at 12 to 36 h post-passage, depending on the seeding density, are the best for this purpose. Second, isotonic media at neutral pH are preferable for use in these assays in order to maintain structural integrity of the particulate organelles. However, when cells are difficult to lyse in isotonic medium, hypotonic media can be used. In these cases, care must be taken to transfer the samples to isotonic medium as soon as possible.

8.2.2.1   Density Gradient Centrifugation

Protocol 8-4: Assay of Receptor Internalization Using Sucrose Gradient Centrifugation (Method 1)[5]

1. Grow cells expressing the receptor to semiconfluence in 15-cm tissue culture dishes.
2. Treat with the desired concentration(s) of the agonist to induce receptor internalization for the appropriate time at 37°C.
3. Transfer the plates onto ice and wash the cell monolayer with ice-cold buffer, as described in Protocol 8-1.
4. Incubate the washed cells with serum-free medium containing 50 µg/ml concanavalin A for 20 min on ice.
5. Wash cells with 10 ml of lysis buffer (1 mM Tris, 2 mM ethylene diamine tetraacetic acid [EDTA], pH 7.4).
6. Incubate the cells in 10 ml of lysis buffer for 20 min on ice to hypotonically lyse the cell.
7. Remove the lysis buffer and collect the lysed cells in a small volume of lysis buffer using a rubber policeman.
8. Layer the lysate (1 ml) on top of a sucrose density gradient comprising of 55, 50, 47.5, 45, 42.5, 40, 37.5, 35, 32.5, and 30% sucrose solution (0.9 ml each) buffered with 2 mM Tris, pH 8.0.
9. Centrifuge the gradient in a SW27 rotor at 25,000 rpm for 2.5 h at 4°C.
10. Collect 20 0.5-ml fractions from the bottom of the gradient.
11. Use an aliquot from each fraction for binding assay to determine the receptor content of each fraction using the appropriate labeled ligand and binding conditions.
12. Plot the results of the binding experiments as fraction number vs. the extent of ligand binding.
13. In the control or untreated cells, the receptor will lie primarily at the cell surface, thus the peak receptor binding will be associated with the heavier membrane fraction at the bottom of the gradient tube. The internalized receptor will be in the lighter vesicular fraction. Thus, in the fractions from the cells treated with agonist, two possible peaks will be observed. The peak associated with the heavier membrane fraction will be the receptors remaining at the cell surface after agonist treatment. The peak associated with the lighter vesicular fractions will correspond to the internalized receptors.
14. It is possible that even in the absence of agonist treatment, an intracellular population of the receptors exist. In this case, two peaks will be seen even in the control cells. If any internalization of the receptor occurs following agonist treatment, a shift in the distribution of the receptors between the two peaks must be observed. The peak representing the cell surface receptors should decrease and the one representing the internalized receptors should increase.

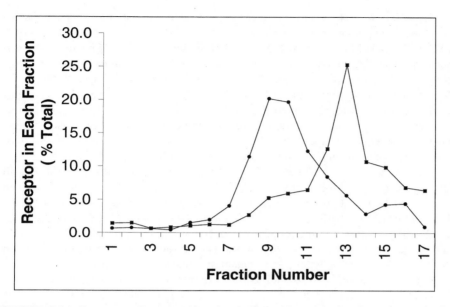

FIGURE 8.2 Sucrose gradient centrifugation: cells stably expressing human muscarinic receptor were treated with or without agonist for 20 min and washed and prepared for sucrose gradient as described in the text. Fractions were collected from the bottom of the gradients. The first fractions (heavy fractions) contain the membrane receptors and the latter fractions (light fractions) contain the vesicular receptor population. The circles are the untreated cells and the squares are the cells treated with the agonist. (Data provided courtesy of Dr. Hirofumi Tsuga, Pharmaceutical Research Laboratories of Ajinomoto Co. Inc., Kanagawa, Japan.)

Figure 8.2 shows the separation by sucrose gradient centrifugation of muscarinic receptors in the heavier membrane and the lighter vesicular fractions.

Protocol 8-5: Assay of Receptor Internalization Using Sucrose Gradient Centrifugation (Method 2)[12]

1. Treat cells to induce internalization in 100-mm dishes.
2. Rinse with ice-cold buffer.
3. Wash twice with ice-cold lysis buffer (20 mM HEPES, pH 7.4, 2 mM EDTA).
4. Allow to swell on ice for 10 min.
5. Scrape cells in 0.8 ml lysis buffer and layer lysate on top of discontinuous sucrose density gradient consisting of 1.7 ml of 15% sucrose, 5.0 ml of 30% sucrose, and 2.5 ml of 60% sucrose.
6. Centrifuge samples at 28,000 rpm for 60 min at 4°C in an SW41 rotor in an ultracentrifuge.
7. Collect fractions from the top of the tubes.
8. Determine the receptor content of each fraction by binding assay.

There are several variations in sucrose gradient centrifugation methods. Cells can be lysed in hypotonic medium containing proteinase inhibitors (50 mM Tris-HCl, pH 8.0, 2.5 mM MgCl$_2$, 1 mM EDTA, 1 mM dithiothreitol, 1 mM phenylmethylsulfonyl fluoride, 5 µg/ml aprotinin, and 10% glycerol).[25] Cells can be lysed by passage through a 27-gauge needle. Nuclei are removed by centrifugation at 3000 rpm for 5 min. The resulting supernatant (the soluble fraction containing the soluble proteins in the cell) is centrifuged for 20 min at 150,000 × g in ultracentrifuge. The pellet (containing the particulate fraction or the membrane population of the cells) is resuspended in an equal volume of lysis buffer. The samples are processed by SDS-PAGE and the receptor content of each fraction quantitated by Western blotting.

The cells can also be lysed by freeze-thawing cycle. The cells are washed with PBS in the plate and transferred to a dry ice-ethanol bath for 2 min. Plates are then transferred to a 37°C water bath and cells thawed in the presence of 0.5 M NaCl lysis buffer (20 mM HEPES, pH 7.2, 0.5 M NaCl, 2.5 mM MgCl$_2$, 1 mM EDTA, 1 mM dithiothreitol, 5 µg/ml leupeptin, 5 µg/ml aprotinin). Broken cells are rapidly scraped and transferred to centrifuge tubes. The suspension is centrifuged for 20 min at 150,000 × g. The pellet and supernatant are treated as described above.

After incubation in lysis buffer, the cells can be homogenized in a Dounce homogenizer and the nuclei removed by centrifugation at 400 × g for 10 min.[26] The cell homogenate can be fractionated on a stepwise sucrose cushion (4 ml each of 60% and 35% sucrose). Centrifugation can be at 150,000 × g for 90 min at 4°C. The fraction from the sucrose gradient can be further centrifuged at 150,000 × g after dilution of the pellets resuspended in Tris-EDTA buffer and protein precipitated with 20% trichloroacetic acid for 60 min on ice. The remaining trichloroacetic acid is removed by centrifugation for 15 min at 14,000 × g. The pellet is resuspended in SDS sample buffer and separated on SDS-PAGE.

In addition to sucrose gradient, other reagents can be used for the subcellular fractionation. The cell lysate is loaded on a 10% Percoll containing 0.25 M sucrose with a 65% sucrose cushion, and is centrifuged at 18,000 rpm in a Beckman 70 Ti rotor for 100 min.[27]

8.2.2.2 Subcellular Fractionation by Sequential Centrifugation[24]

Protocol 8-6: Assay of Receptor Internalization Using Sequential Centrifugation (Method 1)

1. Carry out binding with radioactive agonist.
2. Scrape cells from the plates in lysis buffer containing protease inhibitors (see Protocol 8-5 above).
3. Homogenize the cells in a tissuemizer by two bursts (5 s each, maximum setting).
4. Centrifuge the homogenate at 13,000 × g for 12 min in a microcentrifuge to yield the high density fraction.
5. Centrifuge the supernatant in a Ti-70.1 rotor at 200,000 × g for 80 min to yield the low density microsomal fraction.

6. Further fractionate the supernatant from the above centrifugation on a sucrose density step gradient (10 to 40%) in a SW41 rotor at $260,000 \times g$ for 4.5 h.
7. Dilute the fractions from the sucrose gradient five-fold with 50 mM Tris-HCl, pH 7.4, and further sediment them at $100,000 \times g$ for 60 min in a Ti-70.1 rotor.
8. Each fraction will represent one subcellular compartment: endocytic vesicles, 1.11 to 1.13 g/ml; plasma membranes, 1.15 to 1.19 g/ml; lysosomes, 1.19 to 1.22 g/ml.
9. Determine radioactivity of each fraction to quantitate the relative presence of the receptor in each fraction.

Protocol 8-7: Assay of Receptor Internalization Using Sequential Centrifugation (Method 2)[7]

1. Harvest cells in buffer and incubate with agonist.
2. Terminate the reaction by addition of 30 ml ice-cold buffer and centrifuge the cells at $300 \times g$ for 3 min at 4°C. Aspirate the supernatant and resuspend the pellet in ice-cold buffer without glucose containing 2.5 mg/ml concanavalin A.
3. Incubate the cells on ice for 20 min and recentrifuge as above.
4. Lyse the cells hypnotically by gentle homogenization in 10 ml TE buffer (10 mM Tris-HCl, pH 7.4, 2 mM EDTA, 2 µg/ml aprotinin, 1 µg/ml leupeptin, and 0.7 mg/ml pepstatin).
5. Centrifuge the lysate at $30,000 \times g$ for 10 min to obtain a crude plasma membrane fraction.
6. Centrifuge the supernatant at $200,000 \times g$ for 90 min to obtain a "light" membrane fraction.

8.2.3 Fluorescent Activated Cell Sorter

This method is based on separation of the cells that are bound to the fluorescently labeled antibody or ligand from those that are not bound.

Protocol 8-8: Fluorescent Activated Cell Sorter Using Fluorescent Ligand[16]

1. Harvest the cells and add 10^6 cells per tube in medium containing 10% fetal calf serum.
2. Incubate cells with fluorescently labeled agonist for 60 min at 4°C.
3. Wash cells and resuspend in 200 µl of the buffer containing 0.3% fetal calf serum and 1.5 µg/ml of propidium iodide.
4. Analyze samples using a FACS Star[PLUS] (Becton-Dickinson, Mountain View, CA) equipped with dual argon and helium-neon lasers.
5. Determine cell viability by exclusion of propidium iodide.

6. The difference between the control and treated cells is a measure of receptor internalization.

Protocol 8-9: Fluorescent Activated Cell Sorter Using Antibody to Receptor (Method 1)[28–30]

1. Incubate the cells with agonist for desired time.
2. Fix 2×10^6 cells in suspension with 1% paraformaldehyde on ice for 30 min.
3. Wash cells and resuspend in staining buffer (PBS, 0.02% sodium azide, 0.2% bovine serum albumin, pH 7.4).
4. Add primary antibody at appropriate dilution (e.g., undiluted hybridoma culture supernatant or optimal dilution of ascites, 1:400 or 1:1000).
5. After incubation for 30 min, wash cells again and incubate with secondary antibody for an additional 30 min.
6. Fix cells in 4% paraformaldehyde for analysis on a FACScan flow cytometer (Becton-Dickinson, Mountain View, CA).

Figure 8.3 shows an example for fluorescent activated cell sorter analysis of epitope tagged mu opioid receptors which were treated with or without an agonist.

Protocol 8-10: Fluorescent Activated Cell Sorter Using Antibody to Receptor (Method 2)[29]

1. Incubate cells with primary antibody (15 µg/10^6 cells/ml) to the receptor on ice for 60 min.
2. Wash twice in medium and dilute cells five-fold into 37°C medium containing agonist.
3. Incubate at 37°C and at appropriate times take aliquots, chill on ice by dilution into ice-cold PBS, and centrifuge.
4. After centrifugation, resuspend cells in ice-cold PBS, label with fluorescently labeled secondary antibody, and analyze by flow cytometry.

8.2.4 FLUORESCENT CONFOCAL MICROSCOPY[5,31,32]

Confocal microscopy is a specialized form of fluorescent microscopy. A confocal microscope can take optical horizontal sections of the sample by focusing on one section of the sample at a time and blocking away any fluorescence that may be from adjacent sections of the sample. Thus, the advantage of this technique over regular fluorescent microscopy is that the user can determine if the protein of interest is inside the cell or is clustered at the cell membrane. Furthermore, when multiple proteins are labeled, dual labeling of cellular compartments in a given section of the cell will reveal the co-localization of the two proteins in that compartment.

Confocal microscopy is an extremely sensitive method of visualizing internalized receptors. This method allows the visualization of individual cells and thus is very powerful in determining the behavior of a single cell on the slide. The main limitation

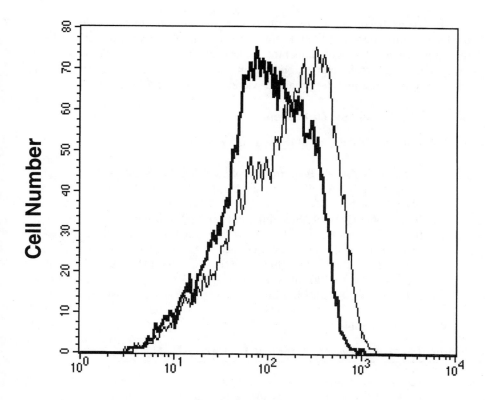

Fluorescence

FIGURE 8.3 Fluorescent activated cell sorter: a representative histogram displaying the fluorescence staining intensity of stably transfected HEK 293 cells expressing an epitope tagged mu opioid receptor (x-axis) vs. cell number (y-axis). Cells were incubated in the presence or absence of 5 μM DAMGO for 30 min. The opioid receptors present in the plasma membrane were specifically stained using a monoclonal antibody recognizing the epitope tag followed by fluorescein-labeled secondary antibody. The right histogram (thin line) shows the distribution of fluorescence intensity of untreated cells. The left histogram (bold line) shows the corresponding distribution after agonist treatment. Each histogram was derived from analysis of 20,000 cells. Mean fluorescence staining intensities (arbitrary units) were used to calculate percent internalization. Mean untreated: 221.68; mean treated: 133.10; internalization: 39.95%. (Graph provided courtesy of Drs. Peter C. Chu and Mark von Zastrow, University of California, San Francisco.)

of this method is the low efficiency of quantitation. While there are computer software programs that allow for quantitation of the extent of receptor internalization, none are yet optimal. The other limitations of this method are threefold. First is the dependence on a cell line expressing at least 100,000 receptor sites per cell. Second is a need for a high affinity antibody for the receptor. Finally, the use of a confocal microscope requires specialized training and understanding of the instrument and its limitations. However, more and more institutions are purchasing this microscope and its use is becoming more common. We recommend that anyone using the confocal microscope spend ample time to get familiarized with the proper use of the equipment and interpretation of the images obtained from it. Furthermore, several experiments have to be carried out and in each case multiple cells on each slide visualized to ensure that the observed phenomena are not artifacts.

8.2.4.1 Tools and Special Reagents

The following tools and special reagents are required:

1. Chamber slide (Lab-Tek, from Applied Scientific, South San Francisco, CA, (catalog no. AS-4804 [4-chambers] or AS-4808 [8-chambers]).
2. Poly-L-lysine.
3. Fluoromount G (Fisher Scientific, Pittsburgh, PA, catalog no. OB100-01).
4. 10% paraformaldehyde—dissolve 4.25 g NaCl in 400 ml distilled water (85% NaCl). Heat solution to just under 70°C. Add 50 g paraformaldehyde while stirring. Add 1 to 2 drops NaOH to get the pH around 7.0. Make up the volume to 500 ml with water. It is best to make this solution in the fume hood. This stock solution can be stored at 4°C for several months.

8.2.4.2 Single Label

Protocol 8-11: Fluorescent Confocal Microscopy[31] (Figure 8.4)

1. Seed 4-well chamber slides with 4 to 5 ×10^4 cells/chamber. We recommend that exponentially growing cells be used for seeding.
2. Allow cells to attach overnight. Ideally the cells have to be spread at low density on the slide. Individual cells are easier to look at under the microscope. On the other hand enough cells have to be on the slide that after the series of washings that follow, enough cells remain on the slide to visualize.
3. Treat cells with agonist in serum-free medium.
4. Aspirate the drug-containing medium and replace with fixing medium (3.7% paraformaldehyde), 10 min at room temperature.
5. Wash the cells with PBS, pH 7.4, containing 0.05% sodium azide.
6. Block the nonspecific fluorescence and permeabilize the cells by incubating with PBS containing 0.25% cold water fish gelatin, 0.04% saponin, and 0.05% sodium azide for 1 h at room temperature.
7. Replace blocking buffer with the same buffer containing primary antibody and incubate 1 h, at room temperature.
8. Wash cells with PBS containing azide four times.
9. Add blocking buffer containing secondary antibody, incubate 30 min at room temperature. Longer incubation with the secondary antibody generally leads to higher background fluorescence.
10. Remove secondary antibody and wash cells four times with PBS and once with water.
11. Dry cells and remove the sides of the chamber slide.

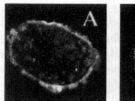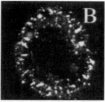

FIGURE 8.4   Single label confocal image of the muscarinic receptor subtype 1 expressed in human embryonic kidney cells. HEK 293 cells were treated with medium alone (A) or agonist, 1 mM carbachol. The staining of the cells was as described in Protocol 8-11. The images shown are from a mid-section of the cells.

12. Add 80 μl/slide of Fluoromount G containing a dash of *p*-phenylenedi-amine. Lay the cover slip on gently and squeeze out and aspirate the excess mounting medium. Seal the cover slip with nail polish.
13. Observe the slides with the confocal microscope, using the appropriate wavelength of the laser beam, depending on the secondary antibody used.

The following are variations in fluorescent confocal microscopy technique:

1. Cells can be grown on coverslips (20 mm^2) treated with poly-L-lysine that is placed on the bottom of a 12- or 6-well tissue culture plate;
2. Methanol:acetone (50:50) solution can be used for cell fixation;[28]
3. Blocking solution can vary—0.2% PBS-gelatin-10% goat serum,[32] 5% non-fat dry milk in 50 mM Tris-HCl, pH 7.6, or PBS[33] or 1% Triton X-100;[25,34]
4. Cells can be permeabilized using 0.1% Triton X-100 in PBS[32] or 0.3% Triton in PBS on ice for 20 min;[34]
5. Other mounting media for fluorescent samples can be used, such as Slow Fade (Molecular Probes, Eugene, OR)[16] or other such reagents available from Kirkengaard and Perry Laboratories, Gaithersberg, MD.

8.2.4.3   Dual Label

Dual labeling is used when it is of interest to determine the location of the receptor with respect to another protein. In this case both the receptor and the other protein can be labeled with their respective primary antibodies and different secondary antibodies with non-overlapping excitation and emission wavelengths.

Protocol 8-12: Fluorescent Confocal Microscopy—Dual Labeling[31]

1. The basic method is similar to that for single staining.
2. The cells are seeded, fixed, and blocked as described above.
3. The two primary antibodies can be added to the cells simultaneously.
4. After the three washes, the two secondary antibodies are added simultaneously as well.

5. For visualization of the fluorescence, protocols for the specific confocal microscope used must be followed.
6. Caution must be taken to determine the bleed through, if any, that may occur between the different channels used for visualization of the two dyes.
7. We have had success using secondary antibodies conjugated to Cy3 (indocarbocyanine) and Cy5 (indodicarbocyanine).
8. If the two primary antibodies are both from the same species, the secondary antibodies can cross-react, resulting in false signal. In this case, one of the primary antibodies can be directly conjugated to a fluorescent dye, thus eliminating the need for a secondary antibody. We have used the labeling kit from Amersham Life Sciences Inc., Arlington Heights, IL, catalog no. PA 3200 to successfully label monoclonal antibodies.

8.2.4.4 Fluorescent Confocal Microscopy—Quantitation

It is possible to quantitate the levels of fluorescence obtained from a confocal microscope. To do this, caution should be taken to normalize the intensity of fluorescence between different cells and different samples. Our experience shows that even among different samples that are treated identically, or even on a given slide, cells can be differentially stained and have variable fluorescence. Thus, for any quantitation to be meaningful, each cell has to serve as its own control.

Using various software, the extent of fluorescence at the cell surface and that in the intracellular vesicles can be measured and compared to determine what percentage of the receptors are intracellular. The BIO-RAD confocal microscope model 1024 is equipped with specialized software that allows for quantitation of the images. Other software that allows for the quantitation is the NIH IMAGE software that is available free of charge from the National Institutes of Health (NIH) online. NIH IMAGE is specific for Macintosh computers. Another program that is available in both Macintosh and Windows versions is Adobe Photoshop™ (Adobe, Mountain View, CA). We recommend that any one using any of these programs be completely familiarized with the program and its limitations before relying heavily on the results obtained from it. We suggest that until more accurate methods are available, any quantitative result obtained from the confocal images be used in conjunction with other quantitative techniques discussed elsewhere in this chapter.

Protocol 8-13: Quantitation Method for Fluorescent Confocal Microscopy[34]

1. Store a computerized image of the optical sections of the cells as a TIFF file.
2. Load the images into Adobe Photoshop™ and split it into green and red channels for independent quantification of two different stains if appropriate.
3. SigmaScan Pro™ (Jandel Scientific, San Rafael, CA) can then be used for quantitation. Make two separate measurements of the cell, one from the total cellular area and the other from the total intracellular area that is fluorescent (a measure of internalized receptor).

4. Outline the surface of the cell using the cursor and use the pixel counting feature of the software to calculate the total fluorescence at the cell surface.
5. To determine the total intracellular receptor, each area containing the fluorescent receptor must be outlined separately and its pixel content calculated. The sum of all the intracellular regions will be a measure of the intracellular receptor content.
6. To improve the accuracy of the calculations, for each data point, 20 to 50 cells need to be assessed, each with 3 to 10 optical sections, and the data plotted with appropriate error bars.

8.2.5   FLUORESCENT MICROSCOPY

When a confocal microscope is not available, a regular fluorescent microscope can be used. The main disadvantage of this method is that the actual location of the stained protein cannot be resolved. The only measure of receptor internalization will be the presence of vesicles, but whether they are at the cell surface or in fact inside the cell cannot be determined. The methods of seeding and staining the cells are identical to that described above for confocal microscopy.

8.2.6   ELECTRON MICROSCOPY[33]

Electron microscopy (EM) has been used in some studies of internalization of GPCRs. The use of this technique has been limited, due mainly to the fact that the instrument is not easily accessible to most of the researchers in the field. Furthermore, with the advent of the confocal microscope, the studies can be carried out more easily.

The main advantage of EM is the high resolution of the instrument. The cells can be studied in extreme detail and the location of the receptor can be visualized with submicromolar resolution. However, the disadvantage of this technique is the fact that only a small region of the cell can be visualized at a time; thus, fewer cells can be analyzed at one time.

Protocol 8-14: Electron Microscopy[32]

1. Grow cells in 35-mm culture dishes.
2. Fix cells with 0.25% glutaraldehyde in medium for 10 min at room temperature.
3. Wash cell with PBS containing 50 mM ammonium chloride buffer.
4. Sequentially incubate the cells with primary antibody, secondary antibody conjugated to horseradish peroxidase, and tertiary antibody reagent reacting with horseradish peroxidase. Each incubation is for 1 h at room temperature and each is followed by three washes with PBS-gelatin buffer.
5. Fix cells with 2.5% gluteraldehyde in PBS for 10 min.
6. Wash cells with PBS and 50 mM Tris-HCl buffer.
7. Incubate in 0.3% diaminobenzidine with 0.01% hydrogen peroxide for 20 min in the dark.
8. Postfix the cells in 1% osmium tetroxide for 45 min.

9. Dehydrate in ethanol and embed the cells *in situ* in epon-araldite.
10. Take ultrathin sections of the samples with a diamond knife (Diatome) and an LKB-Ultratome III (Bromma, Sweden).
11. View the sections with an electron microscope with proper settings.

Protocol 8-15: Quantitative Electron Microscopic Autoradiography[35]

In this method the cells are fixed and processed with a procedure designed to extensively cross-link cellular proteins and enhance preservation of membrane phospholipids.

1. Fix cells in 2% glutaraldehyde, 1% paraformaldehyde, 2 mM $CaCl_2$, and 1% tannic acid in 0.1 M sodium cacodylate buffer (pH 7.4) at room temperature for 1 h.
2. Rinse cells four times in 0.1 M sodium cacodylate buffer containing 2 mM $CaCl_2$.
3. Postfix the cells in 1% osmium tetroxide in cacodylate buffer, rinse twice in 0.8% sodium acetate buffer.
4. Stain en bloc with 0.5% uranyl magnesium acetate.
5. Rinse twice in 0.8% sodium acetate buffer.
6. Dehydrate in sequential changes of 70%, 95%, and two changes of 100% ethanol.
7. Infiltrate the cells with Spurr's low density resin and polymerize the resin at 65°C for 24 h.
8. Section about 100 nm thick and mount on collodium-coated slides, covered with monolayers of Ilford L4 emulsion.
9. Incubate at 4°C for 1 or 2 d.
10. Develop the emulsion in 1.1% *p*-phenylenediamine and 12.6% sodium sulfite and fix in 30% sodium thiosulfate.
11. Analyze grain distributions using the maximum likelihood method.[36]

8.2.7   ENZYME-LINKED IMMUNOSORBENT ASSAY OF RECEPTOR ANTIGEN (ELISA)[33,37]

For this assay, a fluorescent plate reader is required. This method allows quantitative measurement of receptor internalization with the use of a high-affinity antibody to the receptor.

Protocol 8-16: Assay of Receptor Internalization Using the ELISA Method

1. Seed 1000 cells per well in a 96-well microtiter plate pretreated with 0.0005% poly-L-lysine. Larger plates can be used if necessary to improve signal and if the appropriate plate reader is available. For 24-well plates, seed 5×10^4 cells/well.
2. Seed one row of the plate with cells not expressing the receptor (e.g., untransfected cells) for background measurement.

3. Allow cells to grow for at least 4 d.
4. Aspirate the medium and replace with serum-free medium containing the agonist to induce internalization. Incubate for an appropriate time.
5. Remove medium and fix the cells with 3.7% paraformaldehyde at room temperature for 30 min.
6. Wash cells once with PBS and block with medium containing 10% newborn calf serum for 1 h at room temperature (or PBS containing 0.05% Tween 20, 1% BSA, pH 7.4).
7. Replace blocking medium with medium containing primary antibody. Incubate at room temperature for 1 h. The concentration of antibody required needs to be empirically determined. Generally we have used 1:500 to 1:2000 dilutions of the primary antibodies.
8. Remove primary antibody and wash the cell monolayer four times with PBS. For each wash incubate the cells with the buffer for 5 min.
9. Reblock the cells with blocking solution for 15 min.
10. Add appropriate secondary antibody at the suggested concentration. Incubate 1 h at room temperature.
11. Wash cells four times with PBS and once with water as above.
12. Allow the plate to air dry.
13. Add 100 µl per well of PNPP (Pierce®).
14. Block with 50 µl of 2 N NaOH.
15. Read the fluorescence at 405 nM or the appropriate wavelength for the secondary antibody.

8.2.8   INHIBITORS OF INTERNALIZATION

Several different agents have been used to inhibit the internalization of cell surface receptors. Some of these reagents are specific inhibitors of the clathrin-mediated endocytosis, while others nonspecifically interrupt the cytoskeleton and thus inhibit any process that is mediated by the cytoskeleton. Results obtained from the use of inhibitors have to be interpreted with caution, since many of these treatments can affect multiple cellular processes.

8.2.8.1 DEPLETION OF INTRACELLULAR ATP

Requirement of ATP has long been established for receptor-mediated endocytosis of cell surface receptors such as transferrin receptors and β_2 adrenoceptors.[38]

Protocol 8-17: Inhibition of Receptor Internalization by Depletion of Intracellular ATP (Method 1)[39]

1. Deplete cellular ATP by incubating the cells in N_2 atmosphere. Fit 5 ml glass vials (catalog no. 9717-T27; Thomas Scientific, Swedesboro, NJ) with gas-tight rubber septum (catalog no. 1780-J22; Thomas Scientific). Flush N_2 for 30 to 45 min using two 6-port Mino-vap manifolds (catalog no. C5438-12; Baxter Scientific, San Diego, CA) fitted with 20-gauge needle.

2. Introduce into sealed vials on ice a 50 μl suspension of approximately 10^6 cells in PBS containing 5 mM 2-deoxyglucose using a 2.5-ml gas-tight syringe (Hamilton Industries, Two River, WI).
3. Transfer to 37°C for 10 to 15 min to allow ATP depletion and then return to ice.
4. Treat cells with agonist to induce internalization.
5. Assess the extent of receptor internalization by one of the binding methods described above.

Protocol 8-18: Inhibition of Receptor Internalization by Depletion of Intracellular ATP (Method 2)[38,40]

1. When cells are incubated in a medium depleted of substrates for glycolysis and glucogenesis, inhibition of mitochondrial ATP synthesis causes a rapid reduction in cellular ATP content. Thus, this method can be used for depletion of cellular ATP.
2. Rinse cells with PBS and incubate in glucose-free buffer (20 mM HEPES, 140 mM NaCl, 5 mM KCl, 2 mM CaCl$_2$, 1 mM MgSO$_4$, pH 7.4).
3. Treat cells with antimycin A (30 to 300 nM, in ethanol) for 20 min at 37°C.
4. Add agonist directly to the medium to induce internalization.
5. Quantitate internalization using the binding assay described above.

8.2.8.2   Hypertonic Medium—Sucrose

Treatment with hypertonic medium such as sucrose has been shown to disrupt formation of clathrin coated pits. Internalization of several GPCRs has been shown to be clathrin mediated.[16,23,31,41–44] Thus, disruption of this pathway of receptor-mediated endocytosis can result in inhibition of internalization of GPCR.

Protocol 8-19: Inhibition of Receptor Internalization by Hypertonic Medium[41,43,45,46]

1. Grow cells in plates or slides depending on the type of assay to be used for assessing internalization.
2. Treat cells with 0.45 to 0.75 M sucrose for 20 min or more prior to addition of agonist.
3. Treat with agonist in the presence of sucrose.
4. Wash agonist and assay internalization with binding assay or confocal microscopy.

8.2.8.3   Acidification of the Cytosol

Acidification of the cytosol to pH less than 6.5 has been shown to reduce endocytic uptake of transferrin and epidermal growth factor[47] by inhibiting the pinching of the coated vesicles. Acidification is attained by treating cells with 5 mM acetic acid in HEPES buffer, pH 5.0, for 5 min prior to addition of agonist.[31,47]

8.2.8.4   Use of Inhibitors

1. Concanavalin A will disrupt the cellular cytoskeleton. Cells are pretreated with 0.25 mg/ml ConA for 15 to 30 min prior to addition of the agonist.[1]
2. Phenylarsine oxide (PAO) is extremely disruptive to the cells and the monolayer will come off after treatment with it. Treatment with this agent may necessitate washing the cells using the centrifugation technique described above in Protocol 8-1. Dissolve phenylarsine oxide in dimethylsulfoxide and dilute to a final concentration of solvent of 0.1%. Treat cells with 10 to 100 μM for 5 min.[3,48] Add agonist to medium and continue with the desired internalization assay.

8.2.8.5   Low Temperature

Low temperatures have been shown to inhibit receptor-mediated endocytosis. The cells are precooled and agonist treatment is carried out on ice.[34,49]

8.2.8.6   Potassium Depletion

Depletion of intracellular potassium also inhibits clathrin coated pit formation.[50,51] Using this protocol, internalization of GPCRs can also be inhibited.[7]

Protocol 8-20: Inhibition of Receptor Internalization by Potassium Depletion[15,50]

1. Seed cells in appropriate vessel for internalization assay.
2. Wash once with potassium-free buffer (140 mM NaCl, 20 mM HEPES, 1 mM CaCl$_2$, 1 mM MgCl$_2$, 1 mg/ml D-glucose at pH 7.4).
3. Incubate with hypotonic medium (medium:water, 1:1 or above buffer diluted 1:1 with water) for 5 min.
4. Rinse with potassium-free buffer.
5. Incubate with potassium-free buffer above for 15 min at 37°C.
6. Treat with agonist and carry out internalization assay.

Depletion of potassium can also be attained by treatment of cells with 4 μM nigericin, an ionophore that promotes electroneutral K$^+$/H$^+$ exchange, for 40 min.[7]

8.3   DOWNREGULATION

Receptor downregulation is the process by which the receptors are degraded following agonist treatment. This will result in a decrease in the total receptor content of the cells. Generally, the process of receptor downregulation is slower than that of internalization and occurs in a few hours. In some cases the process may not reach maximum until 24 to 48 h after agonist treatment. In order to correctly assess the extent of receptor downregulation, reduction in the total receptors needs to be measured.

8.3.1 BINDING ASSAYS

One method involves the use of lipophilic ligands that can cross the membrane and get access to intracellular compartments. If no lipophilic radioactive ligand is available for the receptor under study, the receptor content in the cell homogenate can be determined.

8.3.1.1 Use of Hydrophobic Ligand in Intact Cells[41,52,53]

The method for the binding assay is similar to that explained above for internalization assay. The main factor to keep in mind is that, generally, higher nonspecific binding is associated with hydrophobic ligands. Thus, careful measurements of nonspecific binding will be required.

Protocol 8-21: Assay of Receptor Downregulation Using Hydrophobic Ligands

1. First determine the saturating concentration of the radioactive ligand.
2. Seed cells and treat with the agonist for long periods of time, such as hours.
3. Wash cells and carry out binding assay with the lipophilic ligand in intact cells.
4. Wash monolayer and determine radioactivity.
5. If desired the binding reaction can be carried out in intact cells in suspension.
6. The extent of receptor downregulation can be expressed as the percent of ligand binding to the receptor in treated cells compared to the untreated cells.

Figure 8.5 shows downregulation of muscarinic receptors.

8.3.1.2 Other Ligands[54,55]

If radioactive lipophilic ligands are not available for the receptor under study, other radioactive ligands can also be used. After agonist treatment, the cells are washed and crude cell homogenate is prepared using a polytron (one 30 s burst at maximum setting). Receptor binding is then carried out in cell homogenate using saturating concentrations of the radioactive ligand.

8.3.2 MEASURING mRNA

Long treatment with agonists can result not only in a decrease in the receptor protein content of the cells, but also in a decrease in the messenger RNA of the receptor. Thus, various techniques can be used to measure the changes in the receptor mRNA.

8.3.2.1 Northern Blotting

This is the simplest method for measuring receptor message levels.

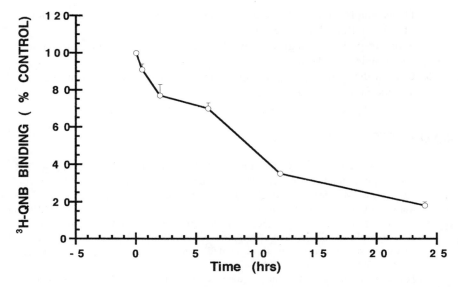

FIGURE 8.5   Time course of downregulation: cells stably expressing human muscarinic receptors were treated with agonist for different times. After washing of the cells, hydrophobic ligand was used to assess total receptors remaining in the cells. The graph shows the percent of the total receptors remaining in the cells at each time point.

Protocol 8-22: Determination of Receptor mRNA Levels by Northern Blotting Analysis[55]

1. Separate RNA from the cells by standard techniques[56] or by using commercially available RNA isolation kits.
2. Load about 10 µg total RNA on a 1% denaturing formaldehyde agarose gel and subject it to electrophoresis.
3. Carry out Northern blot analysis according to standard methods.

8.3.2.2   RNase Protection Assay

Another method for assessing the extent of downregulation of the receptor message is by RNase protection assay. This method takes advantage of the fact that RNA-DNA hybrid is resistant to the nuclease activity of RNases.

Protocol 8-23: Determination of Receptor mRNA Levels by the RNase Protection Method[57]

1. Isolate total RNA from the cells as described above.
2. Generate probes to the receptor using *in vitro* transcription with T3 or SP6 RNA polymerases in the presence of [^{32}P]-nucleotide.
3. Anneal the RNA and the probe at 45°C.

4. Digest single-stranded RNA at 30°C for 45 min using final concentrations of 2.5 U/ml of RNase I and 100 U/ml RNase T_1.
5. Separate protected fragments by electrophoresis on 6% polyacrylamide gel containing 7 M urea.
6. Cut out the bands corresponding to the protected fragments from the gel and measure their radioactivity.

REFERENCES

1. Pippig, S., Andexinger, S., and Lohse, M. L., Sequestration and recycling of β_2-adrenergic receptors permit receptor resensitization, *Mol. Pharmacol.*, 47, 666, 1995.
2. Krueger, K. M., Daaka, Y., Pitcher, J. A., and Lefkowitz, R. J., The role of sequestration in G protein-coupled receptor resensitization, *J. Biol. Chem.*, 272, 5, 1997.
3. Griendling, K. K., Delafontaine, P., Rittenhouse, S. E., M.A. Gimbrone, Jr., and Alexander, R. W., Correlation of receptor sequestration with sustained diacylglycerol accumulation in angiotensin II stimulated cultured vascular smooth muscle cells, *J. Biol. Chem.*, 262, 14555, 1987.
4. Daaka, Y., Luttrell, L. M., Ahn, S., Rocca, G. J. D., Ferguson, S. S. G., Caron, M. G., and Lefkowitz, R. J., Essential role for G protein-coupled receptor endocytosis in the activation of mitogen-activated protein kinases, *J. Biol. Chem.*, 273, 685, 1998.
5. Tsuga, H., Kameyama, K., Haga, T., Honma, T., Lameh, J., and Sadée, W., Internalization and down-regulation of human muscarinic acetylcholine receptor m2 subtype, *J. Biol. Chem.*, 273, 5323, 1998.
6. Koenig, J. A. and Edwardson, J. M., Kinetic analysis of the trafficking of muscarinic acetylcholine receptors between the plasma membrane and intracellular compartments, *J. Biol. Chem.*, 269, 17174, 1994.
7. Sloweijko, D. M., McEwen, E. L., Ernst, S. A., and Fisher, S. K., Muscarinic receptor sequestration in SH-SY-5Y neuroblastoma cells is inhibited when clathrin distribution is perturbed, *J. Neurochem.*, 66, 186, 1996.
8. Pals-Rylaarsdam, R., Xu, Y., Witt-Enderby, P., Benovic, J. L., and Hosey, M. M., Desensitization and internalization of the m2 muscarinic acetylcholine receptor are directed by independent mechanisms, *J. Biol. Chem.*, 270, 29004, 1995.
9. Liles, W. C., Hunter, D. D., Meier, K. E., and Nathanson, N. M., Activation of protein kinase C induces rapid internalization and subsequent degradation of muscarinic acetylcholine receptors in neuroblastoma cells, *J. Biol. Chem.*, 261, 5307, 1986.
10. Lameh, J., Philip, M., Sharma, Y. K., Moro, O., Ramachandran, J., and Sadée, W., Hm1 muscarinic cholinergic receptor internalization requires a domain in the third cytoplasmatic loop, *J. Biol. Chem.*, 267, 13406, 1992.
11. Hausdorff, W. P., Campbell, P. T., Ostrowski, J., Yu, S. S., Caron, M. G., and Lefkowitz, R. J., A small region of the β-adrenergic receptor is selectively involved in its rapid regulation, *Proc. Natl. Acad. Sci. U.S.A.*, 88, 2979, 1991.
12. Wang, J., Zheng, J., Anderson, J. L., and Toews, M., A mutation in the hamster α_{1B}-adrenergic receptor that differentiates two steps in the pathway of receptor internalization, *Mol. Pharmacol.*, 52, 306, 1997.
13. Roettger, B. F., Ghanekar, D., Rao, R., Toledo, C., Yingling, J., Pinon, D., and Miller, L. J., Antagonist-stimulated internalization of the G protein-coupled cholecystokinin receptor, *Mol. Pharmacol.*, 51, 357, 1997.

14. Roettger, B. F., Rentsch, R. U., Hadac, E. M., Hellen, E. H., Burghardt, T. P., and Miller, L. J., Insulation of a G protein-coupled receptor on the plasmalemmal surface of the pancreatic acinar cell, *J. Cell Biol.*, 130, 579, 1995.

15. Roettger, B. F., Rentsch, R. U., Pinon, D., Holicky, E., Hadac, E., Larkin, J. M., and Miller, L. J., Dual pathways of internalization of the cholecystokinin receptor, *J. Cell Biol.*, 128, 1029, 1995.

16. Grady, E. F., Slice, L. W., Bran, W. O., Walsh, J. H., and Payan, D. G., Direct observation of endocytosis of gastrin releasing peptide and its receptor, *J. Biol. Chem.*, 270, 4603, 1995.

17. Arora, K. K., Sakai, A., and Catt, K. J., Effects of second intracellular loop mutations on signal transduction and internalization of the gonadotropin-releasing hormone receptor, *J. Biol. Chem.*, 270, 22820, 1995.

18. Benya, R. V., Fathi, Z., Kusui, T., Pradhan, T., Battey, J. F., and Jensen, R. T., Gastrin-releasing peptide receptor-induced internalization, down-regulation, desensitization, and growth: possible role for cyclic AMP, *Mol. Pharmacol.*, 46, 235, 1994.

19. Laporte, S. A., Servant, G., Richard, D. E., Escher, E., Guillemette, G., and Leduc, R., The tyrosine within the NPXnY motif of the human angiotensin II type 1 receptor is involved in mediating signal transduction but is not essential for internalization, *Mol. Pharmacol.*, 49, 89, 1996.

20. Tseng, M.-J., Coon, S., Stuenkel, E., Struk, V., and Logsdon, C. D., Influence of second and third cytoplasmic loops on binding, internalization and coupling of chimeric bombesin/m3 muscarinic receptors, *J. Biol. Chem.*, 270, 17884, 1995.

21. Slice, L. W., Wong, H. C., Sternini, C., Grady, E. F., Bunnett, N. W., and Walsh, J. H., The conserved NPX_nY motif present in the gastrin-releasing peptide receptor is not a general sequestration sequence, *J. Biol. Chem.*, 269, 21755, 1994.

22. Yu, R. and Hinkle, P. M., Effect of cell type on the subcellular localization of the thyrotropin-releasing hormone receptor, *Mol. Pharmacol.*, 51, 785, 1997.

23. Ghinea, N., Hai, M. T. V., Groyer-Picard, M.-T., Houllie, A., Schoëvaërt, D., and Milgrom, E., Pathways of internalization of the hCG/LH receptor: immunoelectron microscopic studies in leydig cells and transfected L-cells, *J. Cell Biol.*, 118, 1347, 1992.

24. Munoz, C. M. and Leeb-Lundberg, L. M. F., Receptor-mediated internalization of bradykinin, *J. Biol. Chem.*, 267, 303, 1992.

25. Wedegaertner, P. B., Bourne, H. R., and Zastrow, M. V., Activation-induced subcellular redistribution of $G_{s\alpha}$, *Mol. Biol. Cell*, 7, 1225, 1996.

26. Ménard, L., Ferguson, S. S. G., Zhang, J., Lin, F.-T., and Lefkowitz, R. J., Synergistic regulation of β_2-adrenergic receptor sequestration: intracellular complement of β-adrenergic receptor kinase and β-arrestin determine kinetics of internalization, *Mol. Pharmacol.*, 51, 800, 1997.

27. Haraguchi, K. and Rodbell, M., Carbachol-activated muscarinic (M1 and M3) receptors transfected into Chinese hamster ovary cells inhibit trafficking of endosomes, *Proc. Natl. Acad. Sci. U.S.A.*, 88, 5964, 1991.

28. Hoxie, J. A., Ahuja, M., Belmonte, E., Pizarro, S., Parton, R., and Brass, L. F., Internalization and recycling of activated thrombin receptors, *J. Biol. Chem.*, 268, 13756, 1993.

29. Morrison, K. J., Moore, R. H., Carsrud, N. D. V., Trial, J., Millman, E. E., Tuvim, M., Clark, R. B., Narner, R., Dickey, B. F., and Knoll, B. J., Repetitive endocytosis and recycling of the β_2-adrenergic receptor during agonist-induced steady redistribution, *Mol. Pharmacol.*, 50, 692, 1996.

30. Barak, L. S., Tiberi, M., Freedman, N. J., and Al, E., A highly conserved tyrosine residue in G protein coupled receptors is required for agonist-mediated β2-adrenergic receptor sequestration, *J. Biol. Chem.*, 269, 2790, 1994.
31. Tolbert, L. M. and Lameh, J., Human muscarinic cholinergic receptor Hm1 internalizes via clathrin-coated vesicles, *J. Biol. Chem.*, 271, 17335, 1996.
32. Raposo, G., Dunia, I., Marullo, S., Andre, C., Guillet, J.-G., Strosberg, A. D., Benedetti, E. L., and Hoebeke, J., Redistribution of muscarinic acetylcholine receptors on human fibroblasts induced by regulatory ligands, *Biol. Cell*, 60, 117, 1987.
33. Daunt, D. A., Hurt, C., Hein, L., Kallio, J., Feng, F., and Kobilka, B., Subtypes-specific intracellular trafficking of α_2-adrenergic receptors, *Mol. Pharmacol.*, 51, 711, 1997.
34. Berry, S. A., Shah, M. C., Khan, N., and Roth, B. L., Rapid agonist-induced internalization of the 5-hydroxytryptamine$_{2A}$ receptor occurs via the endosome pathway *in vitro*, *Mol. Pharmacol.*, 50, 306, 1996.
35. Saffitz, J. E. and Ligget, S. B., Subcellular distribution of β_2-adrenergic receptors delineated with quantitative ultrastructural autoradiography of radioligand binding sites, *Circul. Res.*, 70, 1320, 1992.
36. Miller, M. I., Larson, K. B., Saffiz, J. E., Snyder, D. C., and Thomas, L. J., Maximum-likelihood estimation applied to election microscopic autoradiography, *J. Electron Microsc. Tech.*, 2, 611, 1985.
37. Hein, L., Ishii, K., Coughlin, S. R., and Kobilka, B. K., Intracellular targeting and trafficking of thrombin receptors, *J. Biol. Chem.*, 269, 27719, 1994.
38. Hertel, C., Coulter, S. J., and Perkins, J. P., The involvement of cellular ATP in receptor-mediated internalization of epidermal growth factor and hormone-induced internalization of β-adrenergic receptors, *J. Biol. Chem.*, 261, 5974, 1986.
39. Schmid, S. L. and Carter, L., ATP is required for receptor-mediated endocytosis in intact cells, *J. Cell Biol.*, 111, 2307, 1990.
40. Liao, J.-F. and Perkins, J. P., Differential effects of antimycin A on endocytosis and exocytosis of transferrin also are observed for internalization and externalization of β-adrenergic receptors, *Mol. Pharmacol.*, 44, 364, 1993.
41. Shockley, M. S., Burford, N. T., Sadee, W., and Lameh, J., Residues specifically involved in down-regulation but not internalization of the m1 muscarinic acetylcholine receptor, *J. Neurochem.*, 68, 601, 1997.
42. vonZastrow, M. and Kobilka, B. K., Ligand-regulated internalization and recycling of human β_2-adrenergic receptors between the plasma membrane and endosomes containing transferrin receptor, *J. Biol. Chem.*, 267, 3530, 1992.
43. Fonesca, M. I., Button, D. C., and Brown, R. D., Agonist regulation of α_{1B}-adrenergic receptor subcellular distribution and function, *J. Biol. Chem.*, 270, 8902, 1995.
44. Garland, A. M., Grady, E. F., Lovett, M., Vigna, S. R., Frucht, M. M., Krause, J. E., and Bunnett, N. W., Mechanisms of desensitization and resensitization of G protein-coupled neurokinin1 and neurokinin2 receptors, *Mol. Pharmacol.*, 49, 438, 1996.
45. Daukas, G. and Zigmond, S. H., Inhibition of receptor-mediated but not fluid-phase endocytosis in polymorphonuclear leukocytes, *J. Cell Biol.*, 101, 1673, 1985.
46. Heuser, J. E. and Anderson, R. G. W., Hypertonic media inhibit receptor-mediated endocytosis by blocking clathrin-coated pit formation, *J. Cell Biol.*, 108, 389, 1989.
47. Sandvig, K., Olsnes, S., Petersen, O. W., and Deurs, B. V., Acidification of the cytosol inhibits endocytosis from coated pits, *J. Cell Biol.*, 105, 679, 1987.
48. Hertel, C., Coulter, S. J., and Perkins, J. J., A comparision of catecholamine-induced internalization of β-adrenergic receptors and receptor-mediated endocytosis of epidermal growth factor in human astrocytoma cells, *J. Biol. Chem.*, 260, 12547, 1985.

49. von Zastrow, M. and Kobilka, B. K., Antagonist-dependent and -independent steps in the mechanism of adrenergic receptor internalization, *J. Biol. Chem.*, 269, 18448, 1994.

50. Larkin, J. M., Brown, M. S., Goldstein, J. L., and Anderon, R. G. W., Depletion of intracellular potassium arrests coated pit formation and receptor-mediated endocytosis in fibroblasts, *Cell*, 33, 1983.

51. Larkin, J. M., Donzell, W. C., and Anderson, R. G. W., Potassium-dependent assembly of coated pits: new coated pits form as planar clathrin lattices, *J. Cell Biol.*, 103, 2619, 1986.

51. Valiquette, M., Bonin, H., and Bouvier, M., Mutation of tyrosine-350 impairs the coupling of the β2-adrenergic receptor to the stimulatory guanine nucleotide binding protein without interfering with receptor down-regulation, *Biochemistry*, 32, 4979, 1993.

52. Wei, H.-B., Yamamura, H. I., and Roeske, W. R., Down-regulation and desensitization of the muscarinic M1 and M2 receptors in transfected fibroblast B82 cells, *Eur. J. Pharmacol.*, 268, 381, 1994.

53. Rousell, J., Haddad, E.-B., Mak, J. C. W., Webb, B. L. J., Giembycz, M. A., and Barnes, P. J., β-adrenoceptor-mediated down-regulation of m2 muscarinic receptors: role of cyclic adenosine 5'-monophosphate-dependent protein kinase and protein kinase C, *Mol. Pharmacol.*, 49, 629, 1996.

54. Lassegue, B., Wayne, R., Nickenig, G., Clark, M., Murphy, T. J., and Griendling, K. K., Angiotensin II down-regualtes the vascular smooth muscle AT1 receptor by transcriptional and post-transcriptional mechanisms: evidence for homologous and heterologous regulation, *Mol. Pharmacol.*, 48, 601, 1995.

55. Chomczynski, P. and Sacchi, N., Single step method of RNA isolation by acid guanidium thiocyanate-phenol-chloroform extraction, *Ann. Biochem.*, 162, 156, 1987.

56. Toth, M. and Shenk, T., Antagonist-mediated down-regulation of 5-hydroxytryptamine type 2 receptor gene expression: modulation of transcription, *Mol. Pharmacol.*, 45, 1095, 1994.

9   Use of Electrophysiology to Monitor and Study Receptor Desensitization

Mark R. Boyett, Zhigang Shui, and Iftikhar A. Khan

CONTENTS

9.1   INTRODUCTION

Desensitization of G protein-coupled receptors and the underlying mechanisms have been studied principally by biochemists. This is because the biochemist has the most appropriate tools for studying the phenomenon. G protein-coupled receptors couple to a range of effectors and the biochemist is equipped to study most of these, e.g., adenylate cyclase. However, G protein-coupled receptors also couple to ion channels and, in this case, the electrophysiologist can make an important contribution. There is a range of receptor–ion channel interactions: (1) the receptor (in this case not a G protein-coupled receptor) and ion channel may be the same protein. For example, in skeletal muscle, acetylcholine (ACh) binds to and activates the nicotinic receptor-channel. (2) The receptor may affect the channel by a second messenger. For example, in the heart, ACh binds to the G protein-coupled muscarinic M_2 receptor. This results in inhibition of adenylate cyclase via an inhibitory G protein and a fall in cAMP. The fall in cAMP either directly or indirectly results in changes in the L-type Ca^{2+} channel, delayed rectifier K^+ channel, and hyperpolarization-activated (i_f) channel. (3) The receptor may activate an ion channel directly via a G protein. In the heart and brain, a wide range of G protein-coupled receptors activate inward-rectifying K^+ channels in this way.[1-4] Desensitization to agonist occurs in all three types of receptor-channel system.[5-7] One of the best-studied cases is the muscarinic M_2 receptor–muscarinic K^+ channel system in the heart (an example of the third type of receptor–channel interaction above). We have been involved in the study of desensitization of this system and this work forms the basis of this chapter.

In the study of desensitization of G protein-coupled receptors, electrophysiology has two advantages over other techniques: (1) it allows the continuous monitoring of desensitization and recovery from it with millisecond time resolution in a single living cell and; (2) it allows the monitoring of a major effector linked to receptors (ion channels) the activity of which cannot be readily measured with other techniques.

9.1.1   DESENSITIZATION OF THE HEART TO ACH

The heart is controlled by parasympathetic nerves. Parasympathetic nerve stimulation leads to the release of ACh. ACh binds to muscarinic M_2 receptors in the heart and, thereby, slows the heart rate, slows conduction through the atrioventricular node, and decreases the strength of contraction of the heart (negative chronotropic, dromotropic, and inotropic effects, respectively). All of these effects of ACh fade during continued parasympathetic nerve stimulation—this is attributed to desensitization to ACh.[8-10] In the heart, ACh affects a number of ion channels, but the negative chronotropic and inotropic effects at least are thought to be primarily the result of the activation of the muscarinic K^+ channel.[10,11] In the presence of ACh, the ACh-activated K^+ current ($i_{K,ACh}$) flowing through the muscarinic K^+ channel fades, i.e., declines as a result of desensitization; this is shown in Figure 9.1A, which shows a whole cell recording of $i_{K,ACh}$ from a rat atrial cell. The decline in $i_{K,ACh}$ as a result of desensitization is thought to be primarily responsible for the fade of the negative chronotropic and inotropic effects of ACh at least.[10,12]

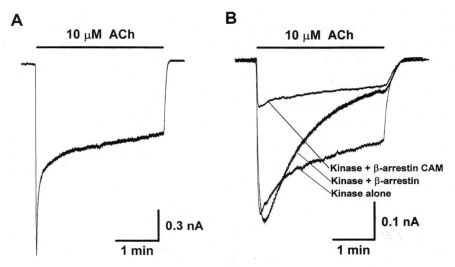

FIGURE 9.1 Desensitization of the muscarinic M_2 receptor-activated K^+ channel: whole cell current recordings on application of 10 μM ACh for 3 min. Holding potential, −60 mV. (A) Recording from a rat atrial cell; (B) superimposed recordings from CHO cells transfected with receptor kinase (GRK2) alone, receptor kinase (GRK2) and β-arrestin 2, and receptor kinase (GRK2) and a constitutively active β-arrestin 2 mutant (β-arrestin CAM). All the CHO cells were also transfected with the human M_2 receptor, Kir 3.1, and Kir 3.4 (the two muscarinic K^+ channel proteins) and GFP.

9.1.2 Role of the Muscarinic M_2 Receptor, Receptor Kinase, and Arrestin in Desensitization

During exposure to ACh, at least three phases of desensitization of $i_{K,ACh}$ are evident: a fast phase developing over ~20 s (Figure 9.1A), an intermediate phase developing over ~5 min (Figure 9.1A), and a slow phase developing over hours and days.[7,13] The fast phase is a channel phenomenon and is perhaps the result of dephosphorylation of the channel.[7,14] However, the intermediate and slow phases are thought to be receptor phenomena.[7,13] In rat atrial cells, we observed the intermediate phase of desensitization in configurations of the patch clamp technique in which the cytoplasm was retained (whole cell, cell attached, perforated outside-out patch).[15] In configurations in which the cytoplasm was lost (inside out, conventional outside-out), the intermediate phase was lost—in these cases the intermediate phase was restored by adding back receptor kinase G protein-coupled receptor Kinase 2, GRK2, or βARK1 to the membrane patches.[15] This suggested that receptor kinase is involved in desensitization of $i_{K,ACh}$.

To test this further, we reconstructed the muscarinic receptor–K^+ channel system in a Chinese hamster ovary (CHO) cell line.[16–18] This involves transfection of the cells with up to six proteins (receptor M_2 muscarinic receptor, two channel proteins Kir 3.1, Kir 3.4, receptor kinase GRK2, arrestin, green fluorescent protein GFP). The system is proving to be a powerful tool—it is yielding much new information about desensitization of $i_{K,ACh}$ and also the technique is dependable and relatively straight-forward. The work has shown that the intermediate phase of desensitization of $i_{K,ACh}$ involves both the receptor kinase (via phosphorylation-dependent and -independent mechanisms) and β-arrestin.[16,18] It has also been shown that the receptor kinase plays a major role in the long-term desensitization of $i_{K,ACh}$.[17] The reconstructed muscarinic receptor–K+ channel system is powerful, because a particular component can be transfected or not, or, alternatively, mutant forms rather than wild-type forms can be transfected. Two examples will be given: Figure 9.1B shows that when CHO cells were transfected with β-arrestin 2 (as well as the other components—see above), desensitization of $i_{K,ACh}$ during the exposure to ACh was marked and, when the cells were not transfected with β-arrestin 2, desensitization was much reduced. Finally, Figure 9.1B shows that when cells were transfected with a constitutively active β-arrestin 2 mutant, which is expected to bind to the receptor at all times (even in the absence of agonist) and, thus, suppress receptor function, $i_{K,ACh}$ was much reduced.

At the present time, our working hypothesis is that on application of ACh, ACh binds to the muscarinic M_2 receptor. This results in the dissociation of an inhibitory G protein. The G protein βγ subunits then bind to and activate the muscarinic K+ channel. However, the βγ subunits also bind to and activate the receptor kinase. The activated receptor kinase binds to the receptor. This causes some uncoupling from the G protein and, thus, some desensitization of $i_{K,ACh}$ (intermediate phase). The receptor kinase then phosphorylates the receptor and this leads to the further uncoupling and desensitization of $i_{K,ACh}$ (again intermediate phase). Arrestin is also involved in the intermediate phase of desensitization. The phosphorylation of the receptor by the receptor kinase then results in long-term desensitization (presumably via internalization and downregulation).

9.1.3 SCOPE OF THE CHAPTER

The brief summary above shows that our understanding of desensitization has resulted from the use of isolated heart cells as well as the reconstructed system in CHO cells and the whole cell, cell-attached, inside-out, conventional outside-out, and perforated outside-out configurations of the patch clamp technique. This chapter deals with all of these techniques.

9.2 ISOLATION AND CULTURE OF ATRIAL CELLS

We use isolated atrial cells from the heart, because the muscarinic K+ channel is present in the cells at high density (the channel is either absent or present at a low density in ventricular cells). We isolate cells from rats, because of economy and ease of isolation from this species. If we are studying the fast or intermediate phases of desensitization, we use freshly isolated atrial cells from adult rats. However, if we are studying the slow phase (for which the cells have to be exposed to agonist for up to 24 h) we culture the cells. For this we use freshly isolated atrial cells from neonatal rats, because adult rat atrial cells do not survive well for prolonged periods in culture. The isolation of both adult and neonatal rat atrial cells is described below.

FIGURE 9.2   Atrial cell isolation apparatus: (A) three water-jacked reservoirs; (B) heat-exchange column for heating solution to 37°C prior to entering the heart; (C) the area enlarged in the inset; (D) three-way taps controlling which solution flows to the heart via the heat-exchange column; (E) peristaltic pump for pumping the chosen solution from a reservoir to the heart; (F) thermostatic bath and circulator for supplying the water-jackets around the reservoirs and the heat-exchange column; (G) shaker for holding and shaking a conical flask during the later stage of the isolation process; (H) rat heart cannulated via the aorta (the clip is holding the aorta on the cannula).

9.2.1   ISOLATION OF ADULT RAT ATRIAL CELLS

Adult rat atrial cells are isolated using the apparatus shown in Figure 9.2. Three solutions described below are held in water-jacketed reservoirs (**A**). Flow from these is controlled by three-way taps (**D**). Solution from one of the reservoirs is pumped by a peristaltic pump (**E**) through a heat exchanger (**B**) to a cannula (**C**). The solution then retrogradely perfuses a rat heart (**H**) mounted on the cannula. The heart is surrounded by a water-jacketed vessel. This vessel helps to keep the heart warm and also collects the perfusate dripping from the heart—all solutions apart from the enzyme solution are not collected in the vessel and are allowed to drip from the vessel through a three-way tap to waste; the enzyme solution, however, is collected in the vessel (using the three-way tap) and is used for recirculation (see below). The temperature of the water bath feeding the water jackets is set at 37°C so that the temperature of the solution dripping from the cannula is at the same level. The solutions held in the three reservoirs are bubbled with oxygen.

The composition of the basic solution for cell isolation is (in mM): NaCl, 130; KCl, 5.4; MgCl$_2$, 1.4; NaH$_2$PO$_4$, 0.4; HEPES, 5; glucose, 10; taurine, 20; creatine,

10 (pH 7.3 with NaOH at room temperature; all chemicals from Sigma). We make 1 l of the isolation solution on the day of an isolation. For different purposes, the isolation solution contains an additional 750 μM $CaCl_2$ (Tyrode's solution), or 100 μM Na_2EGTA (Ca^{2+}-free solution), or 0.8 mg/ml collagenase (200 U/mg, Worthington, type II, 4194), 0.1 mg/ml protease (Sigma, type 14, P-5147), and 50 μM $CaCl_2$ (enzyme solution). We make 500, 200, and 60 ml of Tyrode's solution, Ca^{2+}-free solution, and enzyme solution, respectively.

Protocol 9-1: Isolation of Adult Rat Atrial Cells

1. Kill an adult male rat (about 300 g) by a blow on the neck. Rapidly excise the heart into Tyrode's solution.
2. Cannulate the aorta to retrogradely perfuse the heart, as shown in the inset in Figure 9.2. We use a glass cannula about 2 mm in diameter and with a slight bulge at the end. The cannula should be inserted into the aorta sufficiently so that it passes the various arteries branching off the aorta, but it does not go through the aortic valve into the left ventricle. A length of silk surgical suture is used to tie the aorta onto the cannula (immediately above the bulge at the cannula tip).
3. Perfuse the heart at constant flow (~8 ml/min) via the aorta and coronary arteries with Tyrode's solution for 2 to 3 min to clear the heart of blood. The heart should be beating vigorously at this time.
4. Perfuse the heart with Ca^{2+}-free solution for 4 min. This will cause the heart to stop beating.
5. Perfuse the heart with enzyme solution for about 8 to 10 min. Just before the enzyme solution in the reservoir runs out, recirculate through the heart the enzyme solution collected in the water-jacketed vessel around the heart.
6. Cut the atria from the heart after perfusion with enzyme solution, and chop the atrial tissue finely.
7. Collect all the enzyme solution remaining in the isolation apparatus in a measuring cylinder. Add 1% bovine serum albumin (Sigma, A6003) to the collected enzyme solution.
8. To a 50-ml conical flask, add 10 ml of the supplemented enzyme solution and the chopped atria. Shake for 15-min periods at 37°C. For this we use a shaker—this holds and shakes the conical flask in the water bath (Figure 9.2G).
9. Filter the mixture through fine nylon gauze after each 15-min period. Return the undigested atrial tissue to the conical flask for further enzyme treatment. Sediment the dissociated cells by centrifugation at 60 g for 1 min. Remove the supernatant, and resuspend the cells in Tyrode's solution.
10. This process should be repeated three times. The cells collected after each digest should be stored separately in 5-cm Petri dishes at 4°C.

9.2.2 ISOLATION OF NEONATAL RAT ATRIAL CELLS

We use Hanks' balanced salt solution (Sigma, H4385) for cell isolation.

Protocol 9-2: Isolation of Neonatal Rat Atrial Cells

1. Kill about 10 neonatal rats (1 to 3 d old) by cervical dislocation. Remove both right and left atria with fine forceps under a microscope. Place the atrial tissue in Hanks' balanced salt solution without Ca^{2+}.
2. Incubate the atrial tissue in Hanks' balanced salt solution containing 0.05% collagenase (200 U/mg Worthington, type II, 4194) and 0.03% trypsin (199 U/mg protein, Worthington, 3703) by weight for 10-min periods at 37°C.
3. Filter the mixture through nylon gauze, and sediment the dissociated cells by centrifugation at 145 g for 2 min. Remove the supernatant, and resuspend the cells in Tyrode's solution.
4. This process should be repeated five times. The cells collected after each digest should be mixed and stored in a Petri dish at room temperature.

9.2.3 CULTURE OF NEONATAL RAT ATRIAL CELLS

Cultured neonatal rat atrial cells are grown in Dulbecco's Modified Eagle's Medium (DMEM, Life Technologies, 32430-027) supplemented with 10% horse serum (Life Technologies, 26050-047) and Antibiotic-Antimycotic (containing 100 U/ml penicillin G, 100 µg/ml streptomycin sulfate, and 0.25 µg/ml amphotericin B final concentrations; Life Technologies, 15240-062). Cells are grown in an incubator in 95% air, 5% CO_2 mixture, at 37°C. All manipulations are carried out in a class 2 laminar flow hood, using sterile techniques, and all chemicals and media used are sterilized either by autoclaving or by passing through a 0.22-µm filter.

Cells are grown on sterilized fragments of coverslip glass, cut into squares, which have been precoated with poly-L-lysine.

Protocol 9-3: Culture of Neonatal Rat Atrial Cells

1. Cut 24 × 24 mm borosilicate glass coverslips (BDH, thickness No.1, 406/0187/51) into four pieces (12 × 12 mm) using a diamond cutter. Prepare several dozen glass coverslip fragments in this manner.
2. Sterilize the glass coverslip fragments by autoclaving at 141°C dry heat for 30 min.
3. After the glass fragments have cooled, immerse them for 1 min in a 0.1 mg/ml solution of poly-L-lysine (Sigma, P-6282) before rinsing with 10 mM phosphate-buffered saline (PBS, Sigma, P-3813).
4. Remove excess PBS with a vacuum line and place the glass fragments in each well of a 6-well multidish.
5. Assess the density of the cells in suspension, although a hemocytometer is not always necessary for this purpose; with experience, one can transfer the ideal number of cells without the time-consuming counting procedure. Aim to place at least 5×10^3 cells onto each square fragment of glass, in approximately 0.5 ml of Hanks' balanced salt solution. If the density of the cell suspension is low, then centrifuge at 50 g for 2 min, remove

supernatant, and gently resuspend in a lower volume of Hanks' balanced salt solution.

6. Add 0.5 ml of cell suspension onto each poly-L-lysine coated glass fragment and place multiwell dish in incubator for 2 to 3 h.

7. Gently add 2 ml of warmed DMEM incubation medium to each well.

8. Change medium daily thereafter for the 3 to 4 days that these cells can be experimented on. Gently aspirate old medium and replace with fresh warmed medium.

9.3 RECONSTRUCTION OF THE MUSCARINIC RECEPTOR–K⁺ CHANNEL SYSTEM IN CHO CELLS

CHO cells are derived from the ovary of the Chinese hamster, *Cricetulus griseus*. The small chromosome number of the nuclei of these cells have made CHO cell lines an ideal system for the study of genetic alterations in cultured mammalian cells. Electrophysiological studies involving ion channel expression in CHO cells are made easier by the absence of native voltage-gated currents.

All the CHO cell lines used are grown in Ham's F12 medium (Life Technologies, 31765-027) supplemented with 10% fetal bovine serum (Life Technologies, 10108-074), Antibiotic-Antimycotic and, depending upon the presence of the resistance gene, 0.2 mg/ml final concentration G418 sulfate (Geneticin®, Life Technologies, 11811-023) for antibiotic selection. Cells are grown in 75-cm² tissue culture flasks for routine growth, in 95% air, 5% CO_2 mixture, at 37°C. All manipulations are carried out in a class 2 laminar flow hood, using sterile techniques, and all chemicals and media used are sterilized either by autoclaving or by passing through a 0.22-μm filter.

We use the calcium phosphate transient transfection technique to introduce various DNA plasmids into wild-type cells, or cells already stably transfected. This technique and variations on it are shown in *Molecular Cloning: A Laboratory Manual.*[19]

Our experiments utilize a number of transfection strategies. The protocols below outline the methods used to transiently transfect wild-type CHO cells with, for example, the human M_2 muscarinic receptor, the muscarinic K⁺ channel (a heteromultimer of Kir 3.1 and Kir 3.4), receptor kinase (GRK2, also known as βARK1), β-arrestin 2, and green fluorescent protein (GFP).

9.3.1 CULTURE OF CHO CELL LINES

Protocol 9-4: Culture of CHO Cells

1. Remove all existing medium from a 75-cm² flask, and wash cells with 10 ml cold PBS. Gently rock flask back and forth for 1 min and then remove PBS.

2. Add 1 ml of 1× Trypsin-EDTA (Life Technologies, 35400-019), and wait 5 to 10 min, or until adherent layer of cells dislodges from the surface of the flask upon gently rocking. Do not remove Trypsin-EDTA.

3. Add 10 ml of warmed Ham's F12 incubation medium (containing 10% foetal bovine serum and Antibiotic-Antimycotic), and use to remove and suspend adherent cell layer from bottom of flask.
4. Remove approximately 9 ml of cell suspension from flask and discard.
5. Add 20 ml of warmed Ham's F12 incubation medium to flask.
6. Add G418 sulfate to flasks for final concentration of 0.2 mg/ml (0.1 ml of 40 mg/ml stock solution added to 20 ml of medium). *Note:* Do not add G418 to "blank" wild-type cells containing no G418 resistant gene.
7. Continue culture in same flask for no more than 1 month. Culture each cell line in duplicate.

9.3.2 PREPARATION OF CHO CELLS FOR TRANSIENT TRANSFECTION

Protocol 9-5: Preparation of CHO Cells for Transient Transfection

1. Remove all existing medium, wash cells with PBS and use 1× Trypsin-EDTA to remove and resuspend adherent cells, as in steps 1 to 3 of Protocol 9-4.
2. Count cells in an Improved Neubauer hemocytometer.
3. Seed 5 to 10×10^5 cells into 10 ml of warmed Ham's F12 incubation medium into a 10-cm plastic tissue culture dish. Discard the remainder of the cell suspension, and add a further 20 ml of Ham's F12 medium to a 75-cm^2 flask to continue culture with residual adherent cells.
4. Incubate the dish of cells in incubator overnight.

9.3.3 TRANSIENT TRANSFECTION OF CHO CELLS

Protocol 9-6: Transient Transfection of CHO Cells

1. Prepare the following buffers (all chemicals from Sigma):
   a. *2× HBS (HEPES buffered saline)*
      HEPES   1 g
      NaCl   1.7 g
      Make up to 100 ml with MilliQ (Millipore) water. Adjust pH to 7.15–7.25 (the pH of this solution can be quite critical for production of calcium phosphate-DNA precipitate). Filter and store at 4°C.
   b. *100× PO$_4^-$*
      Na$_2$HPO$_4$·2H$_2$O   1.246 g
      Make up to 100 ml with MilliQ water (to give 70 mM final concentration). Filter or autoclave at 120°C for 20 min, and store at 4°C.
   c. *2.5 M CaCl$_2$*
      CaCl$_2$·2H$_2$O   36.76 g
      Make up to 100 ml with MilliQ water. Filter and store in 1-ml aliquots at –20°C.
   d. *30% glycerol*
      Glycerol   30 g
      Make up to 100 ml with MilliQ water. Filter and store at 4°C.

2. Mix the following in a 1-ml Eppendorf tube:
   Plasmids   10 to 15 μg (1 mg/ml)
   100× PO$_4^-$ 10 μl
   2× HBS   500 μl
   Make up to 950 μl with sterile MilliQ water.
3. Slowly add 50 μl of 2.5 M CaCl$_2$ with stirring over 30 to 60 s (take your time; thorough mixing is essential for efficient calcium phosphate precipitate production).
4. Incubate transfection mixture at room temperature for 20 min.
5. Replace Ham's F12 medium in overnight cultured cells in 100-mm dish with 9 ml fresh warmed medium.
6. Mix transfection mixture again thoroughly and pour the 1 ml mixture on to cells in 9 ml of Ham's F12 incubation medium in a 100-mm dish (to give final plasmid concentrations of 400 ng/ml).
7. Incubate cells at 37°C for 4 to 6 hours. After this time, incubation medium will appear slightly milky, and microscope examination will show appearance of a fine precipitate covering cells.
8. Remove transfection mixture and wash the cells with 5 ml of cold Ham's F12 medium (without serum, without Antibiotic-Antimycotic) twice.
9. Mix 30% glycerol, 1 ml; 2× HBS, 1 ml; 100× PO$_4^-$; 20 μl. Pour 1.5 ml of this mixture onto the cells.
10. After 2 min, remove mixture and wash cells again with 5 ml of cold Ham's F12 medium (without serum, without Antibiotic-Antimycotic).
11. Restore 10 ml of normal warmed Ham's F12 incubation medium to cells and incubate for 48 to 72 h. *Note:* Cells can be reseeded (e.g., onto a glass coverslip) 2 to 3 h after restoring the medium.
12. Use for electrophysiology experiments.

Stable transfectants expressing human M$_2$ receptors and receptor kinase were also used in some of our experiments. Only the DNA for the channel and GFP were transiently transfected in these cases. Stable transfectants were constructed by including 2 μg of pEF-neo in addition to the other required plasmids in a calcium phosphate precipitation method. Stable transfectants were selected in the presence of 400 μg/ml G418, and were subcloned by limiting dilution, using 96-well plates to pick individual colonies.

The current records in Figure 9.1B are from cells stably transfected with the human M$_2$ receptor (pEF-myc-hm2) and GRK2 receptor kinase (pEF-GRK2). These cells were then transiently transfected with the plasmid expression vectors for Kir 3.1 and Kir 3.4 (pEF-GIRK1 and pEF-CIR, respectively), which together form the functional channel heteromultimer, and the S65T point mutation of green fluorescent protein (p-GFP-S65T, Clontech). In one case, the cells were also transiently transfected with wild-type β-arrestin 2 (pCMV5 β-arrestin 2), in another case with a constitutively active mutant of β-arrestin 2 (pCDNA3 arrestin 3-(1-393)), and in the third case with no β-arrestin 2.

9.3.4 PREPARATION OF CHO CELL SUSPENSION AND SEEDING CHO CELLS ONTO GLASS COVERSLIPS

We have used two different methods of introducing transfected cells into the bath chamber for electrophysiological studies. Adherent culture cells can be removed from the surface of the culture dish by chemical or enzymatic treatment, suspended in incubation medium, and plated out onto the bottom of the recording chamber. Alternatively, cells in suspension can be plated onto fragments of glass coverslip, and a coverslip fragment placed in the recording chamber after replacing in the incubator for 1 to 2 h.

Protocol 9-7: Preparation of CHO Cells for Electrophysiological Studies

1. Remove all existing medium from dish of transfected cells.
2. Wash cells with 10 ml cold PBS. Gently rock flask back and forth for 1 min and then remove PBS.
3. Add 1 ml 0.02% EDTA in PBS (preparation: add 1 g EDTA to 50 ml MilliQ water to produce a 2% stock solution, using a little, ~0.2 ml, NaOH to dissolve). Sterilize by filtering using a 0.22-μm syringe filter, and store at 4°C. Dilute 2% EDTA 1:100 with PBS.
4. Wait 5 to 10 min or until adherent layer of cells dislodges from the surface of the flask upon gently rocking.
5. Meanwhile, place sterile glass coverslip fragments in clean 100-mm plastic dishes, or in each well of a 6-well multidish.
6. Add 5 ml of warmed Ham's F12 incubation medium, to remove adherent cell layer from bottom of dish and resuspend cells.
7. Remove cell suspension to centrifuge tube and spin at 100 *g* for 3 min. Decant aspirate and add 1 ml of warmed Ham's F12 medium to resuspend cell pellet, by gently shaking tube.
8. Cover glass coverslips with cells in suspension, and replace in incubator.
9. After 2 to 3 h, refeed cells on coverslip by adding 10 ml of warmed Ham's F12 incubation medium.

9.4 ELECTROPHYSIOLOGY

The voltage clamp technique was invented by Marmont[20] and developed by Hodgkin et al.[21] With the technique, the membrane potential is clamped constant (at the "command potential") while the ionic current flowing through the membrane via ion channels is measured. Since the technique was first developed, many different variants of it have been developed. Currently, the most common is the patch clamp technique. The advantage of the patch clamp technique is that with small changes in the experimental protocol it can be used in a number of different configurations. It can be used in the whole cell configuration to measure the sum of the ionic currents flowing through all ion channels in the cell (whole cell current), or it can be used with isolated patches of membrane to measure the ionic currents flowing through the few (perhaps even one) ion channels in the small patch of membrane.

9.4.1   FUNDAMENTALS OF THE PATCH CLAMP TECHNIQUE

The basic patch clamp circuit is shown in Figure 9.3. At the heart of the circuit is an operational amplifier (**A**) with a feedback resistor (**R$_f$**) connected between the output and negative input of the feedback amplifier. A cell is shown in the whole cell configuration. In this configuration, the pipette potential is approximately equal to the potential within the cell, i.e., the membrane potential. The pipette is connected to the negative input and the command potential (**V$_{command}$**) is connected to the positive input of the feedback amplifier. Since the gain of the operational amplifier (**A**) is very high, if the membrane potential measured by the pipette differs from the command potential, a voltage will appear at the output. This will force a feedback current (**i$_f$**) to flow via the feedback resistor to the negative input of the amplifier. This will cause the potential at this point (pipette potential $\approx$ membrane potential) to change towards the command potential. In this way, the membrane potential is controlled and kept approximately equal to the command potential. It is the ionic current flowing through the ion channels in the cell membrane that causes the membrane potential to differ from the command potential. It follows that to keep the membrane potential approximately equal to the command potential, the feedback current should be equal and opposite to the ionic current. The feedback current (equal and opposite to the ionic current) is monitored by measuring the voltage at the output of the feedback amplifier (**V$_{out}$ = –i$_p$R$_f$**). Similar principles apply for the other configurations of the patch clamp technique. For a more detailed analysis of the patch clamp technique see References 22 and 23. Various patch clamp amplifiers are available commercially. We use amplifiers from Axon Instruments Inc.

9.4.2   CONFIGURATIONS OF THE PATCH CLAMP TECHNIQUE

The different configurations of the patch clamp technique are illustrated in Figure 9.4. The cell-attached, inside-out, whole-cell, and outside-out configurations were introduced by Hamill and coworkers.[24] The perforated whole-cell and perforated outside-out configurations were developed by Lindau and Fernaudez,[25] and Rae et al.[26]

In the cell-attached configuration, the cell is intact, and current through a tiny area of cell membrane in the tip of pipette is measured. In the cell-attached condition, if the pipette is withdrawn from the cell surface a patch of membrane from the cell is excised. The original inside surface of the patch of membrane is then exposed to the bath solution. This is called an inside-out patch.

In the cell-attached condition, the membrane patch within the tip of the pipette can be disrupted by applying a pulse of suction or voltage to the pipette interior. The pipette solution is then in contact with the cell interior, and current through the whole cell membrane is measured. This is the whole cell configuration. In the whole cell condition, an outside-out patch can be obtained by withdrawing the pipette from the cell membrane: membrane around the pipette is excised, and this then reseals to form a membrane patch. The original outside surface of cell membrane is exposed to the bath solution. This is referred to as an outside-out patch.

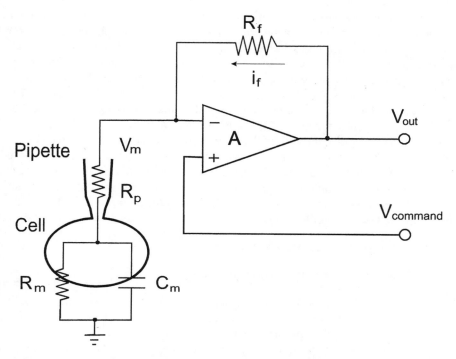

FIGURE 9.3   Patch clamp electronics: A, operational amplifier; C_m, membrane capacitance; R_f, feedback resistor; R_m, membrane resistance; R_p, pipette resistance.

Patch Clamp Configurations

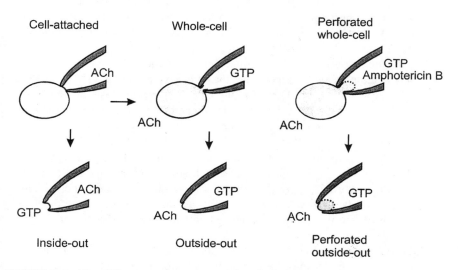

FIGURE 9.4   The different configurations of the patch clamp technique.

From the cell-attached condition, it is possible to obtain the whole cell configuration by including a polyene antibiotic, such as nystatin or amphotericin B, in the pipette: this forms tiny pores in the cell membrane in the pipette tip. This is the perforated whole cell configuration. The tiny pores allow monovalent cations and anions to permeate while excluding multivalent ions and nonelectrolytes. The major cytoplasmic molecules, therefore, are retained in the cell in the perforated whole cell configuration. In contrast, with the conventional whole cell configuration described above, major cytoplasmic molecules can be washed from the cell via the pipette.

The removal of the pipette from the cell membrane under cell-attached condition often causes the formation of a tiny vesicle at the pipette tip. If such a vesicle is formed with amphotericin B in the pipette, the inner membrane in the pipette tip is permeabilized by the antibiotic. Since the vesicle traps a small amount of cytoplasm in its interior, it is equivalent to a tiny cell attached to the pipette tip. This is the perforated outside-out configuration.

9.4.3   MANUFACTURE OF PATCH PIPETTES

Protocol 9-8: Manufacture of Patch Pipettes

1. Use soft glass tube with a diameter of 1 mm (micro-hematocrit tubes, LIP) for patch pipettes. Polish the ends of the micro-hematocrit tubes in the low flame of a Bunsen burner.
2. Pull the tubes by using a vertical pipette puller in two stages. There are many commercial pipette pullers—we use the L/M-3P-A puller from List Medical Electronic. Pull the glass tube to thin the middle to a diameter of 200–400 μm over a narrow region of length about 0.5 cm by the primary pull.
3. Pull the thinned glass tube to two pipettes with tip diameter of about 1 to 2 μm. The diameter of the pipette tip is dependent on the heat current, with lower heat current generating a larger diameter (lower pipette resistance). Pipettes with a resistance of 7 to 10 MΩ are used for single channel recordings and 3 to 5 MΩ for whole cell recording.
4. Coat the tip of the pipette to reduce the capacitance between the conducting solution inside the pipette and the bath. Prepare Sylgard (Sylgard 184 silicone elastomer kit, Dow Corning Corp.) by mixing resin and catalyst in the proportion of 10:1 by weight. Pour the mixed Sylgard into 1-ml plastic beakers, and store the beakers at about −18°C.
5. Apply Sylgard to pipettes with the jig described in Reference 22, chapter 3. Slip the pipette into the jig, and push it through a heated coil. Take a dollop of Sylgard by a glass tube from the beaker to the pipette when the pipette is twirled. Apply Sylgard to cover a pipette surface about 1 cm long from the pipette tip. Do not apply Sylgard to the tip of the pipette. Pull back the pipette into the heated coil to cure the Sylgard in a few seconds.
6. Polish the Sylgard-coated pipette at its tip to create a smooth tip and to burn off the fine film of Sylgard at the tip. Do the polishing on the stage

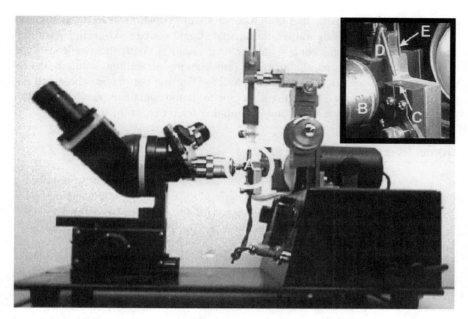

FIGURE 9.5   Microforge: the main photograph shows a Narishige Scientific Instruments microforge. It includes a binocular microscope for watching the pipette tip, a holder for the pipette, a lamp for illuminating the pipette tip, a filament for heating the pipette tip, and the heater power supply. A, the area enlarged in the inset; B, objective lens; C, V-shaped filament; D, metal tube for directing an air stream at the filament; E, pipette.

of a microforge (Narishige Scientific Instruments), as shown in Figure 9.5, under direct observation at a magnification of 300× to 700× with a long-working-distance objective (**B**). Bend the filament in an inverted V shape with an angle of about 60° between the two sides, each about 2 mm long (**C**). Coat the tip of the V-shaped filament with glass to prevent metal atoms from evaporating onto the pipette tip. Direct an air stream at the filament via a thin metal tube (**D**) during polishing to generate a steep thermal gradient. Hold the tip of the pipette (**E**) 5 to 10 μm from the glass coated filament for a few seconds to form a bullet-shaped tip with a clear opening.

9.4.4   PATCH PIPETTE FILLING SOLUTIONS AND FILLING OF PATCH PIPETTES

We fill the pipettes with intracellular or extracellular solution, dependent on the experimental configuration. The intracellular solution contains (in mM): K aspartate, 130; KCl, 20; KH_2PO_4, 1; $MgCl_2$, 5.5 (free Mg^{2+}, 1.8); EGTA, 5; HEPES, 5; Na_2ATP, 3; Na_3GTP, 0.1 (pH 7.4 with KOH at room temperature). The extracellular solution contains (in mM): KCl, 140; $MgCl_2$, 1.8; EGTA, 5; HEPES, 5; with or without 10 μM ACh (pH 7.4 with KOH at room temperature). With the cell attached configuration, both the pipette and the chamber should contain extracellular solution. With the inside-out configuration, the pipette should contain extracellular solution and the chamber should contain intracellular solution. With the whole cell and outside-out configurations, the pipette should contain intracellular solution and the chamber should contain extracellular solution. With the perforated whole cell and perforated outside-out configurations the pipette should contain intracellular solution with amphotericin B and the chamber should contain extracellular solution.

Protocol 9-9: Filling of Patch Pipettes

1. Dip the tip of the pipette into the filling solution. Apply a negative pressure to the back of the pipette with a syringe to suck solution into the pipette tip. Backfill the pipette with the rest of the solution by using another syringe with a 0.22-μm filter, and tap the pipette to displace any air bubbles.
2. Fill the tip of the pipette with an antibiotic-free solution at first when the perforated outside-out configuration is used, and then backfill the pipette with antibiotic-containing solution. Briefly dip the tip of the pipette (for less than 1 s) into the antibiotic-free solution. The column of antibiotic-free solution in the tip of the pipette should not be more than 500 μm. Backfill the pipette with antibiotic-containing solution, and than tap it to displace air bubbles.
3. Keep the fluid level in the pipette as low as possible to avoid additional noise. The solution level should be just sufficient to cover the tip of the AgCl-coated Ag electrode (Axon Instruments Inc.) Prepare the pipettes on the day of an experiment, and store them in a covered box to prevent dust from settling on them.

9.4.5 MAKING A GIGA-SEAL

Protocol 9-10: Making a Giga-Seal

1. Place the resuspended atrial cells in the chamber for about 10 min to allow them to settle onto the glass bottom, and then perfuse the bath for about 2 min. When CHO cells are used, carefully place a cell-covered coverslip fragment in the chamber, and then perfuse the bath for about 2 min.
2. View the cells in the chamber, and choose a good cell. A normal atrial cell is thin and long, about 10 to 15 μm in diameter and 100 to 200 μm in length, and the cell membrane is smooth and transparent. A suitable CHO cell is round with a smooth membrane and a middle level of green fluorescent light.
3. Mount a suitable pipette in a holder (supplied by Axon Instruments Inc. for example), fixed on the headstage of the voltage clamp amplifier—this assembly is held by a micromanipulator (e.g., the three-dimensional hydraulic, Model MW-3, Narishige Scientific Instruments) (Figure 9.6).

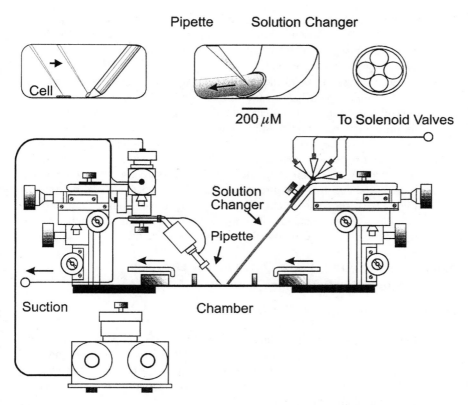

Pipette   Solution Changer

Cell

$\overline{200\,\mu M}$ To Solenoid Valves

Solution
Changer

Pipette

Suction   Chamber

FIGURE 9.6 Chamber: the main panel shows a schematic diagram of the chamber. Not shown is the solution perfusing the chamber, although the pipes through which solution is pumped into the chamber and sucked from the chamber are shown. On the left of the chamber is a hydraulic micromanipulator (controlled by the device at the bottom) holding the voltage clamp headstage. Mounted in the headstage is a pipette holder holding a pipette. Attached to the pipette holder is the suction line (for giga-seal formation). On the right of the chamber is a micromanipulator holding the rapid solution changer. The solution changer consists of a 2-mm glass tube drawn to a ~0.1-mm tip (middle inset). The tube contains 0.5-mm metal tubes that terminate ~2 mm from the tip of the tube (right inset). Each tube is collected to a reservoir via a solenoid valve. Solution from a pump flows via the solenoid (if it is open) to the metal tube and then from the tip of the rapid solution changer. With cell attached, whole cell, and perforated whole cell recordings, the rapid solution changer is moved to the cell of interest. With inside-out, outside-out, and perforated outside-out recordings, the isolated patch is moved from the cell to the mouth of the rapid solution changer (left and middle insets).

4. Apply a slight positive pressure (10 cm H_2O) to the pipette. The pressure is indicated by a glass, ink-filled, U-shaped tube (manometer) connected to the pipette. Move the tip of the pipette through the air–water interface using the micromanipulator.
5. Adjust the current baseline to zero by the manual junction null on the patch clamp amplifier (in voltage clamp mode). Apply a rectangular wave

(−10 mV, 10 Hz) to measure the pipette resistance (and later test the seal resistance). The resistance can be calculated from the current amplitude during the test pulse using Ohm's law. A typical current recording at this stage is shown in Figure 9.7A.

6. Carefully press the tip of the pipette against the cell membrane using the micromanipulator (Figure 9.6). Release the positive pressure when the current during the rectangular wave decreases after touching the pipette tip on the cell membrane. Apply a slight negative pressure (20 cm H_2O) to the back of the pipette. Increase the negative pressure by 100 to 200% if the seal resistance does not reach the expected level within the initial few seconds, and then keep the negative pressure for several seconds. Release the negative pressure soon after a high resistance seal is formed. Figure 9.7B shows a current trace after formation of high resistance seal. The positive and negative spikes are capacitance current due to the charging of the capacitance between the pipette and chamber.

9.4.6 CELL ATTACHED, INSIDE-OUT, OUTSIDE-OUT, AND WHOLE CELL RECORDINGS

Protocol 9-11: Making a Cell Attached Patch

1. Increase the gain of the patch clamp amplifier after a good seal is formed. Cancel the large positive and negative spikes (capacitance transients) on the current waveform by adjusting the fast and slow capacitance compensation controls. Figure 9.7C shows a current trace after compensation of capacitance current.
2. Switch off the rectangular wave, and then turn on an appropriate holding potential (e.g., +60 mV). Record single channel current in the cell-attached configuration.

Protocol 9-12: Making an Inside-Out Patch

1. Withdraw a cell-attached pipette from the cell (the cell must be adhering tightly to the bath bottom) to form an inside-out patch. If the cell is not adhered tightly to the bath bottom, apply another pipette with a large diameter (about 10 μm) to catch the cell and then pull the recording pipette away to form an inside-out patch.
2. Switch off the rectangular wave, and then turn on an appropriate holding potential (e.g., +60 mV). Record single channel current in the inside-out configuration.

Protocol 9-13: Making a Whole Cell Recording

1. Apply a large voltage (1.5 V DC) with a controlled duration (0.1 to 10 ms) to a cell-attached patch by pressing the zap button on the patch clamp

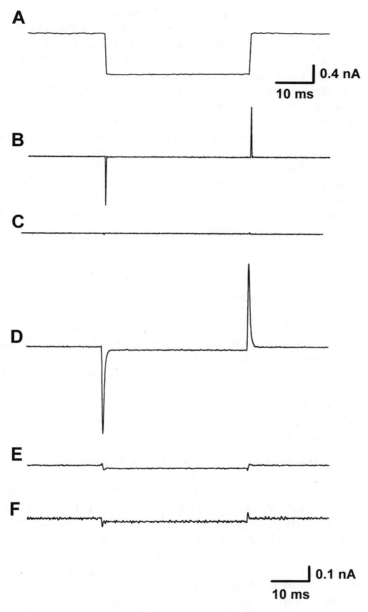

FIGURE 9.7 Typical current recordings during giga-seal formation and formation of the whole cell configuration. All the current recordings were obtained in response to a 40 ms, 10 mV negative voltage clamp step. (A) Pipette in the solution in the chamber; (B) after formation of a giga-seal; (C) after formation of a giga-seal and compensation of the pipette capacity current; (D) after going whole-cell; (E) after going whole-cell and compensation of the whole cell capacity current; (F) after going whole-cell and compensation of both the whole cell capacity current and the series resistance.

amplifier (if it has one) to rupture the membrane in the pipette to obtain the whole-cell configuration. Increase the duration of the voltage pulse, if the patch does not rupture. Rupture of the patch (and, therefore, the whole cell configuration) is indicated by the appearance of whole cell capacitance current as shown in Figure 9.7D.

2. Compensate the whole cell capacitance current. Figure 9.7E shows a current trace after the compensation. Compensate for series resistance (the resistance between the pipette and cell interior; R_p in Figure 9.3). Because whole cell current is a few nA in amplitude and the pipette resistance is several $M\Omega$, the voltage drop across the series resistance could be over 10 mV. It is necessary to compensate this voltage error. Figure 9.7F shows a current trace after compensation of series resistance. The noise of the current trace is a little higher (because the compensation of series resistance is a form of positive feedback). Do not compensate more than 80% of the series resistance to avoid the risk of oscillation.

3. Switch off the rectangular wave, and then set an appropriate holding potential (e.g., –60 mV). Record whole cell current.

Protocol 9-14: Making an Outside-Out Patch

1. In the whole cell configuration, withdraw the pipette from the cell to form an outside-out patch. If the cell does not tightly adhere to the bottom of the chamber, use another pipette with a larger diameter to catch the cell.

2. Switch off the rectangular wave, and then turn on an appropriate holding potential (e.g., –60 mV). Record single channel current in the outside-out configuration.

9.4.7 PERFORATED RECORDINGS

Protocol 9-15: Making a Perforated Whole Cell Recording

1. Use a pipette (about 3 $M\Omega$) filled with antibiotic-containing solution as described before. Apply a negative pressure (20 cm H_2O) soon after the tip of the pipette touches the cell membrane. Decrease the negative pressure to 10 cm H_2O after achieving a giga-seal. Keep the negative pressure at this level for about 10 s to obtain a tiny vesicle that retains a small amount of cytoplasm. Release the negative pressure, and wait for several minutes to allow amphotericin B to make pores in the membrane (during this time, the capacity transients come back and become higher and higher following the rectangular wave command).

2. Compensate the capacitance transients and the series resistance soon after the capacitance transients are stable.

3. Switch off the rectangular wave, and then turn on an appropriate holding potential (e.g., –60 mV). Record current in the perforated whole cell configuration.

Protocol 9-16: Making a Perforated Outside-Out Patch

1. In the perforated whole cell configuration, withdraw the pipette from the cell to form a perforated outside-out patch. If the cell does not tightly adhere to the bottom of the chamber, use another pipette with a larger diameter to catch the cell.
2. Switch off the rectangular wave, and then turn on an appropriate holding potential (e.g., –60 mV). Record single channel current in the perforated outside-out configuration.

9.4.8 RECORDING $I_{K,ACh}$

In the cell-attached configuration, ACh should be present in the patch pipette solution (extracellular solution). Muscarinic K^+ current will be recorded immediately.

In the inside-out configuration, ACh should be present in the patch pipette solution (extracellular solution). If the chamber contains normal intracellular solution, muscarinic K^+ current will be activated immediately. However, we perfuse the chamber with intracellular solution lacking GTP and ATP. We then perfuse the normal intracellular solution to activate $i_{K,ACh}$. The GTP and ATP containing normal intracellular solution should be applied rapidly in about 200 ms (see below).

In the whole cell, outside-out, and perforated whole cell and perforated outside-out configurations, ACh is rapidly applied via the chamber. It is important to apply ACh rapidly (in about 200 ms) in order to faithfully record the time course of activation of $i_{K,ACh}$ and the early phases of desensitization. For this we place the cell or patch at the mouth of a rapid solution changer (Figure 9.6, insets). A description of the rapid solution changer that we use is given in Reference 15.

9.5 ADDRESSES OF SUPPLIERS

- Axon Instruments, Inc., 1101 Chess Drive, Foster City, CA 94408-1102.
- Clontech Laboratories, Inc., 1020 East Meadow Circle, Palo Alto, CA 94303.
- Dow Corning Corporation, Midland, MI 48640.
- Life Technologies Ltd., 3 Fountain Drive, Inchinnan Business Park, Paisley, PA4 9RF, Scotland.
- LIP (Equipment and Services) Ltd., 111 Dockfield Road, Shipley, West Yorkshire, England.
- List-Medical-Electronic, D-6100 Darmstadt-13, Germany.
- Millipore Corporation, 80 Ashby Road, Bedford, MA 01730-2271.
- Narishige Scientific Instrument Lab., 9-28 Kasuya, 4-Chome, Setagaya-ku, Tokyo, Japan.
- Sigma-Aldrich Company Ltd., Fancy Road, Poole, Dorset, BH12 4QH, England.
- Worthington Biochemical Corporation, 730 Vassar Avenue, Lakewood, NJ 08701.

REFERENCES

1. Duprat, F., Lesage, F., Guillemare, E., Fink, M., Hugnot, J.-P., Bigay, J., Lazdunski, M., Romey, G., and Barhanin, J., Heterologous multimeric assembly is essential for K$^+$ channel activity of neuronal and cardiac G protein-activated inward rectifiers, *Biochem. Biophys. Res. Commun.*, 212, 657, 1995.

2. Doupnik, C. A., Davidson, N., and Lester, H. A., The inward rectifier potassium channel family, *Curr. Opinion Neurobiol.*, 5, 268, 1995.

3. Kurachi, Y., Nakajima, T., and Sugimoto, T., Short-term desensitization of muscarinic K$^+$ channel current in isolated atrial myocytes and possible role of GTP-binding proteins, *Pflügers Arch.*, 410, 227, 1987.

4. Bünemann, M., Brandts, B., zu Heringdorf, D. M., van Koppen, C. J., Jakobs, K. H., and Pott, L., Activation of muscarinic K$^+$ current in guinea-pig atrial myocytes by sphingosine-1-phosphate, *J. Physiol.*, 489, 701, 1995.

5. Katz, B. and Thesleff, S., A study of the 'desensitization' produced by acetylcholine at the motor end-plate, *J. Physiol.*, 138, 63, 1957.

6. Pappano, A. J. and Mubagwa, K., Muscarinic agonist-induced actions on potassium and calcium channels in atrial myocytes: differential desensitization, *Eur. Heart J.*, 12 (Suppl. F), 70, 1991.

7. Zang, W.-J., Yu, X.-J. , Honjo, H., Kirby, M. S., and Boyett, M. R., On the role of G protein activation and phosphorylation in desensitization to acetylcholine in guinea-pig atrial cells, *J. Physiol.*, 464, 649, 1993.

8. Martin, P., Secondary AV conduction responses during tonic vagal stimulation, *Am. J. Physiol.*, 245, H584, 1983.

9. Boyett, M. R. and Roberts, A., The fade of the response to acetylcholine at the rabbit isolated sino-atrial node, *J. Physiol.*, 393, 171, 1987.

10. Boyett, M. R., Kirby, M. S., Orchard, C. H., and Roberts, A., The negative inotropic effect of acetylcholine on ferret ventricular myocardium, *J. Physiol.*, 404, 613, 1988.

11. Boyett, M. R., Kodama, I., Honjo, H., Arai, A., Suzuki, R., and Toyama, J., Ionic basis of the chronotropic effect of acetylcholine on the rabbit sinoatrial node, *Cardiovasc. Res.*, 29, 867, 1995.

12. Honjo, H., Kodama, I., Zang, W.-J., and Boyett, M. R., Desensitization to acetylcholine in single sino-atrial node cells isolated from the rabbit heart, *Am. J. Physiol.*, 263, H1779, 1992.

13. Bünemann, M., Brandts, B., and Pott, L., Downregulation of muscarinic M$_2$ receptors linked to K$^+$ current in cultured guinea-pig atrial myocytes, *J. Physiol.*, 494, 351, 1996.

14. Shui, Z., Boyett, M. R., and Zang, W.-J., ATP-dependent desensitization of the muscarinic K$^+$ channel in rat atrial cells, *J. Physiol.*, 505, 77, 1997.

15. Shui, Z., Boyett, M. R., Zang, W.-J., Haga, T., and Kameyama, K., Receptor kinase dependent desensitization of the cardiac muscarinic K$^+$ current in rat atrial cells, *J. Physiol.*, 487, 359, 1995.

16. Shui, Z., Khan, I. A., Tsuga, H., Haga, T., and Boyett, M. R., Role of receptor kinase in short term desensitization of cardiac muscarinic K$^+$ channels expressed in Chinese hampster ovary cells, *J. Physiol.*, 507, 325, 1998.

17. Shui, Z., Khan, I. A., Tsuga, H., Haga, T., Dobrzynski, H., Henderson, Z., and Boyett, M. R., The role of G protein-coupled receptor kinase in long-term desensitization of muscarinic K$^+$ current, *J. Physiol.*, 504, 75P, 1997.

18. Shui, Z., Khan, I. A., Tsuga, H., Haga, T., and Boyett, M. R., Role of β arrestin in modulation of the cardiac muscarinic K$^+$ current, *J. Physiol.*, 511, 78P, 1998.

19. Sambrook, J., Fritsch, E. F., and Maniatis, T., *Molecular Cloning: A Laboratory Manual,* Cold Spring Harbor Laboratory, Cold Spring Harbor, NY, 1989.
20. Marmont, G., Studies on the axon membrane. I. A new method, *J. Comp. Physiol.,* 34, 351, 1949.
21. Hodgkin, A. L., Huxley, A. F., and Katz, B., Measurements of current-voltage relations in the membrane of the giant axon of Loligo, *J. Physiol.,* 116, 424, 1952.
22. Sakmann, B. and Neher, E., *Single-Channel Recording,* Plenum Press, New York, 1983.
23. Sherman-Gold, T., *The Axon Guide for Electrophysiology and Biophysics: Laboratory Techniques,* Axon Instruments, Inc., California, 1993.
24. Hamill, O. P., Marty, A., Neher, E., Sakmann, B., and Sigworth, F. J., Improved patch clamp techniques for high-resolution current recording from cells and cell-free membrane patches, *Pflügers Arch.,* 391, 85, 1981.
25. Lindau, M. and Fernandez, J. M., IgE-mediated degranulation of mast cells does not require opening of ion channels, *Nature,* 319, 150, 1986.
26. Rae, J., Cooper, K., Gates, P., and Watsky, M., Low access resistance perforated patch recordings using amphotericin B, *J. Neurosci. Methods,* 37, 15, 1991.

Section III

3D Structure: Large-Scale
Expression and Electron- and
X-Ray Crystallography

10 Expression in *Escherichia coli* and Large-Scale Purification of a Rat Neurotensin Receptor

Reinhard Grisshammer and Julie Tucker

CONTENTS

10.1 INTRODUCTION

Seven-helix G protein-coupled receptors (GPCRs) are integral components of the cell surface membrane.[1-3] Estimates suggest that there may be several thousand GPCRs in the human genome, of which a substantial proportion could be drug targets. Therefore, it is important to understand the molecular structure and action of GPCRs. Only rhodopsin, the naturally abundant visual pigment, has been crystallized to date. Projection structures at low resolution[4-7] and the three-dimensional arrangement of the transmembrane alpha helices[8,9] have been determined by electron cryomicroscopy using two-dimensional crystals.

In contrast to rhodopsin, most GPCRs do not occur naturally in large quantities. Structural work therefore requires an amenable expression system that permits routine purification of milligram quantities of protein. The potential of various hosts for the expression of membrane proteins has recently been discussed in detail[10,11] (see also Chapter 11). Our work focuses on the high-affinity receptor for neurotensin from rat (NTR)[12] using *Escherichia coli* as the expression system. We describe here the expression of functional NTR inserted in *E. coli* membranes, but do not discuss the production of NTR in nonfunctional, aggregated form.[12a]

For crystallization, purified receptors should be fully functional and homogeneous; thus the purification procedure has to be fast and efficient, as receptors often show reduced stability in the detergents used to extract them from membranes. In this chapter we discuss a two-step purification scheme to obtain functional NTR based on immobilized metal affinity chromatography (IMAC) followed by an agonist ligand column step. We have not yet investigated the refolding of NTR from inclusion bodies, which could also be a route to obtain functional receptors in quantities sufficient for biochemical and structural analyses.[13]

10.2 EXPRESSION OF NTR IN *ESCHERICHIA COLI*

10.2.1 BACKGROUND

The great attraction of *E. coli* as an expression host is the simplicity of its use. The time required for the construction of recombinant expression vectors and the initial expression experiment is very short in comparison to other host systems. This allows rapid optimization of expression yields and testing of affinity purification tags.[14] Low-cost, large-scale fermentation and homogeneous protein expression (few post-translational modifications) may be advantageous for biophysical and structural studies.

E. coli has been used successfully for the expression of GPCRs in functional, membrane-inserted form (see Table 10.1), despite being very different from eukaryotic expression systems.[15] N-linked glycosylation and cholesterol (a major constituent of mammalian plasma membranes but absent from *E. coli* membranes) are often not required for expression of functional receptors. A brief summary of parameters important for the expression of NTR in *E. coli* is given below. The reader is referred to Reference 15 for a more detailed discussion of the folding and membrane insertion of GPCRs related to their expression in *E. coli*.

Expression of the receptor-encoding cDNA alone usually results in very low levels of protein. In contrast, higher expression levels can be achieved by using a fusion protein approach, in which the *E. coli* periplasmic maltose-binding protein (MBP) with its signal peptide is linked to the amino terminus of the receptor. This may facilitate the correct orientation of the receptor in the cytoplasmic membrane with its N terminus in the periplasmic space. The presence of MBP does not seem to interfere with the ligand binding properties of recombinant receptors.[16–20] We found that the pharmacology of NTR expressed in *E. coli* is similar to that of NTR expressed in eukaryotic cells and from native tissues.[14,18,18a]

TABLE 10.1
Expression of GPCRs in *E. coli*

Receptor	Species	Expression as Fusion Protein	Expression Level (Location)[a]	Promoter, Strain and Induction Conditions[b]	Remarks	Ref.
Adenosine A_1	Human	N: first 56 aa of MBP with SP	0.4 pmol/mg (m)	p^{malE}; KS303; LB; 5 h, 30°C	Reconstitution with G protein, altered N improves expression	46
α_2C2-adrenergic	Human	N: CGT without SP C: His tag	n.a. (ib)	p^{lac}; DH5α/F'lacI^q; LB; 6 h, 37°C		Närvä et al., unpublished results
β_1-adrenergic	Human	N: first 279 aa of LamB with SP	50/cell (m)	p^{tac}; pop6510; LB; 4 h, 23°C or 2 h, 37°C	Ligand binding only after *in vivo* proteolytic removal of LamB, reconstitution with G protein	47–49
β_2-adrenergic	Human	N: first 279 aa of LamB with SP	0.45 pmol/mg, 220/cell (m)	p^{tac}; pop6510; LB; 4 h, 23°C or 2 h, 37°C	Ligand binding only after *in vivo* proteolytic removal of LamB, reconstitution with G protein	47–49
β_2-adrenergic	Human		200/cell	$p^{T7\phi10}$; BL21(DE3)/pLysS; M9; 6 h, 23°C	*In situ* filter screening with labeled ligand	50
β_2-adrenergic	Human	N: MBP with SP C: TrxA/His-10	1.9 pmol/mg of total protein, 6 nmol/l (m)	p^{lac}; JLH27C7; 2 × TY; 44 h, 20°C		S. Lin and B. Kobilka, unpublished results
β-adrenergic	Turkey	N: first 279 aa of LamB with SP	33/cell (m)	p^{tac}; pop6510; LB; 2 h, 37°C	Ligand binding only after *in vivo* proteolytic removal of LamB	47
Cannabinoid CB1	Human	N: MBP with SP C: His-6/Flag	Very low level[c] (m)	p^{lac}; DH5α; 2 × TY; 44 h, 20°C	Extensive proteolysis	17
Cannabinoid CB2	Human	N: MBP with SP C: His-6/Flag	16.8–39.5 pmol/mg (m)	p^{lac}; DH5α; 2 × TY; 44 h, 20°C		17
Dopamine D_3	Human	N: MBP with SP	600/cell	p^{lac}; n.a.		20
Endothelin ET_B	Human		41/cell (m)	p^{lac}; BL20; LB; 2 h, 23°C	N: putative own SP[d] expression in protease-deficient strain	51
Muscarinic M_1	Rat	N: MBP with SP C: His-9	50 pmol/mg, 3500/cell (m)	p^{lac}; BL21; minimal; 25°C	Removal of protease-susceptible sequences from i3 loop, addition of antagonist *N*-methylscopolamine to culture medium	16, 52

TABLE 10.1 (CONTINUED)
Expression of GPCRs in *E. coli*

Receptor	Species	Expression as Fusion Protein	Expression Level (Location)[a]	Promoter, Strain and Induction Conditions[b]	Remarks	Ref.
Neurokinin A	Rat	N: MBP with SP C: His-5/cMyc	7.2 pmol/mg (m)	p^{lac}; DH5α; 2 × TY; 40 h, 20°C		53
Neuropeptide Y1	Human	N: MBP with SP C: Flag/His-6	3–9 pmol/mg, 60–250/cell (m)	p^{lac}; TB1; 2 × TY/SB; 1 h, 37°C		19, 54
Neurotensin	Rat	N: with and without enterotoxin B SP	3–11/cell	p^{TSN25}; DH5α; 2 × TY; 1 h, 30°C		18
Neurotensin	Rat	N: MBP with SP C: various tags	10–15 pmol/mg, 450–800/cell (m)	p^{lac} or p^{lac}; DH5α; 2 × TY; 44 h, 20°C	Stabilizing effect of C-terminal tags	14, 18
Olfactory OR5	Rat	N: GST C: His-6	20 mg of purified receptor/l (ib)	p^{lac}; BL21; LB; 4 h, 37°C	Refolding in detergent	13
δ, μ, κ Opioid	Human	N: MBP with SP	0.3–0.8 pmol/mg, 15–30/cell (m)	p^{lac}; JM101; LB; 5 h, 20°C		55, 55a
Opsin	Octopus	N: T7 major capsid protein	0.1–1 mg/l (m+ib)	$p^{T7\Phi10}$; BL21(DE3); NZCYM; 15 h, 30/37°C	Reconstitution with retinal	56
Opsin	Octopus		1–10 mg/l (m+ib)	$p^{T7\Phi10}$; BL21(DE3); NZCYM; 15 h, 30/37°C		56
Serotonin 5-HT$_{1A}$	Human	N: first 56 aa of MBP with SP	120/cell (m)	p^{malE}; KS303; LB; 4.5 h, 30°C		57

Abbreviations: aa, amino acids; C, receptor carboxy terminus; CGT, cyclomaltodextrin glucanotransferase; GST, glutathione S-transferase; ib, inclusion body; m, membrane; MBP, maltose-binding protein; N, receptor amino terminus; n.a., not available; SP, signal peptide; TrxA, *E. coli* thioredoxin.

[a] Expression levels are given in pmol/mg of membrane protein and receptors/cell (ligand binding analysis) or mg of receptor/liter of culture.

[b] Induction conditions are given as medium, time of induction, and temperature; n.a., not available.

[c] CB1 receptor is poorly expressed due to extensive proteolysis (Western blot analysis) and fails to bind ligand.

[d] Endothelin ET$_B$ receptor has its own putative signal peptide, whereas the other receptors listed in this table do not.

Adapted from Reference 15.

No GPCR has yet been overexpressed in *E. coli* in functional form at levels above 100 pmol/mg of membrane protein. The process of membrane insertion and folding into the native receptor conformation seems to be rate-limiting. To avoid overloading of the *E. coli* insertion/translocation machinery, transcription and translation were optimized rather than maximized by using a low copy number plasmid with a weak promoter for expression of the NTR fusion protein (see below). Furthermore, low temperature (20°C) and prolonged induction times (>40 h) gave greater expression. Quantitative data are not yet available on the relative proportions of correctly folded and misfolded receptors in the cytoplasmic membrane of *E. coli*.

The stability of GPCRs can be an important factor for their successful expression, as was found for NTR.[14] When exploring a variety of C-terminal affinity tags for purification, expression levels varied considerably depending on the tag used. Functional expression of NTR was highest when *E. coli* thioredoxin was placed between the C terminus and the tag, indicating a stabilizing effect of the thioredoxin moiety which may be a result of the remarkable stability of this globular domain.[21,22] This observation may be specific for NTR.

10.2.2 METHODS FOR RECEPTOR EXPRESSION

The *E. coli* strain DH5α harbouring the plasmid pRG/III-hs-MBPP-T43NTR-TrxA-H10 is used for the expression of neurotensin receptor. This plasmid is a derivative of pBR322 encoding a truncated neurotensin receptor starting at Thr-43, fused at its N terminus to maltose-binding protein with its signal peptide (MBPP), and fused at its C terminus to thioredoxin (TrxA) followed by a deca-histidine tag (H10). Expression is controlled by the *lac* wild-type promoter. The presence of the *hok/sok* gene cassette (hs), a plasmid maintenance system,[23–25] prevents plasmid loss.[26] The conditions for NTR expression outlined in Protocol 10-1 were established by analyzing receptor levels by ligand binding at different time points after induction and by repeating this time course at various temperatures. As discussed above, the current set-up aims at optimized translation rates allowing the accumulation of functional NTR fusion protein in membranes. These requirements may be specific for NTR.[15] Expression levels of the NTR fusion protein amount to about 1000 receptors per cell (25 pmol/mg of crude membrane protein[18a] or 7 to 8 pmol/mg of total protein obtained after detergent treatment of *E. coli* cells and ultracentrifugation, see Protocol 10-2). This translates into 2 mg of functional, solubilized NTR fusion protein (molecular mass of 96 kDa) from a 9.6 l culture, which is sufficient for purification and two-dimensional crystallization trials.

Protocol 10-1: Expression in E. coli of NTR Fusion Proteins in Functional Form

1. Materials required
   a. *E. coli* strain DH5α carrying the plasmid pRG/III-hs-MBPP-T43NTR-TrxA-H10.
   b. 2× TY medium (per liter: 10 g yeast extract, 16 g bactotryptone, 5 g NaCl)[27] supplemented with ampicillin (100 µg/ml) and glucose (0.2% [w/v]).

  c. 1 M IPTG stock solution.
  d. Orbital shakers with temperature control.
  e. Sorvall centrifuge RC 3B Plus with 1-l buckets or equivalent.
2. Re-streak bacteria on selective agar plate, incubate at 37°C overnight.
3. Grow 12 ml overnight culture at 37°C from a single colony.
4. Inoculate 400 ml of medium in 2-l flask (24 flasks) with overnight culture
    at a dilution of 1:1000.
5. Shake flasks at 300 rpm at 37°C for about 4 h until reaching an OD_{600} of
    0.5.
6. Induce with IPTG at a final concentration of 0.5 mM and decrease tem-
    perature of incubator to 20°C.
7. Monitor OD_{600} during expression experiment.
8. Grow cells for 40 h after induction (OD_{600} should be about 3). Then harvest
    cells by centrifugation (20 min, 5000 rpm, 4°C).
9. Freeze by pouring liquid nitrogen onto cell paste and store at –70°C; 9.6 l
    of culture yield about 54 g of cells (wet weight).

10.3 PURIFICATION OF THE FUSION PROTEIN MBP-T43NTR-TRXA-H10

10.3.1 GENERAL CONSIDERATIONS FOR SOLUBILIZATION AND PURIFICATION

Careful optimization of the detergent and the buffer composition is critical for
obtaining maximum solubilization efficiency and stability of functional receptors in
solution. These two criteria are most important when expression levels are moderate
and when, therefore, a large amount of starting material must be processed to obtain
milligram quantities of pure receptor protein. We have established optimal conditions
for solubilization and purification of NTR by screening a variety of detergents in
combination with various buffers.[14] The highest solubilization efficiency and NTR
stability were achieved when using buffers containing the nonionic detergent *n*-
dodecyl-β-D-maltoside (LM) in combination with the zwitterionic detergent 3-[(3-
cholamidopropyl)dimethylammonio]-1-propanesulfonate (CHAPS), cholesteryl
hemisuccinate (CHS), sodium chloride, and glycerol at pH 7.4 (see Protocols 10-2
and 10-3). Under these conditions, about 85% of functional receptors are solubilized
with a half-life in solution of about 400 h at 4°C (see Reference 14), allowing
subsequent purification.
     Purification of NTR requires two steps. The first step is based on a C-terminal
affinity tag and the second step on neurotensin (NT) binding. The first step has to
be very efficient in order to enrich NTR fusion proteins because of their moderate
expression levels in *E. coli*. We considered the biotin/monomeric avidin system as
an attractive method for purification of *in vivo* biotinylated receptors because *E. coli*
contains only one endogenous biotinylated protein, the soluble biotin carboxyl carrier
protein of acetyl-CoA carboxylase.[28] *In vivo* biotinylation of NTR is achieved by

fusing a 13-amino-acid Bio tag peptide[29] to the receptor C terminus.[14] Crude membrane preparations rather than whole cell lysates have to be used for purification of the biotinylated NTR fusion protein to avoid saturation of the monomeric avidin column by biotin carboxyl carrier protein. This approach results in good purification of NTR at small scale.[14] However, we experienced relatively low recovery rates from the monomeric avidin column and found this system difficult to scale up. In addition, preparation of *E. coli* membranes at large scale is time-consuming. Therefore, we replaced the C-terminal Bio tag peptide with histidine residues so that immobilized metal affinity chromatography could be the first purification step.[26]

IMAC[30] has been used for purification of histidine-tagged membrane proteins in the presence of detergents. Strong binding to the immobilized metal ions is essential for effective purification when expression levels are low. We have investigated the influence of the length of histidine tags and the type of detergent on the purification of NTR fusion proteins[26] and found that hexa-histidine-tagged receptors do bind to Ni^{2+}-NTA resin[31] under conditions that confer maximum stability of functional receptors in solution. However, many *E. coli* protein contaminants also adhere to the column under these conditions. Both hexa-histidine-tagged receptors and contaminants start to elute from a Ni^{2+}-NTA column at an imidazole concentration of about 30 mM which prevents stringent washing of the Ni^{2+}-NTA column without considerable loss of receptor protein. In contrast, the level of purification is substantially improved by extending the histidine tail to ten residues (MBP-T43NTR-TrxA-H10), because this allows stringent washes at high imidazole concentration to remove nonspecifically bound contaminants. This strategy not only results in efficient purification of receptors from crude membranes, but works particularly well for the purification from total cell lysate. Receptors with deca-histidine tails can be directly recovered with high yield (80 to 90% recovery measured by [³H]-NT binding) from cell lysate that has been solubilized with detergent.[26]

The above first purification step is based on the C-terminal histidine tag and selects for NTR fusion protein irrespective of its conformation. Therefore, a subsequent NT affinity column is necessary to separate neurotensin-binding from nonfunctional receptors. The latter arise from possible misfolding during expression and/or from adverse effects of the detergents on receptors leading to loss of ligand binding.

10.3.2 Methods for Receptor Purification

Solubilization and purification must be performed at 4°C or on ice to ensure the integrity of NTR in functional form. Buffers are prepared fresh for each purification and filtered over a 0.2-μm disposable filter (except 2× solubilization buffer). Protease inhibitors (see Protocol 10-2) are present throughout. A GradiFrac system from Pharmacia is used for purification, but other chromatography units are equally suited. Each step is analyzed by [³H]-NT binding[14] and sodium dodecyl sulfate-polyacrylamide gel electrophoresis (SDS-PAGE).[32] The protein content is determined by the method of Schaffner and Weissmann,[33] with bovine serum albumin (BSA) as the standard (see Chapter 5, Protocol 5-16). The recommended timing of solubilization and purification is indicated by [Day one], [Day two], [Day three] in Protocols 10-2 to 10-4.

10.3.2.1 Cell Lysis and Solubilization

Complete resuspension of cells is necessary for successful solubilization and purification of NTR. Chromosomal DNA is degraded by DNase I to reduce the viscosity of the suspension. Ethylene diamine tetraacetic acid (EDTA) is not included as a protease inhibitor because it interferes with the subsequent IMAC step. Mild sonication is necessary to ensure high solubilization efficiencies.

Protocol 10-2: Solubilization of MBP-T43NTR-TrxA-H10 from Intact E. coli Cells

1. Buffers and materials required
   a. Cell paste (about 54 g) from 9.6 l of an *E. coli* suspension (DH5α pRG/III-hs-MBPP-T43NTR-TrxA-H10) (see Protocol 10-1).
   b. 2× solubilization buffer (60% [v/v] glycerol, 400 mM NaCl, 100 mM Tris/HCl, pH 7.4).
   c. Protease inhibitor stock solutions: leupeptin (inhibits serine and thiol proteases) at 1 mg/ml in H_2O, pepstatin (inhibits acid proteases) at 1.4 mg/ml in methanol, phenyl methyl sulfonyl fluoride (PMSF; inhibits serine proteases) at 70 mg/ml in dimethylsulfoxide. The protease inhibitors are used at a final concentration of 1 μg/ml (leupeptin), 1.4 μg/ml (pepstatin), and 70 μg/ml (PMSF), respectively.
   d. 1 M MgCl$_2$ solution.
   e. Deoxyribonuclease I (DNAse I) (Sigma D-4527) at 10 mg/ml.
   f. 6% (w/v) CHAPS (Sigma C-3023) / 1.2% (w/v) CHS (Sigma C-6013) stock solution.
   g. 10% (w/v) LM stock solution (Calbiochem or Sigma D-4641).
   h. 2 M imidazole/HCl, pH 7.4.
   i. Sonicator (Sonicator ultrasonic processor XL, Misonix, 1/2 in. flat tip).
   j. Ultracentrifuge, 45Ti rotor, 45Ti polycarbonate ultracentrifuge tubes (Beckman), or equivalent.
   [Day two]
2. Crush coarsely frozen cell paste between Saran wrap using a cooled metal block.
3. Place crushed cell paste in a 600-ml glass beaker and resuspend, with the aid of a spatula, in 80 ml of chilled 2× solubilization buffer supplemented with protease inhibitors.
4. Add 0.8 ml of a 1 M MgCl$_2$ solution. Add 160 μl of a 10 mg/ml DNase I solution.
5. Determine volume of cell suspension and add H_2O to give a final volume of 128 ml. Transfer material to a 200-ml plastic beaker containing a magnetic stir bar.
6. Add detergent stock solutions *dropwise* with gentle stirring on ice: first 16 ml of CHAPS/CHS, then 16 ml of LM. Stir for 15 min.
7. Place beaker with sample in ice/water bath. Sonicate for 8 min, pulsing 1 s on and 4 s off (level 5).

8. Stir for a further 45 min.
9. Load mixture into three precooled ultracentrifuge tubes. Load tubes into a Beckman 45Ti rotor and spin at 4°C at 40,000 rpm for 2 h.
10. Determine volume of supernatant (about 145 ml). Add protease inhibitors. Add imidazole/HCl to a final concentration of 50 mM.
11. Filter solubilized material over 0.2-μm disposable filter unit with a glass fiber prefilter.
12. Save 1 ml of solubilized material for analysis, use rest for purification.

10.3.2.2   Immobilized Metal Affinity Chromatography (IMAC)

The solubilized material is passed over a Ni^{2+}-NTA column in the presence of imidazole at a concentration of 50 mM. Imidazole at this concentration successfully eliminates binding to the Ni^{2+}-NTA resin of the majority of contaminating *E. coli* proteins, but does not lead to elution of MBP-T43NTR-TrxA-H10. The fusion protein is eluted with imidazole at a concentration of 500 mM (step gradient).

Protocol 10-3: Purification of MBP-T43NTR-TrxA-H10 by IMAC

1. Buffers and equipment required
   a. Buffer A: 50 mM Tris/HCl, pH 7.4, 30% (v/v) glycerol, 50 mM imidazole, 200 mM NaCl, 0.5% CHAPS, 0.1% CHS, 0.1% LM.
   b. Buffer B: 50 mM Tris/HCl, pH 7.4, 30% (v/v) glycerol, 500 mM imidazole, 200 mM NaCl, 0.5% CHAPS, 0.1% CHS, 0.1% LM.
   c. Protease inhibitors (see Protocol 10-2).
   d. GradiFrac system (Pharmacia).
   e. 20 ml Ni^{2+}-NTA column (Ni^{2+}-NTA Agarose, Qiagen) packed into a XK50 column (Pharmacia) (theoretical density of 10 to 13 μmol of Ni^{2+} ions per milliliter of packed resin).

[Day one]
2. Wash Ni^{2+}-NTA column with H$_2$O overnight at a flow rate of 0.3 ml/min.

[Day two]
3. Equilibrate Ni^{2+}-NTA column with 200 ml of buffer A at a flow rate of 0.6 ml/min.
4. Load solubilized material at a flow rate of 0.75 ml/min. Collect flow-through batchwise.
5. Wash Ni^{2+}-NTA column at a flow rate of 1.0 ml/min with 600 ml of buffer A.
6. Elute NTR fusion protein at a flow rate of 0.2 ml/min with 120 ml of buffer B. Collect 1-ml fractions.

[Day three]
7. Combine peak fractions (about 31 ml) and add protease inhibitors. Retain 1 ml for analysis.
8. Wash Ni^{2+}-NTA column with H$_2$O and store in 30% ethanol.

TABLE 10.2
Purification of the NTR Fusion Protein MBP-T43NTR-TrxA-H10

	Protein (mg)	Receptor (pmol)	Specific Binding (pmol/mg)	Purification (-fold)	Recovery (%)
Ni^{2+}-NTA column					
Solubilized material loaded	2883	13892	4.8	1	100
Flow-through	n.d.	1476			10.6
Eluate	6.46	12416	1922	400	89.4
NT column					
Ni^{2+}-NTA eluate loaded	6.27	12208			87.9 (100)
Flow-through	n.d.	305			2.2 (2.5)
Eluate	1.63	6042	3707	772	43.5 (49.5)

Analysis of the purification of MBP-T43NTR-TrxA-H10 (average of ten experiments) according to Protocols 10-2 to 10-4. About 54 g of cells (wet weight, from 9.6 l of cell culture) were used for solubilization. The protein content was measured according to the method of Schaffner and Weissmann.[33] The activity was determined by [^{3}H]-NT (New England Nuclear) binding[58] at a concentration of 1.8 nM using "home-made" Sephadex G 25 Fine columns for separation of bound from free radioligand as described.[14] The specific activities listed in this table have not been corrected by the factors discussed in Section 10.4. Recovery data listed in parentheses refer to the NT column only.

10.3.2.3   Neurotensin Column

The affinity of solubilized receptors for NT is reduced in the presence of sodium ions[34,35] and imidazole at high concentrations (unpublished data). Since the IMAC elution buffer B (Protocol 10-3) contains NaCl and imidazole at concentrations of 200 and 500 mM, respectively, a buffer exchange is required to allow binding of functional receptors to the immobilized agonist. The buffer exchange is achieved by gel filtration using disposable PD10 columns. After desalting, proteins are passed over the NT column, which retains almost all of the NT binding activity. Nonfunctional receptors are found in the flow-through. Proteins retained nonspecifically on the gel by ionic interaction are eliminated by washing the column with buffer containing 200 mM KCl.[36] A small amount of functional receptor protein is lost during this step because potassium ions decrease the NT binding affinity.[34,35] Receptor fusion protein, eluted from the NT column with buffer containing 1 M NaCl,[36] is almost pure, as revealed by SDS-PAGE and silver staining. Purified receptors bind NT in a reversible and saturable manner and with high affinity (see Reference 14). The specific binding is 3707 pmol/mg (Table 10.2) when routinely determined by [^{3}H]-NT binding at a concentration of 1.8 nM using Sephadex G 25 Fine columns for separation of bound from free radioligand as described.[14] More detailed saturation analyses at equilibrium give a specific binding of 6184 pmol/mg (four binding experiments) using prepacked Sephadex G 50 Fine columns (NickSpin columns, Pharmacia). A theoretical value of 10416 pmol/mg is calculated for a protein with a molecular mass of 96 kDa, assuming one ligand binding site per receptor molecule. The above determined value for specific binding may indicate the presence of nonfunctional receptors and minor impurities in the NT-column eluate and/or reflect on the methodology used to determine [³H]-NT binding to detergent-solubilized receptors (see below).

Protocol 10-4: Purification of MBP-T43NTR-TrxA-H10 by Ligand Affinity Chromatography

1. Buffers and equipment required
   a. Buffer A2: 50 mM Tris/HCl, pH 7.4, 30% (v/v) glycerol, 1 mM EDTA, 20 mM NaCl, 0.5% CHAPS, 0.1% CHS, 0.1% LM.
   b. Buffer B2: 50 mM Tris/HCl, pH 7.4, 30% (v/v) glycerol, 1 mM EDTA, 200 mM KCl, 0.5% CHAPS, 0.1% CHS, 0.1% LM.
   c. Buffer C2: 50 mM Tris/HCl, pH 7.4, 30% (v/v) glycerol, 1 mM EDTA, 1 M NaCl, 0.5% CHAPS, 0.1% CHS, 0.1% LM.
   d. Buffer A3: 50 mM Tris/HCl, pH 7.4, 3 mM NaN$_3$, 1 mM EDTA.
   e. Buffer B3: 50 mM Tris/HCl, pH 7.4, 1 mM EDTA, 1 M NaCl.
   f. Protease inhibitors (see Protocol 10-2).
   g. GradiFrac system (Pharmacia).
   h. 4 ml NT column (biotinylated NT(1-13) bound to succinylated tetrameric avidin resin [Sigma A-5150 Lot 42H8290][14] packed into a Pharmacia XK16 column, theoretical ligand density of 120 μM).
   i. PD10 columns (Pharmacia).
   [Day 2]
2. Equilibrate NT column overnight with 60 ml of buffer A2.
   [Day 3]
3. Equilibrate 12 PD10 columns for buffer exchange with 25 ml of buffer A2 per column according to the manufacturer's instructions. Add protease inhibitors to the final 5 ml aliquot.
4. Load Ni^{2+}-NTA eluate (30 ml) onto PD10 columns (2.5 ml per column). Elute protein from PD10 columns with 3.5 ml of buffer A2 per column. Collect PD10 column eluate (approximately 42 ml).
5. Load PD10 column eluate onto NT column at a flow rate of 0.2 ml/min. Collect NT column flow-through batchwise.
6. Wash NT column with 10 ml of buffer A2, then with 16 ml of buffer B2, then with 12 ml of buffer A2 at a flow rate of 0.2 ml/min.
7. Elute with 30 ml of buffer C2 at a flow rate of 0.2 ml/min. Collect 1 ml fractions. Combine peak fractions (about 11 ml) for further use.
8. Prepare NT column for storage. Wash column with 40 ml of buffer B3, and then with 40 ml of buffer A3 at a flow rate of 0.2 ml/min.

10.4   LIGAND BINDING ANALYSIS OF SOLUBILIZED NTR

Progress in purification of NTR is indicated by an increase of the specific binding expressed as "pmol/mg of protein." The amount of functional NTR fusion protein is determined by radioligand binding analysis (for an introduction see References 37

TABLE 10.3
Comparison of Prepacked Sephadex G 50 Fine Columns and "Home-made"
Sephadex G 25 Fine Columns Used for Ligand Binding Analysis of Purified
NTR Fusion Protein

	Prepacked Sephadex G 50 Fine Columns	"Home-made" Sephadex G 25 Fine Columns
Total binding (dpm)	4,814	5,508
Nonspecific binding (dpm)	178	1,955
Specific binding (dpm)	4,636	3,553

Method: 6 ng of purified NTR fusion protein were incubated in 150 µl of "standard" assay buffer at a
[³H]-NT concentration of 2 nM (45,990 dpm per 100 µl of assay mix). Nonspecific binding was determined
in the presence of 2 µM unlabeled NT. Aliquots of 100 µl of the assay mix were loaded onto the columns
and processed as described.[14] The data are the average of 4 experiments performed in duplicate.

to 39) using the agonist [³H]-NT which binds with high affinity to NTR. We determined an equilibrium dissociation constant (K_D) of 0.43 nM under "standard" assay
conditions (incubation on ice for 1 h in assay buffer consisting of 50 mM Tris/HCl,
pH 7.4, 1 mM EDTA, 40 mg of bacitracin per liter, 0.1% (w/v) BSA, and detergent)[14]
(see also References 34, 36, and 40).

Separation of bound from free ligand is achieved by gel filtration assisted by
centrifugation.[14] Using this technique requires that the ligand off-rate is sufficiently
slow that no more than 10% of the specifically bound ligand may dissociate during
the separation.[39] We determined a dissociation rate constant (k_{off}) of 0.043 min^{-1}
(half-life of 16 min) for purified NTR fusion protein assayed under "standard"
conditions (NickSpin columns).[18a] This off-rate makes the above assay method
acceptable for ligand binding analysis of solubilized receptors. However, it remains
to be determined whether a population of purified receptors also shows a very fast
off-rate as we found for membrane-bound NTR (unpublished results). This would
lead to underestimation of the amount of functional NTR in the NT-column eluate.

Specific binding is calculated from total binding minus nonspecific binding. The
performance of the gel filtration column in separating the receptor–ligand complex
from free ligand is in this respect an important parameter which influences the
determination of the B_{max} value. We found that prepacked Sephadex G 50 Fine
columns (NickSpin, Pharmacia) give less background than "home-made" Sephadex
G 25 Fine columns (3 ml of settled resin (Pharmacia), QS-Q columns from Advanced
Laboratory Techniques, U.K.)[14] under otherwise identical experimental conditions
(Table 10.3) and that both column types differ in their separation behavior. The
resulting increase in the specific binding of purified NTR fusion protein by a factor
of 1.3 is thus due to the methodology used rather than due to changes in NTR
binding properties.

The progress of purification is routinely assessed at a radioligand concentration
of 1.8 nM, which suggests a fractional occupancy of receptors with [³H]-NT of 80%

given the equilibrium dissociation constant of 0.43 nM and assuming that the inter-action of the agonist with receptor is a reversible, bimolecular reaction that obeys the law of mass action. Use of higher [^{3}H]-NT concentrations in combination with "home-made" Sephadex G 25 Fine columns leads to unacceptably high background and large scatter of data. Conversion by a factor of 1.3 (Table 10.3) of the data presented in Table 10.2 and consideration of 80% occupancy predict a specific binding of about 6000 pmol/mg for purified NTR. This is confirmed by saturation analysis at equilibrium using Pharmacia NickSpin columns under "standard conditions."

10.5 DETERMINATION OF PROTEIN CONTENT

Commonly used procedures for protein determination are based on the method of Lowry,[41] the bicinchoninic acid method,[42] and dye-binding assays.[43] Each of these methods exhibits some variation in its response toward different proteins and/or suffers from the effect of interfering substances, leading to an inherent inaccuracy in the determination of the protein content (see Reference 15). One major source of error in protein assays is the presence of interfering compounds such as detergents and lipids, major constituents in membrane protein preparations. To help eliminate such interference, the membrane protein sample should be precipitated and washed prior to protein determination.[33,44,45] The Amido Black dye-binding method described by Schaffner and Weissmann[33] is in this respect most useful for the determination of the protein content of membranes and detergent-solubilized receptors and was used throughout for the analysis of NTR purification.

We use BSA as a standard for the above-mentioned Amido Black method assuming that Amido Black binds equally well to BSA and the purified NTR fusion protein. This assumption was confirmed by amino acid analysis of purified receptor fusion protein.[59]

10.6 CONCLUSIONS

The methods described above for the expression in *E. coli* and purification of functional NTR fusion protein have been developed in view of simple procedures to allow the isolation of sufficient receptor protein for crystallization trials and further scale-up. Bacterial expression of NTR is easy and reliable. The use of a deca-histidine tag guarantees efficient purification of receptors by immobilized metal affinity chromatography starting from intact cells, thus avoiding time-consuming membrane preparations. The number of affinity steps is kept to a minimum to make purification as fast as possible. This scheme sets the stage for structural studies of NTR and we are currently in the process of exploring conditions for the formation of two-dimensional crystals.

ACKNOWLEDGMENTS

This work was supported by GlaxoWellcome, Zeneca Pharmaceuticals, and the Medical Research Council (U.K.). We thank David Owen for performing amino acid analyses, and John Berriman, Gebhard Schertler, Chris Tate, and Markus Weiß for critical comments on the manuscript.

REFERENCES

1. Baldwin, J. M., Structure and function of receptors coupled to G proteins, *Curr. Opinion Cell Biol.*, 6, 180, 1994.
2. Schwartz, T. W., Gether, U., Schambye, H. T., and Hjorth, S. A., Molecular mechanism of action of non-peptide ligands for peptide receptors, *Current Pharm. Design*, 1, 355, 1995.
3. Strader, C. D., Fong, T. M., Graziano, M. P., and Tota, M. R., The family of G protein-coupled receptors, *FASEB J.*, 9, 745, 1995.
4. Schertler, G. F. X., Villa, C., and Henderson, R., Projection structure of rhodopsin, *Nature*, 362, 770, 1993.
5. Davies, A., Schertler, G. F. X., Gowen, B. E., and Saibil, H. R., Projection structure of an invertebrate rhodopsin, *J. Struct. Biol.*, 117, 36, 1996.
6. Schertler, G. F. X. and Hargrave, P. A., Projection structure of frog rhodopsin in two crystal forms, *Proc. Natl. Acad. Sci. U.S.A.*, 92, 11578, 1995.
7. Krebs, A., Villa, C., Edwards, P. C., and Schertler, G. F. X., Characterisation of an improved two-dimensional p22121 crystal from bovine rhodopsin, *J. Mol. Biol.*, 282, 991, 1998.
8. Unger, V. M. and Schertler, G. F. X., Low resolution structure of bovine rhodopsin determined by electron cryo-microscopy, *Biophys. J.*, 68, 1776, 1995.
9. Unger, V. M., Hargrave, P. A., Baldwin, J. M., and Schertler, G. F. X., Arrangement of rhodopsin transmembrane α-helices, *Nature*, 389, 203, 1997.
10. Tate, C. G. and Grisshammer, R., Heterologous expression of G protein-coupled receptors, *Trends Biotechnol.*, 14, 426, 1996.
11. Grisshammer, R. and Tate, C. G., Overexpression of integral membrane proteins for structural studies, *Q. Rev. Biophys.*, 28, 315, 1995.
12. Tanaka, K., Masu, M., and Nakanishi, S., Structure and functional expression of the cloned rat neurotensin receptor, *Neuron*, 4, 847, 1990.
12a. Kiefer, H. and Grisshammer, R., unpublished results.
13. Kiefer, H. et al., Expression of an olfactory receptor in *Escherichia coli*: purification, reconstitution, and ligand binding, *Biochemistry*, 35, 16077, 1996.
14. Tucker, J. and Grisshammer, R., Purification of a rat neurotensin receptor expressed in *Escherichia coli*, *Biochem. J.*, 317, 891, 1996.
15. Grisshammer, R., Expression of G protein-coupled receptors in *Escherichia coli*, in *Identification and Expression of G Protein-Coupled Receptors*, Lynch, K.R., Ed., John Wiley & Sons, New York, 1998, 133.
16. Hulme, E. C. and Curtis, C. A. M., Purification of recombinant M1 muscarinic acetylcholine receptor, *Biochem. Soc. Trans.*, 26, S361, 1998.
17. Calandra, B., Tucker, J., Shire, D., and Grisshammer, R., Expression in *Escherichia coli* and characterisation of the human central CB1 and peripheral CB2 cannabinoid receptors, *Biotechnol. Lett.*, 19, 425, 1997.
18. Grisshammer, R., Duckworth, R., and Henderson, R., Expression of a rat neurotensin receptor in *Escherichia coli*, *Biochem. J.*, 295, 571, 1993.
18a. Grisshammer, R. and Averbeck, P., unpublished results.
19. Herzog, H., Münch, G., and Shine, J., Human neuropeptide Y1 receptor expressed in *Escherichia coli* retains its pharmacological properties, *DNA Cell Biol.*, 13, 1221, 1994.

20. Vanhauwe, J., Luyten, W. H. M. L., Josson, K., Fraeyman, N., and Leysen, J. E., Expression of the human dopamine D3 receptor in *Escherichia coli*: investigation of the receptor properties in the absence and presence of G proteins, in *International Symposium on G protein Coupled Receptors: Structure and Function*, Janssen Research Foundation, Beerse, Belgium, 1996, poster abstract 51.

21. Katti, S. K., LeMaster, D. M., and Eklund, H., Crystal structure of thioredoxin from *Escherichia coli* at 1.68 Å resolution, *J. Mol. Biol.*, 212, 167, 1990.

22. Jeng, M.-F. et al., High-resolution solution structures of oxidized and reduced *Escherichia coli* thioredoxin, *Structure*, 2, 853, 1994.

23. Thisted, T., Nielsen, A. K., and Gerdes, K., Mechanism of post-segregational killing: translation of Hok, SrnB and Pnd mRNAs of plasmids R1, F and R483 is activated by 3'-end processing, *EMBO J.*, 13, 1950, 1994.

24. Thisted, T., Sørensen, N. S., Wagner, E. G. H., and Gerdes, K., Mechanism of post-segregational killing: Sok antisense RNA interacts with Hok mRNA via its 5'-end single-stranded leader and competes with the 3'-end of Hok mRNA for binding to the *mok* translational initiation region, *EMBO J.*, 13, 1960, 1994.

25. Thisted, T., Sørensen, N. S., and Gerdes, K., Mechanism of post-segregational killing: secondary structure analysis of the entire Hok mRNA from plasmid R1 suggests a fold-back structure that prevents translation and antisense RNA binding, *J. Mol. Biol.*, 247, 859, 1995.

26. Grisshammer, R. and Tucker, J., Quantitative evaluation of neurotensin receptor purification by immobilized metal affinity chromatography, *Protein Expression Purification*, 11, 53, 1997.

27. Sambrook, J., Fritsch, E. F., and Maniatis, T., *Molecular Cloning: A Laboratory Manual*, 2nd ed., Cold Spring Harbor Laboratory Press, Cold Spring Harbor, 1989.

28. Fall, R. R., Analysis of microbial biotin proteins, *Methods Enzymol.*, 62, 390, 1979.

29. Schatz, P. J., Use of peptide libraries to map the substrate specificity of a peptide modifying enzyme: a 13 residue consensus peptide specifies biotinylation in *Escherichia coli*, *Bio/Technology*, 11, 1138, 1993.

30. Porath, J., Carlsson, J., Olsson, I., and Belfrage, G., Metal chelate affinity chromatography, a new approach to protein fractionation, *Nature*, 258, 598, 1975.

31. Hochuli, E., Döbeli, H., and Schacher, A., New metal chelate adsorbent selective for proteins and peptides containing neighbouring histidine residues, *J. Chromatogr.*, 411, 177, 1987.

32. Laemmli, U. K., Cleavage of structural proteins during the assembly of the head of bacteriophage T4, *Nature*, 227, 680, 1970.

33. Schaffner, W. and Weissmann, C., A rapid, sensitive, and specific method for the determination of protein in dilute solution, *Anal. Biochem.*, 56, 502, 1973.

34. Zsürger, N., Mazella, J., and Vincent, J.-P., Solubilization and purification of a high affinity neurotensin receptor from newborn human brain, *Brain Res.*, 639, 245, 1994.

35. Mazella, J., Chabry, J., Kitabgi, P., and Vincent, J.-P., Solubilization and characterization of active neurotensin receptors from mouse brain, *J. Biol. Chem.*, 263, 144, 1988.

36. Mazella, J., Chabry, J., Zsürger, N., and Vincent, J.-P., Purification of the neurotensin receptor from mouse brain by affinity chromatography, *J. Biol. Chem.*, 264, 5559, 1989.

37. Limbird, L. E., *Cell Surface Receptors: A Short Course on Theory and Methods*, 2nd ed., Kluwer Academic Publishers, Boston, 1996.

38. Motulsky, H. J. and Searle, P., *GraphPad Prism Version 2.0—User's guide*, 1994/1995.

39. Hulme, E. C., Receptor-binding studies—a brief outline, in *Receptor Biochemistry: A Practical Approach*, Hulme, E. C., Ed., IRL Press, Oxford, 1990, 303.

40. Miyamoto-Lee, Y., Shiosaka, S., and Tohyama, M., Purification and characterization of neurotensin receptor from rat brain with special reference to comparison between newborn and adult age rats, *Peptides*, 12, 1001, 1991.

41. Lowry, O. H., Rosebrough, N. J., Farr, A. L., and Randall, R. J., Protein measurement with the folin phenol reagent, *J. Biol. Chem.*, 193, 265, 1951.

42. Smith, P. K. et al., Measurement of protein using bicinchoninic acid, *Anal. Biochem.*, 150, 76, 1985.

43. Bradford, M. M., A rapid and sensitive method for the quantitation of microgram quantities of protein utilizing the principle of protein-dye binding, *Anal. Biochem.*, 72, 248, 1976.

44. Chang, Y.-C., Efficient precipitation and accurate quantitation of detergent-solubilized membrane proteins, *Anal. Biochem.*, 205, 22, 1992.

45. Pande, S. V. and Murthy, M. S. R., A modified micro-Bradford procedure for elimination of interference from sodium dodecyl sulfate, other detergents, and lipids, *Anal. Biochem.*, 220, 424, 1994.

46. Jockers, R. et al., Species difference in the G protein selectivity of the human and bovine A1-adenosine receptor, *J. Biol. Chem.*, 269, 32077, 1994.

47. Chapot, M.-P. et al., Localization and characterization of three different β-adrenergic receptors expressed in *Escherichia coli*, *Eur. J. Biochem.*, 187, 137, 1990.

48. Freissmuth, M., Selzer, E., Marullo, S., Schütz, W., and Strosberg, A. D., Expression of two human β-adrenergic receptors in *Escherichia coli*: functional interaction with two forms of the stimulatory G protein, *Proc. Natl. Acad. Sci. U.S.A.*, 88, 8548, 1991.

49. Marullo, S. et al., Expression of human $β_1$ and $β_2$ adrenergic receptors in *E. coli* as a new tool for ligand screening, *Bio/Technology*, 7, 923, 1989.

50. Breyer, R. M., Strosberg, A. D., and Guillet, J.-G., Mutational analysis of ligand binding activity of $β_2$-adrenergic receptor expressed in *Escherichia coli*, *EMBO J.*, 9, 2679, 1990.

51. Haendler, B., Hechler, U., Becker, A., and Schleuning, W.-D., Expression of human endothelin receptor ET_B by *Escherichia coli* transformants, *Biochem. Biophys. Res. Commun.*, 191, 633, 1993.

52. Curtis, C. A. M. and Hulme, E. C., in *7th International Symposium on Muscarinic Subtypes, Life Sciences*, 60, 1176, 1997.

53. Grisshammer, R., Little, J., and Aharony, D., Expression of rat NK-2 (neurokinin A) receptor in *E. coli*, *Receptors Channels*, 2, 295, 1994.

54. Münch, G., Walker, P., Shine, J., and Herzog, H., Ligand binding analysis of human neuropeptide Y1 receptor mutants expressed in *E. coli*, *Receptors Channels*, 3, 291, 1995.

55. Stanasila, L., Kieffer, B. L., and Pattus, F., Expression of human opioid receptors in *E. coli*, in *25th Silver Jubilee FEBS Meeting*, FEBS '98 Abstracts, Copenhagen, Denmark, 1998, p. 184, poster abstract P35.16.

55a. Stanasila, L., Massotte, D., Kieffer, B. L., and Pattus, F., Expression of δ, κ and μ human opioid receptors in *Escherichia coli* and reconstitution of the high-affinity state for agonist with heterotrimeric G proteins, *Eur. J. Biochem.*, 260, 430, 1999.

56. Harada, Y., Senda, T., Sakamoto, T., Takamoto, K., and Ishibashi, T., Expression of octopus rhodopsin in *Escherichia coli*, *J. Biochem.*, 115, 66, 1994.

57. Bertin, B. et al., Functional expression of the human serotonin 5-HT_{1A} receptor in *Escherichia coli*, *J. Biol. Chem.*, 267, 8200, 1992.

58. Goedert, M., Radioligand-binding assays for study of neurotensin receptors, *Methods Enzymol.*, 168, 462, 1989.

59. Grisshammer, R. and Tucker, J., unpublished results.

11 Large-Scale Expression of Receptors in Yeasts

Helmut Reiländer, Christoph Reinhart and Agnes Szmolenszky

CONTENTS

11.1 INTRODUCTION

The investigation of hormone and neurotransmitter receptors has become one of the most active areas in biochemical and biomedical research. Among the different families of integral receptors, the G protein-coupled receptors (GPCRs) comprise the largest and, from the viewpoint of pharmacology, also the most interesting family. Due to the rapid development of modern molecular biology, over 1000 GPCRs have been cloned to date from a variety of organisms and it is expected that a plethora of novel GPCRs will be cloned in the future. All GPCRs have a similar structure, consisting of a single polypeptide chain, and share a similar hydrophobicity profile consistent with the existence of seven transmembrane spanning segments. Although the receptors share a common mechanism of signal transduction, they transduce a variety of different external signals (light, neurotransmitters, hormones, odorants, etc.) to the interior of the cell.[1,2,3] In the activated state GPCRs trigger GDP to GTP exchange in heterotrimeric G proteins ($G_{\alpha\beta\gamma}$) which after that dissociate (G_α and $G_{\beta\gamma}$) to interact with a multitude of effector proteins (adenylate cyclase, phospholipase C, Ca^{2+}- and K^+-channels, mitogen-activated protein kinase signal transduction pathway).[4] Selectivity in this signal transduction system is achieved by the specificity of ligand–receptor as well as receptor–G protein interaction and the system is modulated at different levels.[5]

The pharmacological relevance of this receptor family also led to increasing interest in the three-dimensional structure of these receptors. Current receptor models are based on structural studies of bacteriorhodopsin (two-dimensional crystallization;[6] three-dimensional crystallization[7,8]), several projection structures of rhodopsins,[9–13] and topographical studies of receptors,[14–16] mostly of the β_2-adrenergic receptor.[17,18] However, knowledge of a "real" three-dimensional structure of a GPCR would significantly add to our understanding of the molecular mechanisms involved in ligand binding as well as signaling and would also be useful in the design of drugs without unwanted side effects. An absolute prerequisite for determination of a three-dimensional structure is the availability of milligram amounts of pure and homogeneous protein.

Traditionally, GPCRs were investigated in, or purified from, animal or human tissue. With the exception of rhodopsin, which can be isolated and purified in high yield from retina, the procedure of isolation and purification from natural sources tended and still tends to be laborious and time consuming, if possible at all. A further problem in the study of receptors isolated from tissues was the discovery of multiple receptor subtypes, many of which had not been anticipated on the basis of pharmacological characterization. To circumvent these problems, heterologous production of GPCRs in defined host systems was established as an important and valuable tool for the pharmacological and biochemical analysis of defined receptor subtypes. Also, analysis of chimeric or mutant receptors is performed in host systems. The analysis of G protein interaction and signal transduction, the detection of "lead" compounds for a GPCR by high-throughput screens (HTS), as well as for the search of the associated ligand for an orphan receptor, are other important tasks that are also discussed in this chapter. First, we provide a general review of production of GPCRs in the yeast system and compare it with other expression systems. Second, we provide some essential recipes for GPCR production using the *Pichia pastoris* expression system.

11.2 THEORETICAL OVERVIEW

11.2.1 Yeast Systems Used for the Production of GPCRs

Production of membrane proteins, especially of GPCRs, has been described from microorganisms and cell lines derived from a variety of species. The choice of expression systems is nowadays, in theory at least, much broader and more complicated than it was 10 years ago. For large-scale production the choice of an expression system now involves consideration of the problem not only from the point of view of the molecular biologist but also from that of the microbial or cell physiologist, the fermenter specialist, and the downstream processing, which includes breakage of the cells and purification of the recombinant protein.

Membrane proteins produced in bacteria, e.g., *Escherichia coli*, if produced at all, are often insoluble and therefore inactive. The high-level production of active protein in *E. coli* is frequently challenging and there are a plethora of publications dealing with tricks to enhance production levels, including the optimization of signals for transcription and translation, codon usage, fusion proteins and growth conditions (see Chapter 10 and Reference 19) and the use of different promoters or plasmids with different copy numbers. Mammalian expression systems comprise a plethora of different cell lines used for the production of recombinant proteins. Most commonly, Chinese hamster ovary (CHO) cells, baby hamster kidney (BHK) cells, human embryonic kidney (HEK) 293 cells, NIH3T3 cells and others have been described for production. Being eukaryotic cells they have the capacity to carry out post-translational modifications such as glycosylation, phosphorylation, and addition of lipid compounds (for review see Reference 20). Recombinant viruses can also be constructed and used for infection of mammalian host cells which then synthesize the desired protein. Here, the Vaccinia system of the *Semliki Forest* virus expression systems has to be mentioned.[21–23]

Production of GPCRs in insect cells by infection with recombinant baculovirus has become a popular and reliable system in the last decade.[24,25] In comparison to mammalian systems production levels of recombinant proteins in infected insect cells in most cases are higher—sometimes $\geq$200-fold higher. GPCRs produced in insect cells from the viewpoint of pharmacology, biochemistry, and also signal transduction probability resemble their native counterpart. The versatility of this combination with good production levels and the possibility for large-scale fermentation, which is much easier than with mammalian systems, accounts for its attractiveness. Alternatives to the systems mentioned above are the yeast-based systems which are very attractive for heterologous protein production due to high level production, easy manipulation, and low costs (for general reviews see References 26 to 29). Conventionally, yeasts have been used in fermentation and food industries for a long time and have proven to be safe. In comparison to yeasts, the higher eukaryotic systems (insect or mammalian cells) are more difficult to handle and large-scale culture is difficult and more costly. Another disadvantage of higher cell systems is that in fermentation only low cell densities can be reached. Contrary to *E. coli*, yeasts have the potential to perform eukaryotic post-translational modifications which are in some cases necessary for active protein.[30] In Table 11.1 an overview is presented of the different yeast systems as well as the different GPCRs that have been tested for production since 1990. More information on the production of GPCRs using other expression systems (bacteria, insect or mammalian cells) can be obtained in this book (Chapter 10) and in the reviews by Grisshammer and Tate,[31] and Tate and Grisshammer.[32]

As can be seen in Table 11.1, beside the classical expression system of *Saccharomyces cerevisiae*, there are only two other yeasts that have been tested for production of GPCR: the fission yeast *Schizosaccharomyces pombe* and the methylotrophic yeast *Pichia pastoris*. Other yeast, such as *Hansenula polymorpha*, *Kluyveromyces lactis*, *Yarrowia lipolytica*, or *Candida utilis*, which are used for the production of a number of commercially useful proteins, have never been tested for production of a GPCR.

11.2.1.1 *Saccharomyces cerevisiae*

In most studies, the well known "brewers" yeast or "bakers" yeast *S. cerevisiae* has been used for the production of GPCRs. *S. cerevisiae* is the most intensively studied and characterized yeast from molecular genetic and cell biological aspects and its complete genome sequence has been published recently.[53-55] A multitude of different promoters and strains makes life easy when working with this yeast. Therefore, *S. cerevisiae* is more amenable to rapid genetic manipulation than many other eukaryotic cell expression systems also allowing other components of mammalian GPCR signal transduction pathways to be coexpressed. Although its genetics and physiology are well known, *S. cerevisiae* has been found to have some limitations as a host for production of recombinant proteins. It is often difficult to obtain high production level, plasmid stability during fermentation can cause problems, and glycoproteins may become hyperglycosylated.

11.2.1.2 *Schizosaccharomyces pombe*

The fisson yeast *Schizosaccharomyces pombe* (*Sz. pombe*) is phylogenetically as distant from budding yeast *S. cerevisiae* as it is from mammalian and has many properties in common with higher eukaryotic organisms. Therefore, *Sz. pombe* is a very attractive host for the production of complex proteins from eukaryotic origin.[56] Nevertheless, the biotechnical application lags far behind the developments achieved in *S. cerevisiae* and *P. pastoris*. Nowadays a range of different expression vectors have become available, some of which are based on the thiamine-repressible *nmt* promoter (*nmt1*, no message in thiamine)[57] and which have been established for the production of recombinant protein.

TABLE 11.1
Production of G Protein-Coupled Receptors in Yeasts

Receptor	Type	Species	Yeast Host	Promoter	Signal-Peptide/Fusion	Tag	Mol wt (kDa)	Glycosylation	Production Level	Activities Measured on Protein Produced	Signal Transduction	Ref.
Adenosine receptor	A_{2A}	Rat	S. cerevisiae	ADH1	+	+	ND	ND	0.45 ± 0.13 pmol/mg	Agonist binding	+	33
Adrenergic receptor	β_2	Human	S. cerevisiae	GAL1	Ste2	+	ND	ND	115 pmol/mg	Antagonist binding	+	34
	β_2	Human	Sz. pombe	adh	Ste2	+	69/46	Possible	7.5 pmol/mg	Antagonist binding	ND	35
	β_2	Human	P. pastoris	AOX1	αF	F, B	55	n.g.	25 pmol/mg	Antagonist binding	ND	36
	β_2	Human	S. cerevisiae	GAL1	Ste2	+	ND	ND	36 ± 7 pmol/mg	Antagonist binding	ND	37
	α_2-C2	Human	S. cerevisiae	GAL1	+	H6	50	ND	7–70 pmol/mg	Ligand binding	ND	37
Muscarinic acetylcholine receptor	M_1	Human	S. cerevisiae	ADH	+	+	ND	ND	0.02 pmol/mg	Antagonist binding	ND	38
	M_3	Rat					66	Detected	ND	ND	ND	#
	M_5	Rat	S. cerevisiae	αF	αF	+	ND	ND	0.13 pmol/mg	Antagonist binding	–	39
Serotonin (5HT) receptor	$5HT_{5A}$	Mouse	S. cerevisiae	PRB1	Bm	m	43	Detected	16 pmol/mg	Agonist binding	ND	40
	$5HT_{5A}$	Mouse	Sz. pombe				38	ND	2–3 pmol/mg	Agonist binding	ND	#
	$5HT_{5A}$	Mouse	P. pastoris	AOX1	αF, aP	m	ND	ND	22 pmol/mg	Agonist binding	ND	41
	$5HT_{5A}$	Mouse	P. pastoris	AOX1	αF	F, m, B	61/41	Detected	40 pmol/mg	Agonist binding	ND	36
Dopamine receptor	D_{1A}	Human	S. cerevisiae	GPD	Ste2	F, H	55	Detected	127 ± 25 fmol/mg	Antagonist	ND	42
	D_{2S}	Human	S. cerevisiae	PMA1	Ste2		40	ND	1–2 pmol/mg	Antagonist binding	ND	43
	D_{2S}	Human	S. cerevisiae	GAL10			40	ND	2.8 pmol/mg	Antagonist binding	ND	44
	D_{2S}	Human	Sz. pombe	nmt1			40	ND	14.6 pmol/mg	Antagonist binding	ND	44
	D_{2S}	Human	P. pastoris	AOX1			51	Detected	100 pmol/mg	Antagonist binding	ND	44a
Neurokinin receptor	NK2	Human	Sz. pombe	nmt1	aP	+	ND	ND	1.16 pmol/mg	Agonist binding	–	45
Opioid receptor	μ	Human	P. pastoris	AOX1	αF/Ste2	+	ND	ND	0.4 pmol/mg	Antagonist binding	ND	46
Rhodopsin		Bovine	S. cerevisiae	GAL1	+	+	40	Detected	2 ± 0.5 mg/10^{10}c	11-cis-retinal binding	+	47
		Bovine	P. pastoris	AOX1	aP	+	37	Detected	0.3 mg/l ($A_{600} = 1$)	11-cis-retinal binding	+	48

TABLE 11.1 (CONTINUED)
Production of G Protein-Coupled Receptors in Yeasts

Receptor	Type	Species	Yeast Host	Promoter	Signal-Peptide/ Fusion	Tag	Mol wt (kDa)	Glycosylation	Production Level	Activities Measured on Protein Produced	Signal Transduction	Ref.
Somatostatin receptor	SSTR2	Rat	S. cerevisiae	GAL1//10	+	+	ND	ND	0.2 pmol/mg	Agonist binding	+	49
Lysophosphatidic acid receptor	Edg2/ Vzg1	Human	S. cerevisiae	GAL	+	+	ND	ND	ND	Agonist binding	+	50
Growth hormone releasing hormone receptor		Human	S. cerevisiae	GAL1//10	+	+	ND	ND	ND	Agonist binding	+	51
α-factor receptor	Ste2	S. cerevisiae	S. cerevisiae	GDP	+	F, H6	60	Detected	350 pmol/mg	Agonist binding	+	52

Bm, Bacillus; aP, acidic phosphatase; F, FLAG-tag; H6, hexa-histidine-tag; m, c-myc-tag; B, biotin-tag; Ste2, α-factor receptor fusion; AOX1, alcohol oxidase 1 promoter; GAL1, GAL1 promoter; GAL10, GAL10 promoter; GPD, glyceraldehyd-3-phosphate dehydrogenase promoter; nmt1, nmt1 promoter (no message in thiamine); PMA1, PMA1 promoter; PRB1, endopeptidase B promoter; ND, not determined; n.g., not given, #, Bach et al., unpublished results.

11.2.1.3 *Pichia pastoris*

In the last decade, the use of yeast began to change when Phillips Petroleum converted its single cell protein organism, the methylotrophic yeast *P. pastoris*, into a producer of recombinant protein. Since 1993 the system has been available from Invitrogen and, therefore, *P. pastoris* has been developed as a host for the efficient production of foreign proteins in the academic world as well.[58] In almost all cases expression of a foreign gene in this yeast was accomplished using the strongly regulated promoter of the *P. pastoris* alcohol oxidase I (*AOXI*). In the meantime there are vectors available that are based on the constitutive glyceraldehyde-3-phosphate dehydrogenase (*GAP*) promoter for heterologous protein production in *P. pastoris*.[59] Here, it will be interesting to compare production under control of the *GAP* promoter to that of the *AOX1* promoter (vectors with different promoters and selections can be obtained commercially from Invitrogen). In *P. pastoris* only vectors for genomic integration are available; after transformation and integration of the linearized vector resulting recombinant *P. pastoris* clones are therefore extremely stable even under conditions of induction (cultures are induced by supplementation of methanol to the growth medium, reviewed in References 60 to 62).

Since *P. pastoris* is the first choice for high-density fermentation, this yeast system comprises an interesting organism for large-scale production of GPCRs. Nevertheless, up to now there are only a few examples of functional production of GPCRs in *P. pastoris*. However, it seems that production of large amounts of at least some receptors will be possible. Here, genetically improved strains and more sophisticated culture conditions should lead to even better production levels.

11.2.2 PRODUCTION OF GPCRs IN YEAST FOR LIGAND SCREENING ASSAYS

Traditionally, drug discovery for GPCRs has relied on the use of animals or tissues isolated from animals to screen compound libraries in biological or pharmacological assays. Nowadays, the implementation of highly sensitive cell-based, high-throughput screening (HTS) assays which facilitate the rapid identification of selective and potent lead compounds from known and orphan GPCRs have replaced the old assay systems.

The modern HTS assays use the functional coupling of reporter genes to heterologously produced GPCRs. Widely used reporter genes for this purpose are the chloramphenicol acetyltransferase, β-galactosidase, β-glucoronidase, alkaline phosphatase and firefly luciferase. Most commonly, engineered mammalian cells[63] were used for these assays (see Chapter 4), but in the last few years *S. cerevisiae* has developed as an alternative and therefore will be described here in more detail (see also Chapter 2).[64] For a long time mating of the two different haploid cell types of baker's yeast *S. cerevisiae* to form diploid cells provided a simple system for studying the biochemical basis of signal transduction and developmental control.

S. cerevisiae haploid cells of different mating-types (**a** or α mating-type; MATa or MATα) secrete complementary tridecapeptide phermones (**a**- or α-factor) which bind to the complementary receptors. These membrane-bound receptors expressed on the cell surface by cells of opposite mating type after stimulation activate the mating-signal transduction pathway which results in cell cycle arrest in G1, morphological changes of the yeast cells, and transcriptional activation of genes involved in pheromone response and mating (reviewed in Reference 65). For production of diploid cells (a/α-cells) a functional response is absolutely essential. The two receptors, the **a**-factor (Ste3p) and α-factor (Ste2p) receptors, after binding of the respective factor, interact with a heterotrimeric G protein composed of Gpa1p (α subunit), Ste4p (β subunit), and Ste18p (γ subunit), catalyzing the exchange of GDP for GTP bound to the α subunit. Contrary to the situation in mammalian cells, in *S. cerevisiae* the α subunit, after dissociation from the βγ subunits, functions as a negative component that prevents activation of the pathway, whereas the βγ subunits activate the downstream pathway which includes a mitogen-activated protein kinase cascade (MAP) (reviewed in Reference 66). The MAP-kinase cascade includes the Ste20 (a p21-activated protein kinase homolog), the Ste11 (MEK kinase), the Ste7 (MAP-kinase kinase or MEK), as well as the MAP kinases Fus3 and Kss1 which activate the transcriptional activator Ste12. The Ste12 in turn activates transcription of several mating factor-inducible genes such as *FUS1*.

The first evidence that a foreign GPCR was interacting with the internal pheromone-response pathway was reported by the Lefkowitz group in 1990.[34] Here, it was possible to activate the yeast downstream signal transduction pathway by the addition of the adrenergic agonist isoproterenol, after coproduction of the human β_2-adrenergic receptor with a mammalian $G_{\alpha s}$ subunit in a *S. cerevisiae* strain where the *GPA1* gene had been inactivated. Signaling resulted in the induction of pheromone-responsive reporter gene expression (*FUS1-lacZ* gene fusion), cell-cycle arrest, and morphological changes of the yeast cells. Nevertheless, the receptor production level required for this response was extremely high (about 100 pmol/mg of protein).

Over the past few years, the *S. cerevisiae* system has been optimized by genetic manipulation (development of *S. cerevisiae* strains with deletions of *STE2*, the Ste2 receptor, *GPA1*, the internal G_α subunit, *SST2* which encodes a GTPase-activating protein for the G_α subunit and mutation of *FAR1*, leading to uncoupling of the MAP kinase activation from cell cycle arrest). The heterologously produced GPCRs were coupled to sensitive reporters that were under the transcriptional control of the *FUS1* promoter (the *FUS1-LacZ* reporter uses the *E. coli* β-galactosidase which can be quantified by a sensitive enzyme assay; *FUS1-HIS3*, here, agonist activation confers histidine prototrophy on selective medium in his3-deficient host backgrounds) (reviewed in Reference 67).

Using this approach it was possible to investigate several GPCRs. The mammalian somatostatin receptor subtype 2 (SSTR2) was cloned under the transcriptional control of the *GAL1/10* promoter. The receptor was detected by radioligand binding and in *sst2Δ* cells the agonist-bound receptor signaled through both the wild-type yeast G_α protein as well as a Gpa1-$G_{\alpha i2}$ fusion protein which was introduced into Gpa1-lacking *S. cerevisiae* cells.[49] Membranes prepared from a recombinant yeast clone producing the rat A_{2a}-adenosine receptor under control of the *ADH1* promoter exhibited saturable binding of the radiolabeled adenosine agonist [³H]-5′-*N*-ethyl-carboxamidoadenosine and also the rank order of potency and binding affinities for some agonists were comparable to those in mammalian systems. Adenosine agonists led to receptor-dependent activation of the signal transduction pathway inducing the pheromone-responsive *FUS1-HIS3* reporter element.[33] Here, for assaying the agonist response adenosine deaminase had to be added to the plates. Contrary to the situation with the somatostatin receptor, the A_{2a} receptor failed to mediate signaling in yeast strains producing the chimeric Gpa1-$G_{\alpha i2}$ protein. Also functional signaling of the human growth hormone releasing hormone (GHRH) receptor to the *FUS1-HIS3* reporter required substitution of the yeast G_α by a Gpa1-$G_{\alpha s}$ fusion protein. A strain bearing the chimeric Gpa1-$G_{\alpha i2}$ exhibited no detectable response to addition of GHRH. As expected the receptor mutant resembling the so-called little mouse mutation (D60G) failed in coupling upon addition of agonist.[51] The power of the system was also demonstrated in the attempt to determine the agonist specifity of the putative lysophosphatidic acid (LPA) receptor Edg-2 (also named Vzg-1). As was determined via the *FUS1-lacZ* reporter, LPA efficiently activated the pheromone transduction pathway in *S. cerevisiae* producing this receptor in a time- and dose-dependent manner. Also here, efficient activation was dependent on inactivation of the *SST2* gene, the yeast GTPase-activating protein for the G_α protein.[50]

The *S. cerevisiae* system, as described here, meets the criteria of being extremely robust and highly sensitive, making it suitable for high-throughput screenings. Therefore, the system will be a valuable tool for the search of specific drugs for GPCRs. Also the search for specific ligands for so-called orphan receptors which emerged and which will continue to emerge due to the ongoing sequencing projects can be started with this system; see Chapter 4 and Reference 68. Nevertheless, the system faces at least two difficulties: it is absolutely dependent on the production of correctly folded receptor as well as on the co-production of the correct G_α subunit or chimeric Gpa1-G_α protein. In case of a negative result, especially for screening of orphan receptors where the ligand is unidentified, it will be difficult to distinguish whether the receptor was not produced at all or produced in a nonfunctional state, or whether the G protein chosen in the assay was the correct one.

A solution to one of the problems may be to use G_α subunits or Gpa1-G_α fusions of the G_q family, the mouse $G_{\alpha 15}$ subunit or the human $G_{\alpha 16}$ subunit which behave promiscuously in respect to their interaction to a variety of GPCRs.[68–70] Furthermore, the development of vectors allowing engineering of a C-terminal fusion of a certain receptor directly to a $G_{\alpha 15/16}$ or a chimeric Gpa1-$G_{\alpha 15/16}$ subunit may be an effective strategy for signaling studies. In case of a Ste2-Gpa1 as well as a Ste2-Gpa1-$G_{\alpha s}$ fusion protein, efficient signal transduction was promoted in yeast cells devoid of the endogenous *STE2* and *GPA1* genes.[71] A similar approach has also been described recently for a mammalian system.[72]

11.2.3 PRODUCTION OF GPCRs IN YEASTS FOR PHARMACOLOGICAL, BIOCHEMICAL, AND STRUCTURAL STUDIES

11.2.3.1 Production of Adrenergic Receptors in Yeast

Adrenergic receptors, also named adrenoceptors, bind the catecholamines adrenaline and noradrenalin. Both catecholamines act as neurotransmitters in the sympathetic nervous system. Noradrenalin also plays an important role in the regulation of the

TABLE 11.2
Production of Adrenergic Receptors

Receptor	Type	Species	Host	Signal-Peptide/Fusion	Tag	Promoter	Mol. Wt. (kDa)	Glycosylation	Production Level	Ref.
Adrenergic receptor	$\beta_{2/1}$	Human	E. coli	—	—	T7	46	ND	200 receptors per cell	91
receptor	β_2	Human	E. coli	LamB (cleaved)	—	tac	45	ND	220 receptors per cell; 450 fmol/mg	77
	β_2	Human	E. coli	LamB (cleaved)	—	tac	43	ND	224 receptors per cell	78
	β_2	Human	S. cerevisiae	Ste2	—	GAL1	ND	ND	115 pmol/mg	34
	β_2	Human	Sz. pombe	Ste2	—	adh	64/46	Possible	7.5 pmol/mg	35
	β_2	Human	P. pastoris	F	F, B	AOX1	55	ND	25 pmol/mg	36
	β_2	Human	S. cerevisiae	Ste2	—	GAL1	ND	ND	36±7 pmol/mg	37
	β_2	Turkey	Sf9	—	—	pol	45	Possible	5–30 pmol/mg	82
	β_2	Human	Sf9	—	—	IE1	46	Possible (mannosidic type)	$1-4 \times 10^5$ receptors/cell	85
	β_2	Human	Sf9	—	—	pol	46	Possible	1×10^6 receptors/cell	83
	β_2	Hamster	Sf9	—	—	pol	46/48	Possible	12×10^6 receptors/cell; 16–20 pmol/mg	81
	β_2	Human	Sf9	—	c-myc		40–50	Detected	120 receptors/extracellular virions; 55.1 ± 1.8 pmol/mg cell membrane	92
	β_2	Avian	Murine cell line 293	—	—		45		0.81–0.27 pmol/mg	93
	β_2	Human	CHO	—	—	CMV	67	N-linked glycosylation	200 pmol/mg	84
	β_2	Human	HeLa	—	—				14.8–59.2 pmol/mg	84
	β_2	Human	Murine L-cell line B-82	—	—		68	Detected	420–900 fmol/mg	94
	α_2C2	Human	S. cerevisiae	—	his6	GAL1	50	ND	60–70 pmol/mg	37
	α_2C2	Human	S. cerevisiae	—	his6		43	—	3 pmol/mg	89
	α_2C2	Human	Sf9	—	—		43	—	11 pmol/mg	89
	α_2C2	Human	S115 mouse mammary tumor cells	—	—		43	—	5 pmol/mg	89

F, FLAG-tag; his6, hexa-histidine-tag; B, biotin-tag; Ste2, α-factor receptor fusion; ND, not determined. pol, polyhedrin promoter; tac, tac promoter; adh, alcohol dehydrogenase promoter; IE1, immediate-early baculovirus promoter; CMV, cytomegalovirus promoter; AOX1, alcohol oxidase 1 promoter; GAL1, GAL1 promoter; T7, T7 promoter.

cardiovascular system in the periphery. Based on pharmacology, signaling mechanisms, and sequence data, three main receptor types, the α_1, α_2, and β receptors, can be distinguished in the adrenergic receptor family.[73] These different receptor types can further be subdivided in subtypes (α_{1A}-, α_{1B}-, α_{1D}-, α_{2A}-, α_{2B}-, α_{2C}-, β_1-, β_2-, and β_3-adrenergic receptors). Activation of the α_1 receptors leads to hydrolysis of phosphatidylinositol by phospholipase C. The β receptors activate adenylyl cyclase via G_s, whereas the α_2 receptors mediate their action upon activation of G_i protein resulting in inhibition of adenylyl cyclase.[74,75]

The β_2-adrenergic receptor is probably the best characterized hormone and neurotransmitter receptor, and therefore the production of this receptor will be discussed in more detail (Table 11.2). Before the first production of an adrenergic receptor in a yeast was described in 1990,[34] the β_2 receptor had already been produced in other expression systems, including E. coli, Xenopus oocytes, insect cells, and several mammalian cell lines. In first attempts in E. coli up to 450 fmol receptor/mg was produced.[76,77] Later on the receptor was N-terminally fused to the outer membrane protein LamB, the maltose binding protein MalE, and the production level could be enhanced to the low pmol range.[78,79] The halophilic bacteria Halobacterium salinarium and Haloferax volcanii were also investigated for production of the β_2-adrenergic receptor.[80] In Haloferax volcanii, receptor production of about 180 fmol/mg was achieved with a vector carrying the ferredoxin promoter. Nevertheless, in comparison to the production levels of β_2-adrenergic receptors from different origin (turkey, hamster, human) in other systems like insect cells after infection with recombinant baculovirus (up to 30 pmol/mg[81–83]) or CHO cells (up to 200 pmol/mg[84]), this production level for the β_2-adrenergic receptor was rather poor. Large-scale production using mammalian cells tends to be cumbersome, due to biotechnological hurdles. Stably transformed insect cells which were tested as an alternative to mammalian cells exhibited the production of maximally 0.3 to 0.4 × 10^6 recombinant receptors per cell.[85] This production level was lower than that of CHO cells and also was about one magnitude lower than that in insect cells after infection with an appropriate baculovirus.

Purification of the receptor to homogeneity has been reported by two groups.[86,87] In both cases the receptor was purified from infected insect cells by a similar approach. In a first step, the tags which had been engineered to the amino and/or carboxyl termini of the receptor (KT3-tag,[86] His6-tag, and Flag-tag[87]) were used for affinity purification. In a second step, ligand affinity chromatography was applied to separate functional from nonfunctional receptor. Alprenolol affinity chromatography has already been used for purification of the β_2-adrenergic receptor.[88]

Subsequently several yeasts were examined for high-level production of adrenergic receptors. Production of the human β_2 receptor in S. cerevisiae resulted in yield of about 100 pmol/mg.[34] To reach this extremely high level of production, the receptor was engineered as follows: first the transcriptional transactivator gene (LAC9) of the GAL1 promoter which was used to drive production was co-produced; second, the N-terminal region of the receptor was replaced by the coding region of the first 14 amino acids of the α-factor receptor. Additionally, the recombinant yeast cells were induced at late exponential growth and during induction the growth medium was supplemented with an adrenergic ligand. Also the successful coupling of the receptor to a co-produced $G_{s\alpha}$ subunit which in a *GPA1*-deficient strain led to partial activation of the yeast pheromone response pathway, was reported (see above). Later on, the conditions for large-scale production of the receptor in fermenters were developed.[37] Some crucial factors for successful production in 1- to 15-l fermenters involved the use of selective glucose-free medium, the regulation of the pH in the culture medium at 7.2 to 7.5 during growth, and the presence of an antagonist in the growth medium. Here, production levels of about 40 pmol receptor per mg of protein in a 15-l fermenter, which corresponded to 20–30 mg of functional protein per fermenter, were obtained. In the same paper, a similar approach was described also for the production of a N-terminally His₅-tagged human α_2-C2-adrenergic receptor. This receptor has an extremely short N terminus which lacks sites for N-linked glycosylation and, in a previously published comparative study from the same group, was found to be produced in mammalian (S115 cells, 5 pmol/mg total protein), insect cells (11 pmol/mg), and *S. cerevisiae* (3 pmol/mg).[89] Contrary to the β_2-adrenergic receptor, the α_2-C2 receptor was not N-terminally fused to the Ste2p. The production of the receptor was traced by immunoblot analysis using a receptor-specific antiserum as well as by ligand binding analysis with the specific ligand [³H]-rauwolscine. After growth of the recombinant yeast clone in the presence of the antagonist phentolamine about 35×10^3 receptors per cell were measured. Subsequent fermentations that were performed under identical conditions described for the β_2 receptor revealed varying production levels ranging from 7 to 70 pmol/mg of receptor. As stated by the authors the level of 70 pmol/mg of protein corresponded to about 10 to 15 mg of receptor in a 15-l fermenter. Nevertheless, in the case of the β_2 receptor initial scaled-up protocols for cell disruption and isolation of the yeast membranes resulted in a dramatic loss of receptor (up to 60%).

The fission yeast *Sz. pombe* and the methylotrophic yeast *Pichia pastoris* were used also for the heterologous production of the human β_2-adrenergic receptor. For production in *Sz. pombe* two different receptor constructs were tested: one construct without any fusion, the second construct with a N-terminal fusion to the first 14 amino acids of the *S. cerevisiae* α-factor receptor.[35] In both constructs the receptor gene was placed under the control of the constitutive *Sz. pombe* alcohol dehydrogenase (*adh*) promoter. Recombinant *Sz. pombe* cells revealed no production of the unmodified receptor whereas the STE2/β₂AR receptor fusion was traced by either immunoblot analysis using antiserum against the STE2 part of the fusion (apparent molecular mass 64 kDa) or by binding assays performed with [¹²⁵I]-cyanopindolol. The production level of the fusion protein was estimated to be about 7.5 pmol receptor/mg protein. From a pharmacological point of view the fusion receptor exhibited the same ligand specificities as that of the "native" receptor.

The β_2-adrenergic receptor was the second receptor produced in the methylotrophic yeast *P. pastoris*. The receptor was produced in a tagged form: N-terminally the FLAG-tag and C-terminally a Bio-tag (biotinylation domain of the *Pb. shermanii*) was fused.[36] Additionally, the α-factor signal pre pro peptide which is cleaved by the internal Kex 2 endopeptidase was engineered to the N terminus. Here, although the receptor contained a motive for N-glycosylation (N-X-S/T), the receptor was found to be unglycosylated and the α-factor leader was correctly processed by the Kex2p. In general, production levels of receptors were improved by the use of protease-deficient strains, the selection of clones bearing multiple integrations on G418-plates,[90] and the fusion to the α-factor leader from *S. cerevisiae*. Nevertheless, clones with copy numbers ≥3 showed comparable production levels which indicates that post-translational events, e.g., folding, membrane insertion, etc., are the bottleneck for production of functional receptors. Supplementation of ligands into the culture medium which had been described for *S. cerevisiae* to enhance production level here also resulted in a significant increase of receptor yield. Using these optimized conditions, binding assays with [^{3}H]-CGP12177 revealed that approximatly 25 pmol of β_2-adrenergic receptor per milligram of crude membrane was produced. The pharmacological profile for the receptor was comparable to that in mammalian or in insect cells. In initial fermentation attempts about 8 mg receptor per liter were produced. Localization studies showed that the receptor was mainly targeted to internal membranes in recombinant yeast.[36]

11.2.3.2 Production of Muscarinic Acetylcholine Receptors in Yeast

The muscarinic acetylcholine receptors (mAChR) modulate a variety of intracellular events, such as opening of K$^+$ channels, and the activation or inhibition of enzymes that control levels of intracellular second messengers such as cyclic AMP and inositol triphosphate by interaction with respective G proteins.[95,96] By cloning and nucleotide sequencing, five distinct subtypes of muscarinic receptors (M_1–M_5) have been identified from mammalian sources.[97,98] The different types of mAChRs regulate different transduction pathways, although functional specifity seems to be affected by cell type and receptor levels.[99,100]

The M_1 mAChR was the first muscarinic receptor produced in yeast (Table 11.3).[38] Like in the case of the β_2-adrenergic receptor, *Xenopus* oocytes[101] and different mammalian cell lines have been used for the production of the M_1 receptor, including CHO cells,[102–104] A9, COS-7,[97,105] mouse Y1 adrenal cells,[106] and human embryonic kidney cells,[99] with more or less success. The receptor yields obtained in these expression systems excluded their use for large-scale production and purification of the receptor. Enhanced production levels were reached in insect cells after infection with respective recombinant baculovirus.[107–109]

These systems are of value for examination of pharmacological properties and signal transduction (reconstitution with various G proteins[107]). For production of the M_1 receptor in *S. cerevisiae* the receptor gene was cloned under the transcriptional control of the constitutive *ADH* promoter.[38] Nevertheless, the production level of the receptor in the yeast was significantly lower than that in mammalian cells (20 to 40 receptors per cell [20 fmol/mg], compared to 10,000–200,000 receptors per cell [0.2 to 2 pmol/mg]),[99,110] and in insect cells after infection with recombinant baculovirus,[108,109] and the number of receptors per cell stayed well below the number of endogenous yeast STE2 receptors (3000 to 8000 receptors per cell). The K_Ds determined for several antagonists correlated with data obtained in rat and human cerebral cortex membranes,[111,112] whereas the K_Ds for two agonists revealed that the M_1 receptor was present in its low-affinity state. Therefore, coupling to endogenous yeast G protein was not expected and also not detected. Also the M_3 receptor subtype

TABLE 11.3
Production of Muscarinic Receptors

Type	Species	Host	Promoter	Production Level	Vector	Ref.
M_1	Human	Sf9	pol	0.1–1 pmol/mg	pLucGRBac1	109
M_1	Human	HEK		121,000–199,000 receptors/cell	pF8CIS9080	99
M_1	Human	CHO		41.5 ± 5.2 fmol	pCMV-3	104
				6.6 ± 2.7 fmol		
M_1	Porcine	*Xenopus*		2 ± 0.8 fmol receptors/cell		101
M_1	Mouse	Mouse Y1 adrenal cells	Zinc-inducible metallothionein	300 fmol/mg 4,000 fmol/mg	pZEM228	106
M_1	Mouse	Mouse Y1	Zinc-inducible metallothionein	500–1000 fmol/mg	pZEM 228	117
M_1	Rat	COS-7	SV40	2×10^4–10^5 receptors/cell	pcD	97
M_1	Rat	A9		1×10^4, 2×10^4 receptors/cell		97
M_1	Human	*S. cerevisiae*	ADH	1–20 fmol/mg	pVTUHM1	38
M_1		CHO		1.05–2.06 pmol/mg	pSVL	103
M_1	Porcine	*Xenopus*		1.9–0.5 fmol/cell		118
M_1		A9 L fibroplast		100–200 fmol/mg		119
M_1	Rat	CHO-K1		Wild type: 1,530 fmol/mg ser mutant: 410–2,240 fmol/mg	pSVL	102
M_1	Human	Sf9		5–30 pmol/mg		107
M_1	Human	A9L		369 ± 73 fmol/mg		120
M_1	Human	CHO		824 ± 59 fmol/mg		120

pol, polyhedrin promoter; ADH, alcohol dehydrogenase promoter.

from rat could be produced in *S. cerevisiae*, as was revealed by immunoblot analysis of membrane preparations of recombinant clones. This receptor previously was functionally produced in insect cells after infection with recombinant baculovirus (0.5 to 1×10^6 receptors per cell).[109,113–116] Nevertheless, in yeast, although the receptor was present and found to be glycosylated and located in the plasma membrane, ligand binding was not successful, neither on intact cells nor on membranes (unpublished results).

Contrary to that, another muscarinic acetylcholine receptor, the M_5 subtype, was functionally produced in *S. cerevisiae* in 1992.[39] In the first attempt, the receptor was produced under control of the constitutive yeast alcohol dehydrogenase (*ADH1*) promoter. Specific [^{3}H]-*N*-methyl scopolamine (NMS) binding sites were detected in crude membranes but not on intact cells which might be due either to incorrect targeting or incorrect folding of the receptor. In a second attempt, the receptor gene for expression was placed under the control of the yeast α-factor promoter and, additionally, the coding region of the receptor was N-terminally fused to the α-factor leader sequence to ensure proper translocation within the recombinant cells. With this construction binding on intact cells was possible and further analysis revealed a K_D for NMS which was comparable to that of crude membrane extracts. Also here, despite the receptor being N-terminally fused to the α-factor leader sequence, the production level reported with only 45 receptors per cell, corresponding to about 135 fmol/mg protein, excluded this for large-scale preparation of the protein. Also in baculovirus-infected insect cells the yields for this receptor were moderate.[109] Growth inhibition after addition of the agonist carbachol to recombinant yeast clone was not observed, indicating that interaction to the pheromone signal pathway did not occur.

11.2.3.3   Production of Dopamine Receptors in Yeast

Dopamine receptors play a key role in central nervous system (CNS) neurotransmission. Dopamine has been implicated in several severe disorders of the CNS, including Parkinson's disease and schizophrenia.[121,122] Originally, two dopamine receptor subtypes, D_1 and D_2, were recognized based on their opposing effects on adenylyl cyclase activity and their disparate pharmacological profiles.[123] Molecular cloning has identified five receptor subtypes to date, designated D_{1A}, D_2, D_3, D_4, and D_5 (D_{1B}), thus pointing to the complexity of the dopaminergic system.[124,125] The D_2 subtype is of particular interest, as it is thought to be the target for many antipsychotic drugs, known to be dopamine antagonists,[126,127] but it also exists in two alternatively spliced isoforms. The two splice variants, D_{2S} and D_{2L}, differ by an insert of 29 amino acids in the third intracellular loop, which is a major site of interaction with G proteins. At the molecular level, activation of D_2 dopamine receptors results in inhibition of adenylyl cyclase and in a variety of cell-type specific responses reviewed in Civelli et al.[124]

In our laboratory, we were able to produce the D_{2S} dopamine receptor in a variety of hosts: *E. coli* (3 to 5 pmol/mg; unpublished results), stably transformed insect cells (3 to 6×10^5 receptors per cell; unpublished results) and baculovirus-infected insect cells ($10–11 \times 10^6$ receptors per cell[128]), as well as in several yeast systems. In *S. cerevisiae* the D_{2S} receptor was produced under the control of the regulatable *GAL10* promoter,[129] as well as under the constitutive *PMA1* promoter.[43] No significant differences in production levels were noted between the native receptor and a chimeric STE2/D2S receptor fusion. In each case production leveled at about 1 to 2 pmol/mg protein. Bound [^{3}H]-spiperone was displaced by a variety of unlabeled dopamine-specific ligands, resulting in a typical D_2 subtype-specific pharmacological fingerprint. No significant differences were noted between the fingerprint of the native receptor and that of the STE2 receptor chimera. Worthwhile mentioning is that the absolute K_i values, which were calculated from the displacement curves, differed from published values derived from both receptor native environment and heterologous higher eukaryotic host systems. As revealed by immunoblot analysis the receptor and the STE2 receptor fusion exhibited an apparent molecular mass of about 40 and 43 kDa, respectively. Protease-deficient *S. cerevisiae* strains were tested and proved to be the best candidates for the heterologous production. As was seen on the immunoblots comparing wild-type and different protease-deficient strains, the receptor was severely degraded in protease containing cells.[129] Post-embedding immunostaining of ultrathin sections from yeast cells producing the receptor revealed that most of the recombinant protein was targeted to internal membranes although the receptor is a resident of the plasma membrane.

Encouraged by the successful production of the seven transmembrane protein bacteriorhodopsin in *Sz. pombe*,[113–132] this yeast was also tested for the production of the D_{2S} receptor.[44] The receptor gene was cloned under the control of the tightly regulatable and strong *nmt1* promoter in the high-copy number plasmid pREP1 and after transformation and selection of recombinant clones, receptor production was traced by immunoblot analysis and radioligand binding using [^{3}H]-spiperone on crude membranes. In accordance with the results in *S. cerevisiae*, an apparent molecular mass of about 40 kDa was determined for the recombinant receptor using the D_{2S}-specific polyclonal antiserum. The comparative study also revealed that the production level in *Sz. pombe*, about 14 pmol/mg, was significantly higher than that in *S. cerevisiae* bearing an equivalent receptor construct under control of the *GAL10* promoter (2.8 pmol/mg protein). The rank order of potencies for several dopaminergic ligands in displacement measurements was the same as those for "native" receptor, whereas the individual ligand affinities differed by a factor of 2 to 3. An extremely high production level for this receptor was detected in recombinant *P. pastoris*.[44a] The cloning strategy for construction of the vector was comparable to that of the β_2-adrenergic receptor reported above. The receptor was tagged with the Flag-tag and additionally fused to the pre propeptide of *S. cerevisiae*. After transformation into the protease-deficient *P. pastoris* strain SMD1163 up to 100 pmol receptor per mg of membrane protein was measured by radioligand binding. Nevertheless, the recombinant receptor suffered severe proteolytic digestion as was revealed by immunoblot analysis.[44a]

Recently, the D1 receptor was produced in *S. cerevisiae*.[42] According to the results obtained for the D_{2S} receptor, this receptor was also produced in a protease-deficient *S. cerevisiae* strain. In comparison a STE2 deleted strain which was wild-type in respect to its protease content was used and in both strains no degradation was detected. Production of a receptor fusion protein (N-terminal fusion to the first amino acids of the Ste2p) revealed twofold higher production in comparison to an unmodified receptor construct. In both cases the receptor was Flag- and His6-tagged at the C terminus. As was revealed by immunoblot analysis, the Ste2/D1 fusion protein was glycosylated whereas the unmodified receptor was not. Purification of the receptor was achieved by affinity chromatography making use of the His6- and the Flag-tag. The authors claimed the purification of 0.2 to 0.5 mg of active receptor from 60 g cells (corresponds to 20 l of culture). In comparison to insect cells which produce the receptor after infection with a recombinant baculovirus (up to 33 pmol/mg; unpublished results and Reference 133) the yields in the yeast system with 76 or 127 fmol/mg protein was low. Nevertheless, the receptor was functional as was demonstrated by reconstitution into phospholipid vesicles.

11.2.3.4 Production of Serotonin Receptors in Yeast

Serotonin (5-hydroxytryptamine; 5-HT) is a biogene amine that functions as both a neurotransmitter and a hormone in the mammalian nervous system (CNS) and in the periphery.[141] The neuromodulator serotonin elicits and modulates a wide range of behaviors, such as sleep, appetite, locomotion, sexual activity, and vascular contraction. This pleiotropy can be explained both by the wide distribution of 5-HT and

TABLE 11.4
Production of Dopamine Receptors

Type	Species	Host	Promoter	Mol. Wt. (kDa)	Glycosylation	Production Level	Vector	Signal Peptide Fusion	Ref.
D_{2S}	Human	Sf9	pol	39–40	Yes	2×10^6 receptors/cell	ACPolD2S	Pol	128
D_{2S}	Human	Sf 21	pol	39–40	Yes	1.2×10^6 receptors/cell	ACPolD2S	Pol.	128
D_{2S}	Human	Tn368	pol	39–40	Yes	4×10^6 receptors/cell	ACPolD2S	Pol.	128
D_{2S}	Human	Mb	pol	39–40	Yes	4×10^6 receptors/cell	ACPolD2S	Pol.	128
D_{2S}	Human	Sf9	pol	38–41	Yes	4–5×10^6 receptors/cell	VLMelIMycD2S	Prepromelittin	128
D_{2S}	Human	Sf9	pol	38–41	Yes	4–5×10^6 receptors/cell	VLMeID2SH6	Prepromelittin	128
D_{2S}	Human	Tn368	pol	38–41	Yes	10–11×10^6 receptors/cell	VLMelIMycD2S	Prepromelittin	128
D_{2S}	Human	Tn368	pol	38–41	Yes	10–11×10^6 receptors/cell	VLMeID2SH6	Prepromelittin	128
D_{2S}		S. cerevisiae	GAL10	40/85	No	2.8 pmol/mg	Yep51D2		44
D_{2S}	S. pombe		nmt1	40/85	No	14.6 pmol/mg	pREP1D2		44
D_{2S}		S. cerevisiae	GAL10	40	No	1–2 pmol binding sites/mg	Yep51D2		129
D_{2S}		S. cerevisiae	GAL10	42	No	1–2 pmol binding sites/mg	Yep51D2S	Ste2	129
D_{2S}		S. cerevisiae	PMA1	43	ND	1–2 pmol binding sites/mg	pRS421D2S		43
D_{1A}	Human	S. cerevisiae	GPD	55	Yes	127 ± 25 fmol/mg	pG3	Ste2p	42
D_{2S}	Human	Ltk-cells		—	ND	4.05 ± 0.3 pmol/mg			134
D_{2L}	Rat	Sf9	pol	46	No	15 pmol/mg			135
D_{2S}	Rat	Sf9	pol	44	No	15 pmol/mg			135
D_{2S}		Cos-7		—	ND	100–150 fmol/mg	pMAMneo		136
D_{2L}		Cos-7		—	ND	150–200 fmol/mg	pMAMneo		136
D_{2L}	Rat	Ltk⁻	Metallothionein	—	ND	945 fmol/mg			137
D_2	Rat	Mouse L cells		—	ND	3,000 receptors/cell			138
D_2	Human	Sf9		54–60	Yes	2.9 ± 1.1 pmol/mg			139
D_{2L}	Human	293 kidney cells	CMV			20–1,000 fmol/mg	pRc/CMV		140
D_{2S}									
D_{2a}									
D_2	Human	Tn5		54–60	Yes	6.9 ± 1.0 pmol/mg			139

Ste2, α-factor receptor fusion; ND, not determined; pol, polyhedrin promoter; Pol, polyhedrin fusion; GAL10, GAL10 promoter; GPD, glyceraldehyd-3-phosphate dehydrogenase promoter; nmt1, nmt1 promoter (no message in thiamine); PMA1, PMA1 promoter; CMV, cytomegalovirus promoter.

by the multiplicity of 5-HT receptors. Within the mammalian brain serotonergic neurons are concentrated in the raphe nuclei of the brainstem and from here 5-HT-releasing fibers project to most areas of the CNS. In the periphery, serotonin is involved in diverse functions. Pharmacological studies of serotonin receptors have revealed considerable heterogeneity and it became clear that multiple 5-HT receptors exist. The receptors that have been cloned so far belong to either the ligand-gated ion channel family (5-HT$_3$ receptor) or to the large family of G protein-coupled receptors (5-HT$_{1A}$, 5-HT$_{1B}$, 5-HT$_{1D-F}$, 5-HT$_{2A-C}$, 5-HT$_{5A-B}$, 5-HT$_6$, 5-HT$_7$).[142,73]

For production of the mouse 5HT$_{5A}$ serotonin receptor in *S. cerevisiae*, the receptor cDNA was N-terminally engineered to the *Bacillus macerans* (1-3, 1-4)-β-glucanase signal sequence. To permit immunological detection of the recombinant protein, the receptor was fused to a c-*myc* tag. The vacuolar endopeptidase B (*PRB1*) promoter which gets activated when cells have exhausted the glucose supply in the medium was used for production of the receptor in a protease-deficient yeast host strain. The receptor was traced by immunoblot analysis using the c-*myc* tag-specific monoclonal antibody 9E10. Treatment of membranes with *Endo*H revealed that the recombinant receptor was glycosylated in the host. Binding studies performed with the serotonergic ligand [^{3}H]-lysergic acid diethylamine (LSD) revealed a production level of about 16 pmol/mg protein corresponding to about 12,000 receptors per cell. The pharmacological profile displayed a rank order of potencies which was comparable to data obtained for the receptor after production in mammalian cells. Interestingly, after a heat shock the yield of functional receptor could be doubled on average. The receptor was mainly located inside the producing cell as was revealed by immunogold electron microscopy.[40]

The mouse 5HT$_{5A}$ receptor was the first GPCR whose production was tested in *P. pastoris*.[41] Here, two N-terminal fusions, one to the *P. pastoris* acid phosphatase signal peptide, the other to the *S. cerevisiae* α-factor prepropeptide, were analyzed in comparison to the unaltered receptor. Additionally, these two constructs had been c-*myc*-tagged for immunological detection of the recombinant receptor. Best production was achieved with the α-factor receptor fusion in the protease-deficient strain SMD1163 (22 - 32 × 10^3 receptors per cell, corresponding to about 22 pmol/mg of membrane protein, 24 to 48 h after induction with methanol). In a subsequent paper, biochemical analysis of the 5HT$_{5A}$ receptor was provided.[36] Worthwhile mentioning is, that as revealed by immunoblot analysis, the α-factor leader was not removed from the myc-tagged 5HT$_{5A}$ receptor, nor from the FLAG- and Bio-tagged receptor by the Golgi resident Kex2 protease. This failure in processing resulted in an about 20-kDa increase in the molecular mass. The Kex2-dependent processing was found to be hindered by glycosylation of the receptor; correct processing could be achieved by either addition of tunicamycin during culturing of the yeast cells or by enzymatic deglycosylation of membrane preparations. By radiologand binding with [^{3}H]-LSD it was estimated that approximately 40 pmol of 5HT$_{5A}$ receptor per milligram of crude membrane was produced. From a pharmacological point of view the receptor exhibited a profile comparable to that in mammalian or insect cells. In initial fermentation attempts about 10 mg 5HT$_{5A}$ receptor per liter of fermenter culture was obtained. As was revealed by localization studies, the receptor was mainly targeted to internal membranes in the recombinant yeast clones.[36]

11.2.3.5   Production of the Opsins in Yeast

Rhodopsin is the only GPCR that can be isolated in high yield and purity from retina. Two-dimensional crystallizations with bovine and frog rhodopsin led to the first receptor maps (see introduction). However three-dimensional crystals which frequently could be obtained were not usable for structure determination by X-ray crystallography. This was mainly due to microheterogeneity of the protein preparations. To circumvent this problem, the opsin has been produced and reconstituted in mammalian cells. Alternatively, a synthetic bovine opsin gene was produced also under the control of the inducible *GAL1* promoter in *S. cerevisiae*.[47] After galactose induction 2 ± 0.5 mg opsin per liter of cell culture (about 10^{10} cells) were produced (0.4% of total protein) as estimated by enzyme-linked immunosorbent assay (ELISA). After spheroplasting recombinant cells were treated with 11-*cis*-retinal, lysed, and the membrane fraction solubilized. Only 2 to 4% of the total opsin produced could be reconstituted and purified to homogeneity. This protein which was homogeneously glycosylated was indistinguishable from its natural counterpart by several functional criteria. However, as has been already described for other glycosylated heterologously produced proteins, glycosylation of the recombinant rhodopsin was distinct from that of bovine rhodopsin.[47] Production of opsin in *P. pastoris* resulted in levels of up to 0.3 mg per liter of cell culture ($A_{600} = 1.0$). Here, for optimal production the opsin gene had been fused to the acid phosphatase signal sequence. As was revealed by subsequent analysis of the recombinant protein, it showed an apparent molecular mass of 37 kDa and high mannosidic glycosylation. Nevertheless, only about 4 to 15% of the recombinant opsin could be reconstituted with 11-*cis*-retinal to form functional rhodopsin which was immunoaffinity-purified.[48]

11.2.3.6   Production of Neurokinin Receptors in Yeast

Tachykinins are peptides that are widely distributed in the CNS, as well as in peripheral tissues. They are involved in the regulation of many complex physiological functions, like contraction of vascular and nonvascular smooth muscles. Three receptors, NK_1-, NK_2-, and NK_3-neurokinin receptors, are known to bind with different affinities to their targets, substance P, neurokinin A (previously known as substance K, neurokinin α, neurokinin L), neurokinin B (previously known as neurokinin β, neuromedin K), neuropeptide K, and neuropeptide γ (N-terminally extended forms of neurokinin A).[73]

The human NK_2 receptor was the first tachykinin receptor to be produced in a yeast system.[45] Earlier, this receptor had been produced in fusion to the maltose-binding protein at a level of 2.5 pmol/mg protein in *E. coli*.[143] The production level obtained in *E. coli* was higher than that in insect cells as well as in different stable or transiently transfected mammalian cells.[144-147] Nevertheless, mammalian cells infected with a recombinant *Semliki Forest* virus revealed production levels of about 6.5 pmol/mg protein.[148] Contrary to the production in *E. coli*, the mammalian cells allowed examination of the signal transduction pathway after agonist binding.[145-147] The yeast *Sz. pombe* contains a signal transduction chain resembling the systems found in mammalian cells.[149] Contrary to the situation in *S. cerevisiae*, the G_α subunit (*gpa1*) in *Sz. pombe* positively transmits the signal from the activated mating factor receptor (M-factor and P-factor are secreted by h⁻ or h⁺ type of *Sz. pombe*, respectively) to the effectors. Therefore, *Sz. pombe* was tested for production of the receptor and also for functional coupling of the receptor to the internal signal chain.[45] To ensure correct receptor trafficking and translocation to the cell surface the NK_2 receptor had been N-terminally fused to the signal peptide from the endogenous yeast cell surface phosphatase *pho1*. A construct without signal peptide did not work in the production of functional receptor. Binding assays performed on intact recombinant yeast cells with the nonpeptide antagonist [³H]-SR48968 revealed the presence of about 430 ± 315 receptors per cell whereas binding to membranes indicated 5514 ± 1599 receptors per cell (1.6 pmol/mg protein). Therefore, one can conclude that only a certain portion of the receptor population was targeted to the cell surface. Coupling to different co-produced mammalian G protein subunits ($G_{\alpha q}$, $G_{\alpha 16}$, $G_{\beta 1}$, $G_{\beta 2}$, and $G_{\gamma 3}$) and subsequent activation of phospholipase C-$\beta 1$ (PLC$\beta 1$) was undetectable and the authors argue that this is most probably due to inactive G_{α} subunits.

11.2.3.7   Production of the μ-Opioid Receptor in Yeast

Opioid receptors are part of a neuromodulatory system that plays a major role in the control of nociceptive pathways. The system is a key element for pain perception and is also involved in the modulation of affective behavior as well as neuroendocrine physiology. Furthermore it also controls respiration, gastrointestinal motility, blood pressure, and thermoregulation. The receptor is the main target for exogenous narcotic drugs including drugs of abuse. After cloning of the mouse δ opioid receptor in 1992, today the sequences of all three major opioid receptors, μ-, δ-, and κ-receptor from several species are available.[150]

   The *AOX1* promoter was used to drive production of the human μ-opioid receptor in *P. pastoris*.[46] Here, two plasmids have been constructed, one bearing the receptor gene deleted in its first 32 N-terminal codons fused to the *S. cerevisiae* α-mating factor signal and the other bearing the receptor gene deleted in its first 11 N-terminal codons in fusion to the 18 amino-terminal codons of the *S. cerevisiae* STE2 gene. Only the α-factor signal receptor fusion recombinant clones which were selected after transformation of the *P. pastoris* strain GS115 and the protease-deficient strain SMD1168 displayed [³H]-diprenorphine binding in a saturable mode on whole cells (maximum of 810 receptors per cell) as well as in membranes. With relation to the *P. pastoris* strain used for transformation and production no difference in the production level was reported. In comparison to the production in insect cells via the baculovirus expression system the production level of about 0.4 pmol/mg in *S. cerevisiae* was low. Maximally about 1×10^6 receptors per infected insect cell (High Five insect cells revealed higher production levels than the commonly used Sf9 insect cell line) were measured which resulted in the production of up to 1.7 nmol recombinant receptor per liter of insect cell culture.[151] Nevertheless, the pharmacological profile for antagonists and agonists of the receptor produced in *S. cerevisiae* was comparable. The K_i values of the agonists, however, were higher than those of "native" μ-opioid receptors (also in Sf9 cells the K_D for diprenorphine was reported to be slightly higher compared to that in COS cells). The addition of Gpp(NH)p to the binding assays which, in mammalian cells producing the receptor results in a shift from high-affinity to the low-affinity state, had no such effect in the yeast cells, most probably due to the absence of interacting protein.

11.2.3.8   Production of the STE2 Receptor in Yeast

An efficient production of the Ste2p α-factor receptor from *S. cerevisiae* has been achieved under the control of the glyceraldehyde-3-phosphate dehydrogenase (*TDH3*) promoter on a high-copy number plasmid in receptor-deficient *S. cerevisiae* strains.[52] The receptor was C-terminally tagged with a FLAG- and a His$_6$-tag for immunological detection and/or purification, respectively. These tags appended to the receptor were shown not to compromise either signal transduction or ligand binding characteristics. In a protease-deficient yeast strain carrying the recombinant expression plasmid about 80-fold more receptor than in a wild-type strain could be detected by radioligand binding with [^{3}H]-α-factor. Effectiveness of receptor production was also monitored by immunoblot analysis which revealed several closely spaced bands in the molecular mass range of about 55 to 60 kDa. Only a portion of the receptor was present on the cell surface, about 79-fold more receptor was detected in membranes in comparison to whole cell measurements. Due to this extremely high production level (>140,000 receptors per cell; approximately 350 pmol/mg of membrane protein), purification after solubilization was achieved in only two steps. Yields as high as 1 mg of nearly homogeneous receptor were obtained from about 60 g of frozen cells (corresponded to approximately 20 l of cell culture grown to A_{600} = 3). Nevertheless, the protein isolated revealed some microheterogeneity due to glycosylation and the receptor had to be reconstituted into proteoliposomes for functional assays. The pharmacological profile of the receptor in membrane preparations as well as of the receptor after purification and reconstitution into proteoliposomes was as expected, indicating that the receptor maintained its integrity.

11.2.4   CONCLUSIONS

Production of GPCRs in yeasts is yet at the beginning. Screening systems based on yeasts have the advantage of fast and easy handling. Here, the use of genetically modified yeasts for rapid and convenient high-throughput screening of compound libraries from diverse sources is one important task. *S. cerevisiae* is suitable for this purpose since this yeast is well characterized and the components of the pheromone signal transduction pathway are known. Interesting results have been obtained with specially constructed strains that should have a broad applicability for different receptors,[67] and so it may not be necessary to establish similar systems using other yeasts.

Another task is the high-level production of GPCRs for structural studies. For revealing the three-dimensional structure of a GPCR, large quantities of pure and homogeneous protein will be needed. *S. pombe* and *P. pastoris* were the first and remain the only non-*Saccharomyces* yeasts that have been tested for production of GPCRs and acceptable production levels achieved for at least some receptors. Here, *P. pastoris*, due to its ability to grow to extreme densities and its ease of fermentation,[152] has much better properties than either *S. cerevisae* or *Sz. pombe*. This was also demonstrated for many soluble proteins produced in non-*Saccharomyces* yeasts.[153] Nevertheless, coupling of the heterologously produced receptors to the internal signal pathway in these yeasts have not yet been demonstrated. Also posttranslational modifications like palmitylation or glycosylation of the receptors have not yet been analyzed, indicating that a lot of basic work has still to be done.

A general disadvantage of all yeast-based expression systems is that they are faced with the severe problem of protein degradation. Unfortunately, the vacuoles of yeast cells, the homolog to lysosomes of higher cells, are filled with highly active proteases.[154] Therefore, addition of protease inhibitor cocktails during cell breakage is absolutely essential. Nevertheless, large-scale disruption of yeast cells and subsequent preparation of membranes due to proteolytic digestion of the recombinant protein is still a major obstacle which has to be solved. Here, from a biotechnological point of view, protease-deficient yeast strains are of special interest and in some cases have already been found to have positive effects on receptor production and therefore should be examined for heterologous receptor production. Also the addition of receptor-specific ligands to the culture has been found to improve receptor production significantly. Another trick to enhance receptor production seems to be to fuse the N terminus of the given receptor to either a yeast signal peptide (acid phosphatase or the α-factor prepropeptide from *S. cerevisiae*) or to the first amino acids of the Ste2p. However, many receptors produced in yeasts remain in internal membrane compartments even when N-terminal fusions with leader peptides were constructed and in some cases ligand binding affinities are slightly reduced when compared to mammalian based systems. For the slightly reduced affinities two reasons are discussed: one is the lipid composition of the yeast membrane and especially the different sterols found in yeast and mammalian cell membranes, the other is the lack of a fitting G protein for the receptor. Another commonly used technique, the attachment of tags to receptors, will be of great value too. The approaches described above for the *S. cerevisiae* Ste2p receptor as well as for the $5HT_{5A}$ receptor produced in *S. cerevisiae* and *P. pastoris* can easily be adapted to other GPCRs. Subsequent detection of production of a tagged receptor by specific monoclonal antibodies (for example the M_1 and M_2 mAbs in the case of the FLAG-tag; mAB 9E10 in the case of the c-*myc*-tag, etc.) is easy and will provide important information about production maxima, apparent molecular mass (glycosylation state), localization within the cell, proteolysis of the recombinant protein, etc. Initial purification attempts will be facilitated by some of these tags (FLAG-, $His_{6/10}$-tag).

In conclusion, yeast-based expression systems have had and will have a great impact on revealing the secrets of G protein-coupled receptors. Much more work will have to be invested in the future to understand these proteins and to find compounds that act specifically on them.

11.3   METHODS

11.3.1   *Pichia Pastoris* Vectors

In contrast to *S. cerevisiae*, for the production of recombinant proteins in *P. pastoris*, no episomal plasmids are available and therefore vectors that can be integrated into the host genome are used. A selection of commonly used expression vectors and

TABLE 11.5
P. pastoris Expression Vectors

Vector	Selection Marker	Characteristics	Ref.
Intracellular			
pHIL-D2	*HIS4*	*AOX1* promoter	Invitrogen manual
pAO815	*HIS4*	*AOX1* promoter; *Bam*HI and *Bgl*II restriction sites for *in vitro* multimerization of the expression casette	157
pPIC3.5K	*HIS4* and *kan*ʳ	*AOX1* promoter; multiple cloning site; G418 selection for clones with multiple insertions of the expression casette	90
pPICZ	*ble*ʳ	*AOX1* promoter; multiple cloning site; zeocin selection for clones with multiple insertions of the expression casette; *Bam*HI and *Bgl*II restriction sites for *in vitro* multimerization of the expression casette; His6 and c-*myc* epitope tag (optional)	160
pHWO10	*HIS4*	*GAP* promoter	59
pGAPZ	*ble*ʳ	*GAP* promoter; multiple cloning site; Zeocin selection for clones with multiple insertions of the expression casette; *Bam*HI and *Bgl*II restriction sites for *in vitro* multimerization of the expression casette; His6 and c-*myc* epitope tag (optional)	Invitrogen manual
Secretion			
pHIL-S1	*HIS4*	*AOX1* promoter; *PHO1* signal sequence	Invitrogen
pPIC9K	*HIS4* and *kan*ʳ	*AOX1* promoter; *S. cerevisiae* α-factor prepro-sequence	90
pPICZα	*ble*ʳ	*AOX1* promoter; *S. cerevisiae* α-factor prepro-sequence; multiple cloning site; zeocin selection for clones with multiple insertions of the expression casette; *Bam*HI and *Bgl*II restriction sites for *in vitro* multimerization of the expression casette His6 and c-*myc* epitope tag (optional)	160
pGAPZα	*ble*ʳ	*GAP* promoter; *S. cerevisiae* α-factor prepro-sequence; multiple cloning site; zeocin selection for clones with multiple insertions of the expression casette; *Bam*HI and *Bgl*II restriction sites for *in vitro* multimerization of the expressions casette; His6 and c-*myc* epitope tag (optional)	Invitrogen manual

Most of the vectors listed above can be obtained from Invitrogen; sequences and maps of them are available from the internet (http://www.invitrogen.com).

Modified after Cregg.[62]

their characteristics are given in Table 11.5. Almost all the vectors listed in the table can be obtained from Invitrogen (for further information about the vectors, including sequence, restriction maps, etc., contact the following internet address: http://www.invitrogen.com). Most of the vectors allow the cloning of the foreign gene under the control of the strong, highly inducible *AOX1* promoter. Recently, vectors allowing production under the constitutive *GAP* promoter have been described.[35] The major difference between the two promoters is that for heterologous production under the GAP promoter the growth medium does not have to be changed. In general, all *P. pastoris* vectors are shuttle vectors, for propagation in *E. coli* the plasmids contain a ColE1 replication origin and an ampicillin resistance gene for selection. In most vectors the histidine dehydrogenase gene (*HIS4* gene) is utilized as a selection marker. Additionally some vectors contain a kanamycin resistance gene for the selection of *P. pastoris* clones with multiple genomic insertions.[155,156] Nowadays, a new vector generation has been developed containing only one resistance marker, the *Streptoalloteichus hindustanus* bleomycin/phleomycin resistence-gene (*ble*; zeocin resistance). This marker can be utilized for selection in both *E. coli* and *P. pastoris*.[34] Therefore, vectors bearing this marker have a dramatically reduced size (3600 bp) in comparison to vectors of the older generation (7000 to 9000 bp) and also have an optimized multiple cloning site. Furthermore, recombinant yeast clones with multiple insertions can be obtained directly by selection on plates with increasing concentrations of zeocin. For production of secreted proteins two different signal sequences, the *P. pastoris PHO1* signal sequence and the *S. cerevisiae* α-factor prepro leader can be tested.

11.3.2 INTEGRATION OF RECOMBINANT VECTORS INTO THE *P. PASTORIS* GENOME

In order to produce recombinant *P. pastoris* clones, the vector containing the foreign gene has to be integrated into the host genome by homologous recombination. Prior to transformation the *HIS4*-based vectors have to be linearized by restriction at a unique restriction site either in the *HIS4* gene (e.g., *Sal*I in pPIC9K) or the *AOX1* promoter sequence (e.g., *Sac*I in pPIC9K). The zeocin vectors can only be integrated into the 5′-*AOX1* region after restriction with either *Sac*I, *Pme*I, or *Bst*XI. Alternatively, the whole expression cassette can be used for replacement of the *AOX1* after resriction at appropriate sites (e.g., *Bgl*II in pPIC9K). After transformation of the resulting fragment, approximately 10 to 20% of the His transformants have the *AOX1* gene replaced by the expression cassette. These strains, which are called mut^S (methanol utilization sensitive), in contrast to mut$^+$ strains metabolize methanol only at a very reduced rate because they have to rely on the Aox2p for growth on methanol.

11.3.3 *P. PASTORIS* HOST STRAINS

Table 11.6 shows the commonly used *P. pastoris* host strains. Most of the strains are *his4* auxotrophs. Following transformation, the auxotrophy will be complemented by recombination with the corresponding linearized *His4* vector. For transformation with a zeocin vector the strain need not be auxotroph, and therefore, prototroph strains can be utilized. The main strain used for transformation of *His4*-based vectors is GS115. Sometimes it is advantageous for production of recombinant protein to use the strains SMD1168 and SMD1163 which are protease-deficient, bearing mutations either in the *PEP4* or the *PEP4* and *PRB1* genes, respectively. As protease-deficient *S. cerevisiae* strains, also the protease-deficient strains of *P. pastoris* have to be handled with care. The strains are not as robust as the wild-type strains. The strain KM71 strain has an insertion in the *AOX1* locus, therefore resulting recombinant clones are all Mut^s.

TABLE 11.6
***P. pastoris* Host Strains**

Strain	Genotype	Phenotype	Ref.
GS115	*his4*	Mut+ His−	155
KM71	*aox1Δ::SARG4 his4 arg4*	Mut^s His−	158
SMD1168	*pep4Δ his4*	Mut+ His−; protease deficient	159
SMD1163	*pep4Δprbhis4*	Mut+ His−; protease deficient	41

All the strains listed above can be obtained from Invitrogen: http://www.invitrogen.com.
Adapted from Cregg (1997).[62]

11.3.4 EXPERIMENTAL PROCEDURES

11.3.4.1 Preparation of the Recombinant Vector

The following experimental protocols already require a recombinant *His4*-based or a zeocin-based vector. Here, the receptor gene most probably N-terminally fused to one of the two signal sequences was cloned under the transcriptional control of the *AOX1* promoter. It should be noted that in the case of usage of a zeocin vector, the zeocin concentration in the agar plates for selection in *E. coli* should be 25 µg/ml. It should also be considered that a low-salt LB medium has to be used, because with the salt concentration present in ordinary LB plates one faces problems at zeocin selection. In other "high-salt" media, like SOB, 2× YT or TB, the zeocin concentration has to be raised up to 50 µg/ml. After linearization of the recombinant vector with an appropriate restriction enzyme, it has to be purified by a phenol/chloroform extraction followed by an ethanol precipitation. The plasmid DNA that is to be used for transformation should have a concentration of about 1 µg/µl either in 1× TE or in sterile distilled water.

11.3.4.2 Transformation of *P. Pastoris*

Four methods have proven to be successful for the transformation of *P. pastoris*: (1) transformation by electroporation; (2) transformation after spherophlasting preparation; (3) transformation using PEG; and (4) transformation using LiCl. In the following part the electroporation and LiCl transformation methods are described. For the spheroplast method a versatile kit from Invitrogen is available. The advantage of electroporation is that this is the only method in which cells do not have to be treated in a special way in advance. The efficiency of the electroporation and the spheroplast methods are similar but are always dependent on the given *P. pastoris* strain. The spheroplast method cannot be utilized for zeocin vectors. For these vectors the electroporation or the LiCl method is recommended. The LiCl method is the modified version of the same method in *S. cerevisiae* and is easy to handle in any lab. The efficiency of the method with about 10^2 to 10^3 transformants per microgram of DNA is similar to that of the PEG method, but is not as good as the electroporation or the spheroplast methods.

Protocol 11-1: Preparation of Materials for Transformation of P. pastoris

1. Agar plates with the *P. pastoris* strains GS115, KM71, SMD1186, or SMD1163.
2. 15 and 50 ml sterile tubes (e.g., Sarstedt, Greiner).
3. 2 l baffled flasks (sterile).
4. Shaking incubator (≥250 rpm).
5. YPD medium and agar plates (yeast extract, peptone, dextrose medium): 10 g yeast extract and 20 g peptone are dissolved in 900 ml water and autoclaved at 121°C for 20 min; add 100 ml 10× D solution (200 g D-glucose dissolved in 1 l water either autoclaved at 121°C for 15 min or sterile filtered). For agar plates, add 15 g agar.
6. Sterile deionized water.
7. 1 M sorbitol (sterile).
8. Recombinant, linearized plasmid DNA.
9. Electroporation instrument: Bio-Rad Gene Pulser or Electroporator II with sterile electroporation cuvets.
10. MD agar plates (minimal dextrose medium): dissolve 15 g agar in 800 ml water and autoclave. After cooling down to 60°C, add 100 ml of 10× YNB (134 g filter sterilized), 2 ml of 500× biotin (20 mg biotin dissolved in water and filter sterilized), and 100 ml of 10× D solution (200 g D-glucose dissolved in 1 l water and autoclaved at 121°C for 15 min).
11. YPDS + zeocin agarose plates (yeast extract, peptone, dextrose, sorbitol medium): 20 g agar, 10 g yeast extract, 182 g sorbitol, 20 g peptone are dissolved in 900 ml water and autoclaved for 20 min at 121°C. Cool down to 55°C, add 100 ml 10× D solution and 1 ml of a 100 mg/ml Zeocin solution. Keep plates with Zeocin cool and dark (plates can be stored for 2 to 3 weeks).
12. TE buffer: 10 mM Tris-HCl (pH 7.4), 1 mM ethylene diamine tetraacetic acid (EDTA).
13. 1 M LiCl in deionized water, filter sterilized.
14. 50% PEG3350 in deionized water, filter sterilized.
15. 20 mg/ml denatured, fragmented herring sperm DNA in TE buffer (boil solution for 5 min and then keep on ice).
16. YPD + zeocin agar plates: prepare YPD as described above and after autoclaving cool down to 55°C and add 1 ml of a 100 mg/ml zeocin solution. Keep plates with zeocin cool and dark (plates can be stored at 4°C for 2 to 3 weeks).

Protocol 11-2: Transformation of P. pastoris by Electroporation

1. Inoculate a 5-ml YPD culture from a fresh agarose plate with a single *P. pastoris* colony of the strain that has to be transformed, and incubate at 30°C overnight with shaking at 250 rpm.

2. Inoculate 500 ml YPD medium in a 2-l Erlenmeyer flask with 500 µl overnight culture and shake with 250 rpm at 30°C until culture reaches an OD_{600} of 1.3 to 1.5 (overnight).
3. Wash the culture after centrifugation at 1800 g for 5 min at 4°C with 500 ml ice-cold sterile water, once with 250 ml sterile water, and once with 20 ml 1 M sorbitol.
4. Resuspend the pellet with 1 ml 1 M sorbitol.
5. Add 10 µg linearized plasmid DNA in 5 to 10 µl 1× TE buffer to 80 µl of the cells and mix, apply the sample into a prechilled sterile electroporation cuvet, and incubate for 5 min on ice.
6. Electroporate the cells as described by the producer (e.g., Bio-Rad).
7. Immediately after electroporation add 1 ml ice-cold 1 M sorbitol solution into cuvet.
8. For *his4* vectors:
   a. Transfer the sample from the cuvet into a sterile Eppendorf cup.
   b. Plate 200 to 600 µl onto an MD agar plate.
   c. Incubate the plate until colonies are visible (about 2 d).
9. For Zeocin vectors:
   a. Transfer the sample from the cuvet into a sterile 15-ml tube and incubate at 30°C without shaking for about 1 to 2 h.
   b. Plate 50, 100, and 200 µl samples onto YPDS agar plates supplemented with 100 µg/ml zeocin.

Protocol 11-3: Transformation of P. pastoris by LiCl Treatment

1. Inoculate 50 ml YPD medium with a single colony of a *P. pastoris* strain and incubate at 30°C with shaking until they reach an OD_{600} of 1.0.
2. Wash the cells after centrifugation at 1500 g for 10 min at room temperature with 25 ml sterile water.
3. After the wash, resuspend the cells in 1 ml 100 mM LiCl and transfer the sample into an Eppendorf cup.
4. Centrifuge the cells at maximum speed in an Eppendorf centrifuge for 15 s.
5. Resuspend the pellet in 400 µl 100 mM LiCl.
6. Make 50 µl aliquots of the resuspended cells and transfer to Eppendorf cups.
7. Before transformation, centrifuge the cells and remove the supernatant.
8. For transformation add, in the following order: 240 µl 50% PEG, 36 µl 1 M LiCl, 25 µl 2 mg/ml fragmented herring sperm DNA, and 5 to 10 µg plasmid DNA in 50 µl sterile water.
9. Mix the sample by vortexing the tube until the cells are resuspended.
10. Incubate the samples at 30°C for 30 min and then for 20 to 25 min at 42°C.
11. For *his* vectors:
    a. Centrifuge the cells and resuspend in 100 µl sterile water.
    b. Plate the whole sample onto MD agar plates.
    c. Incubate the plates at 30°C until colonies are visible (about 2 d).

12. For zeocin vectors:
    a. Centrifuge the cells and remove the supernatant.
    b. Resuspend the cells in 1 ml YPD and incubate at 30°C.
    c. After 1 h and 4 h plate 25, 50, and 100 µl samples onto YPD agar plates with 100 µg/ml Zeocin.
    d. Incubate the plates at 30°C until colonies are visible.

11.3.4.3 Determination of the Mut Phenotype

The phenotype of His⁺ P. pastoris clones that grow on MD plates after transformation has to be characterized. Mutˢ transformants underwent recombination into the AOX1 locus, whereas Mut⁺ transformants still have a functional AOX1. In the case of Mutˢ clones the Aox2p allows only slow growth on agar plates which contain methanol as sole carbon source (generation time Mut⁺ clone: 4 to 6 h; Mutˢ clone: 18 to 20 h). It is easy to distinguish these two types of His⁺ transformants from each other. The resulting phenotype depends on both the strain and on the restriction site that was used for restriction of the recombinant plasmid prior to transformation. When the P. pastoris KM71 strain was used for transformation each His⁺ clone is Mutˢ after transfomation. In case of transformation of a zeocin vector into the strains GS115, SMD1168, or SMD1163 most of the resulting recombinant clones are Mut⁺; therefore, here it is not necessary to test phenotype for zeocin vectors.

Protocol 11-4: Determination of the Mut Phenotype

1. Material
   a. MD plates (see above).
   b. MM agar plates: autoclave 15 g agar in 800 ml water, cool down to 60°C, and then add 100 ml 10× YNP (see above) and 100 ml 10× M (5 ml methanol in 95 ml water, sterile filtrated). The solution can be stored for 2 months.
2. pick single colonies from the transformation plate and strike them out onto MM and MD plates.
3. incubate the plates for 2 d at 30°C.
4. screen the colonies by eye: all should grow on MD plates (His⁺). If the colonies grow normally on MM plates they are Mut⁺; if they show a significantly reduced growth they are Mutˢ.

11.3.4.4 Selection of Recombinant P. pastoris Clones with Multiple Vector Insertions

After transformation into P. pastoris, spontaneous multiple insertions of the recombinant plasmid into the host genome can occur (spontaneous insertion in either the AOX1 or the HIS4 locus occurs at a frequency of 1 to 10%). These are welcome because they can lead to a copy number-dependent increase of protein production. Recombinant yeast clones with accidental multiple insertions can be quantitatively analyzed with the help of dot blot or Southern blot hybridizations. Certain vectors, like pAO815, pPICZ, or pGAPZ allow for the *in vitro* production of multiple insertions. Some vectors, like pPIC3K or pPIC9K (see Table 11.5), additionally contain a kanamycin marker (*kan* gene of the transposon Tn903). Therefore, it is possible to enrich and to select recombinant clones with multiple insertions. The level of resistance conferred by the resistance gene is roughly proportional to the copy number present in the genome. In the case of the vectors mentioned above it is essential to select for His$^+$ transformants first and then to screen for G418-resistant clones. For clones that have been transformed by zeocin vectors, the selection for multicopy transformants can be done directly on agar plates with increasing concentrations of zeocin. After the selection and characterization of the multicopy clones the relation between copy number and protein production can be determined by hybridization.

Protocol 11-5: Selection of Recombinant P. pastoris

1. Material
   a. G418 solution: 100 mM G418/ml water, filter sterile, and divided in aliquots (store at −20°C).
   b. YPD G418 agar plates: 2.5 g yeast extract, 5 g peptone, and 5 g agar in 225 ml water, after autoclaving add 25 ml 10× D solution (see above). Cool down to 60°C and add G418 in different concentrations. For strains like GS115, KM71, and SMD1168 the following G418 concentrations are recommended: 0.25, 0.5, 0.75, 1.0, 1.5, 1.75, 2.0, 3.0, and 4.0 mg/ml. For the strain SMD1163 ten times lower concentrations are recommended.
   c. Hemocytometer.
2. After transformation add 1 to 2 ml sterile water onto the plates with the His$^+$ transformants.
3. With a sterile spatula transfer the cells into a 50-ml centrifuge tube.
4. Determine the cell density with a hemocytometer (alternatively measure OD$_{600}$).
5. Plate 200 µl aliquots containing about 1×10^5 cells onto YPD G418 plates with different G418 concentrations. For each concentration several plates should be used (if not all cells are plated onto plates add glycerol to a final concentration of 15% and store at −75°C).
6. Incubate the plates for several days at 30°C.
7. Sometimes only a few G418-resistant strains grow on the plates and some of them may have different sizes, but the morphology should be the same.
8. Pick the G418-resistant colonies with a sterile toothpicker and strike them out onto a new G418-containing plate in order to avoid false positives.
9. Incubate plates at 30°C for several days until colonies get visible.
10. Inoculate 10 ml YPD medium with a single colony and incubate overnight at 30°C with shaking.
11. Also prepare a glycerol stock.

**11.3.4.5   Screening of Recombinant *P. pastoris* Clones for
             Protein Production**

The production level of the recombinant protein can vary significantly from clone
to clone. Factors that influence the level of production are the number of integrated
expression cassettes (single copy or multicopy clones), the Mut phenotype (Mut⁺ or
Mut⁻), the type of strain used for transformation, and some other not yet so clearly
defined parameters. Therefore, it is highly recommended to screen several clones in
order to be able to find the best producing one. It is important to keep in mind that
the production of foreign proteins is also influenced by culture and medium parameters, such as oxygen and methanol concentration. Each parameter has to be checked
and tested to ensure the most appropriate conditions. To determine the time dependence of protein production after induction, samples should be taken at different
time points. For a Mut⁺ clone the best time points for such a time course experiment
include 0 (before induction), 6, 12, 24, 36, 48, 72, and 96 h; for a Mutˢ clone 0, 24,
48, 72, 96, and 120 h are convenient. It is recommended to have a control clone
while doing time course analysis.

Protocol 11-6: Screening of Recombinant P. pastoris

1. Material
   a. Sterile 50-ml tube for Mut⁺ clones and 1-l Erlenmeyer flask for Mutˢ
      clones.
   b. 100-ml baffled flasks (sterile).
   c. Shaking incubator.
   d. BMGY medium: 10 g yeast extract and 20 g peptone dissolved in
      700 ml water and autoclaved at 121°C for 20 min. Cool down, add
      100 ml 10× GY (10% glycerol in water, autoclaved), 100 ml 10× YNB
      (see above), 100 ml 10× phosphate buffer (1 M potassium phosphate,
      pH 6.0, autoclaved), and 2 ml 500× B (see above).
   e. BMM medium: like BMGY, but with 10× M (see above) instead of
      10× GY.
   f. Breaking buffer (50 mM potassium phosphate, pH 7.4, 1 mM EDTA,
      5% [v/v] glycerol).
   g. Membrane buffer (50 mM Tris, pH 7.4, 150 mM NaCl).
   h. 100× PMSF (100 mM phenylsulfonylfluoride in dimethylsulfoxide
      [DMSO]), store at –20°C.
   i. Acid washed glass beads (500 μm diameter).
   j. Methanol, p.A.
   k. Equipment for sodium dodecylsulfate-polyacrylamide gel electro-
      phoresis (SDS-PAGE).
2. Growth of recombinant *P. pastoris* clones
   a. Inoculate 10 ml BMGY medium in a 50-ml flask with a single Mut⁺
      clone or 100 ml BMGY medium in a 1-l Erlenmeyer flask with a single
      Mutˢ clone and incubate overnight at 30°C with shaking at 250 rpm.

    b. If the OD_{600} reaches 2 to 6 centrifuge the cells (5 min, 2000 g at room temperature) and resuspend the pellet in BMMY medium to a final OD_{600} of 1 for Mut$^+$ clones and 20 for Mut^s clones.

    c. Transfer 20 ml of this suspension into a 100-ml baffled flask and cover with a loose aluminum foil (good aeration is very important), shake with 250 rpm at 30°C.

    d. To guarantee good induction, every 24 h add methanol to a final culture volume of 0.5%.

    e. Collect samples after indicated time intervals, centrifuge the samples for 2 to 3 min at 3000 g, and freeze the cell pellets in ethanol/dry ice or liquid nitrogen.

    f. Store the pellets at −80°C until further analysis.

3. Processing of cell probes

    a. Thaw the cell pellets on ice (e.g., cells from 1 ml culture) and mix with 100 μl breaking buffer.

    b. Add one sample volume of glass beads and 1 μl 100× PMSF to the suspension (PMSF is stable only for a short time in aqueous solution).

    c. Mix the sample by vortexing for 30 s and then put on ice for 30 s, repeat these steps eight times.

    d. Centrifuge with maximum speed in a table centrifuge at 4°C for 10 min, transfer the supernatant to an Eppendorf cup (keep on ice).

    e. For analysis by SDS-PAGE and subsequent Western analysis mix a small part of the probe with sample buffer, add PMSF, and apply the sample (5 to 10 μl per lane) directly onto the SDS gel.

    f. The rest of the sample can be used for binding assay.

4. Small-scale membrane preparation

    a. Harvest the yeast cells from 10- to 15-ml cultures 24 to 48 h after induction by centrifugation with 800 g at 4°C for 5 min in 12-ml centrifuge tubes.

    b. Resuspend the cell pellet in 2.5 ml of ice-cold breaking buffer with protease inhibitor, and add 2 ml of cold acid-washed glass beads (always keep the probe on ice).

    c. Vortex the tubes for 30 s, and then cool for about 30 s on ice, repeat these two steps eight times.

    d. Remove unbroken cells and cell debris by centrifugation at 1800 g for 5 min at 4°C.

    e. Transfer supernatant (1.5 to 2 ml) without glass beads to ultracentrifuge tubes, and centrifuge at 100,000 g for 30 min at 4°C.

    f. Resuspend each pellet in 500 μl cold membrane buffer supplemented with protease inhibitors (the protein concentration should be in the range of 9 to 13 mg/ml when a 15-ml culture is used and grown to a final OD_{600} of 10 to 12).

    g. Freeze the samples in liquid N_2 or dry ice/ethanol bath and store at −80°C until use.

11.3.4.6　Large-Scale Production of Crude Membranes from Recombinant *P. pastoris*

After the characterization of a clone reveals good production of the recombinant protein, scaled up experiments can be introduced. If large amounts of recombinant protein will be needed a 4- to 10-l fermenter can be applied. *P. pastoris* is excellently suitable for fermenter culturing. Large shake flask cultures are almost as easy to handle as small ones. Here, adequate aeration is important, especially after methanol induction. Baffled flasks filled up to a fifth of their volume are recommended. For intracellular and membrane proteins larger volumes of cells have to be broken. For this purpose glass mills and microfluidizers are available. In the following section the breakage and membrane preparation of 20 g cells (wet weight) is presented. Normally, about 20 g *P. pastoris* cells can be obtained from 1 l BMMY culture. The advantage of this method is that there is no need for a special machine for breakage, the cells are disrupted by harsh mixing (vortexing) with glass beads.

Protocol 11-7: Large-Scale Production of Crude Membranes from P. pastoris

1. Material
   a. 5-l baffled flasks and 3-l Erlenmeyer flasks (sterile).
   b. 50-ml tubes (sterile).
   c. BMGY and BMMY medium (see above).
   d. Breaking buffer (see above).
   e. 100 mM PMSF solution (in DMSO).
   f. Glass beads (see above).
   g. Membrane buffer (50 mM Tris, pH 7.4, 150 mM NaCl).
   h. Potter (25 ml volume).
2. Inoculate 10 ml BMGY medium in a 50-ml tube with a single clone and incubate at 30°C overnight with shaking at 250 rpm.
3. Determine OD$_{600}$ of this preculture.
4. Inoculate 500 ml prewarmed medium (30°C) in a 3-l Erlenmeyer flask with the preculture so that the culture will have an OD$_{600}$ of 2 to 6 the next day (in complex medium the yeast cells have a generation time of about 2 h).
5. Incubate the 500-ml culture overnight at 30°C with shaking at 250 rpm.
6. Determine the cell density in the BMGY culture (OD$_{600}$ should be between 2 and 6).
7. Calculate the amount of BMGY culture that is necessary to start the 1-l culture with an OD$_{600}$ of 1 (for example: if the OD$_{600}$ of the culture is 4 then centrifuge 250 ml culture and resuspend in 1 l BMMY medium).
8. Harvest the cells by centrifugation (2000 g, room temperature, 5 min).
9. Resuspend cell pellet in BMMY medium and transfer into a 5-l baffled flask.

10. Incubate cells at 30°C and shake with 250 rpm until the maximum protein production is reached (check protein production by convenient method, e.g., binding assay).
11. Harvest the cells by centrifugation (2000 g, 4°C, 5 min) and resuspend in breaking buffer.
12. Divide the suspension into six 50-ml tubes (on ice) and add 10 to 12 ml glass beads and 120 μl 100× PMSF to every tube.
13. Vigorously vortex the tubes for 30 s, and then cool for about 30 s on ice, repeat eight times.
14. Add 10 ml cold breaking buffer and 100 μl 100× PMSF to the suspension and vortex again for 20 s.
15. Centrifuge the tubes at 1800 g, 4°C for 5 min.
16. Transfer the supernatant of each tube into an ultracentrifuge tube and centrifuge for 100,000 g, 4°C for 60 min.
17. Add 250 μl 100× PMSF and resuspend the membrane pellet in 250 ml membrane buffer using a potter (always on ice).
18. Divide the membrane suspension in aliquots (in 1.5-ml Eppendorf cups) and freeze in liquid nitrogen or in an ethanol/dry ice bath, store at −80°C until use (the protein concentration of the membrane suspension should be about 30 mg/ml).

11.3.4.7 Long-Term Storage of *P. pastoris* Strains and Recombinant *P. pastoris* Clones

Transformation plates with the recombinant clones can be stored for some weeks at 4°C. YPD plates can be used to store *P. pastoris* strains for some months. For the long-term storage of *P. pastoris* transformants and strains glycerol stocks have to be prepared.

Protocol 11-8: Long-Term Storage of P. pastoris Strains

1. Material
   a. 50-ml tubes (sterile).
   b. YPD medium (see above).
   c. 86% glycerol (sterile).
2. Inoculate 15 ml YPD medium in a 50-ml tube with a single colony.
3. Incubate the culture at 30°C overnight and determine the OD_{600}.
4. Harvest the cells by centrifugation (2000 g, 5 min).
5. Add glycerol to the YPD medium to a final concentration of 15%.
6. Resuspend the cells in YPD + 15% glycerol, aliquot the suspension, and shock freeze in a dry ice/ethanol or liquid nitrogen (store at −75°C).
7. After a long storage time G418-resistant recombinant cells should be plated on the appropriate selection plates to check His and Mut phenotype.

REFERENCES

1. Strader, C. D., Fong, T. M., Graziano, M., and Tota, M. R., The family of G protein-coupled receptors, *FASEB J.*, 9, 745, 1995.
2. Helmreich, E. J. M., and Hofmann, K.-P., Structure and function of proteins in G protein-coupled signal transfer, *Biochim. Biophys. Acta*, 1286, 285, 1996.
3. Wess, J., G protein-coupled receptors: molecular mechanisms involved in receptor activation and selectivity of G protein recognition, *FASEB J.*, 11, 346, 1997.
4. Gutkind, J. S., The pathways connecting G protein-coupled receptors to the nucleus through divergent mitogen-activated protein kinase cascades, *J. Biol. Chem.*, 273, 1839, 1998.
5. Böhm, S. K., Grady, E. F., and Bunnett, N. W., Regulatory mechanisms that modulate signalling by G protein-coupled receptors, *Biochem. J.*, 322, 1, 1997.
6. Henderson, R., Baldwin, J. M., Ceska, T. A., Zemlin, F., Beckmann, E., and Downing, K. H., Model for the structure of bacteriorhodopsin based on high-resolution electron cryo-microscopy, *J. Mol. Biol.*, 213, 899, 1990.
7. Pebay-Peyroula, E., Rummel, G., Rosenbusch, J. P., and Landau, E. M., X-ray structure of bacteriorhodopsin at 2.5 angstroms from microcrystals grown in lipidic cubic phases, *Science*, 277, 1676, 1997.
8. Luecke, H., Richter, H.-T., and Lanyi, J. K., Proton transfer pathways in bacteriorhodopsin at 2.3 angstrom resolution, *Science*, 280, 1934, 1998.
9. Schertler, G. F. X., Villa, C., and Henderson, R., Projection structure of rhodopsin, *Nature,* 362, 770, 1993.
10. Schertler, G. F. X. and Hargrave, P. A., Projection structure of frog rhodopsin in two crystal forms, *Proc. Natl. Acad. Sci. U.S.A.,* 92, 11578, 1995.
11. Unger, V. M. and Schertler, G. F., Low resolution structure of bovine rhodopsin determined by electron cryo-microscopy, *Biophys. J.*, 68, 1776, 1995.
12. Davies, A., Schertler, G. F. X., Gowen, B. E., and Sabil, H. R., Projection structure of an invertebrate rhodopsin, *J. Struct. Biol.*, 117, 36, 1996.
13. Unger, V. M., Hargrave, P. A., Baldwin, J, M., and Schertler, G. F., Arrangement of rhodopsin transmembrane alpha-helices, *Nature*, 389, 203, 1997.
14. Baldwin, J. M., The probable arrangement of the helices in G protein-coupled receptors, *EMBO J.,* 12, 1693, 1993.
15. Baldwin, J. M., Structure and function of receptors coupled to G proteins, *Curr. Opin. Cell Biol.*, 6, 180, 1994.
16. Baldwin, J. M., Schertler, G. F. X., and Unger, V. M., An alpha-carbon template for the transmembrane helices in the rhodopsin family of G protein-coupled receptors, *J. Mol. Biol.*, 272, 144, 1997.
17. Collins, S., Lohse, M. J., O'Dowd, B., Caron, M. G., and Lefkowitz, R. J., Structure and regulation of G protein-coupled receptors: the β_2-adrenergic receptor as a model, *Vitam. Horm.*, 46, 1, 1991.
18. Kobilka, B. K., Adrenergic receptors as models for G protein-coupled receptors, *Annu. Rev. Neurosci.*, 15, 87, 1992.
19. Weickert, M. J., Doherty, D. H., Best, E. A., and Olins, P. O., Optimization of heterologous protein production in *Escherichia coli*, *Curr. Opin. Biotechnol.*, 7, 494, 1996.
20. Geisse, S., Gram, H., Kleuser, B., and Kocher, H. P., Eukaryotic expression systems: a comparison, *Protein Exp. Purif.*, 8, 271, 1996.
21. Carroll, M. W. and Moss, B., Poxviruses as expression vectors, *Curr. Opin. Biotechnol.*, 8, 573, 1997.

22. Lundstrom, K., Alphaviruses as expression vectors, *Curr. Opin. Biotechnol.*, 8, 578, 1997.
23. Possee, R. D., Baculoviruses as expression vectors, *Curr. Opin. Biotechnol.*, 8, 569, 1997.
24. Luckow, V. A. and Summers, M. D., Trends in the development of baculovirus expression vectors, *Bio/Technology*, 6, 47, 1988.
25. O'Reilly, D. R., Miller, L. K., and Luckow, V. A., *Baculovirus Expression Vectors: A Laboratory Manual*, W. H. Freeman, New York, 1992.
26. Gellisen, G., Melber, K., Janowicz, Z. A., Dahlems, U. M., Weydemann, U., Piontek, M., Strasser, A. W. M., and Hollenberg, C. P., Heterologous protein production in yeast, *Antonie van Leeuwenhoek*, 62, 79, 1992.
27. Romanos, M. A, Scorer, C. A., and Clare, J. J., Foreign gene expression in yeast, *Yeast*, 8, 423, 1992.
28. Sudbery, P.E., The expression of recombinant proteins in yeasts, *Curr. Opin. Biotechnol.*, 7, 517, 1996.
29. Buckholz, R. G., Yeast systems for the expression of heterologous gene products, *Curr. Opin. Biotechnol.*, 4, 538, 1993.
30. Eckart, M. R. and Bussineau, C. M., Quality and authenticity of heterologous proteins synthesized in yeast, *Curr. Opin. Biotechnol.*, 7, 525, 1996.
31. Grisshammer, R. and Tate, C. G., Overexpression of integral membrane proteins for structural studies, *Q. Rev. Biophys.*, 28, 315, 1995.
32. Tate, C. G. and Grisshammer, R., Heterologous expression of G protein-coupled receptors, *Trends Biotechnol.*, 14, 426, 1996.
33. Price, L. A., Strnad, J., Pausch, M. H., and Hadcock, J. R., Pharmacological characterization of the rat A_{2a} adenosine receptor functionally coupled to the yeast pheromone response pathway, *Mol. Pharmacol.*, 50, 829, 1996.
34. King, K., Dohlman, H. G., Thorner, J., Caron, M. G., and Lefkowitz, R. J., Control of yeast mating signal transduction by a mammalian β_2-adrenergic receptor and $G_{S\alpha}$ subunit, *Science*, 250, 121, 1990.
35. Ficca, A. G., Testa, L., and Tocchini-Valentini, G. P., The human β_2-adrenergic receptor expressed in *Schizosaccharomyces pombe* retains its pharmacological properties, *FEBS Lett.*, 377, 140, 1995.
36. Weiß, H. M., Haase, W., Michel, H., and Reiländer, H., Comparative biochemical and pharmacological characterization of the mouse $5HT_{5A}$ 5-hydroxytryptamine receptor and the human β_2-adrenergic receptor produced in the methylotrophic yeast *Pichia pastoris*, *Biochem. J.*, 330, 1137, 1998.
37. Sizmann, D., Kuusinen, H., Keränen, S., Lomasney, J., Caron, M. G., Lefkowitz, R. J., and Keinänen, K., Production of adrenergic receptors in yeast, *Receptors Channels*, 4, 197, 1996.
38. Payette, P., Gossard, F., Whiteway, M., and Dennis, M., Expression and pharmacological characterization of the human M1 muscarinic receptor in *Saccharomyces cerevisiae*, *FEBS Lett.*, 266, 21, 1990.
39. Huang, H.-J., Liao, C.-F., Yang, B.-C., and Kuo, T.-T., Functional expression of rat m5 muscarinic acetylcholine receptor in yeast, *Biochem. Biophys. Res. Commun.*, 182, 1180, 1992.
40. Bach, M., Sander, P., Haase, W., and Reiländer, H., Pharmacological and biochemical characterization of the mouse $5HT_{5A}$ serotonin receptor heterologously expressed in the yeast *Saccharomyces cerevisiae* strains, *Receptors Channels*, 4, 129, 1996.

41. Weiß, H. M., Haase, W., Michel, H., and Reiländer, H., Expression of functional mouse 5-HT$_{5A}$ serotonin receptor in the methylotrophic yeast *Pichia pastoris*: pharmacological characterization and localization, *FEBS Lett.*, 377, 451, 1995.

42. Andersen, B. and Stevens, R. C., The human D$_{1A}$ dopamine receptor: heterologous expression in *Saccharomyces cerevisiae* and purification of the functional receptor, *Protein Exp. Purif.*, 13, 111, 1998.

43. Sander, P., Grünewald S., Maul, G., Reiländer, H., and Michel, H., Constitutive expression of the human D$_{2S}$ dopamine receptor in the unicellular yeast *Saccharomyces cerevisiae*, *Biochim. Biophys. Acta*, 1193, 255, 1994.

44. Sander, P., Grünewald, S., Reiländer, H., and Michel, H., Expression of the human D$_{2S}$ dopamine receptor in the yeasts *Saccharomyces cerevisiae* and *Schizosaccharomyces pombe*: a comparative study, *FEBS Lett.*, 344, 41, 1994.

44a. Grünewald, S., Haase, W., Molsberger, E., Michel, H., and Reiländer, H., High-level production of the human D$_{2S}$ receptor in the methylotrophic yeast *Pichia pastoris*, in preparation.

45. Arkinstall, S., Edgerton, M., Payton, M., and Maundrell, K., Co-expression of the neurokinin NK2 receptor and G protein components in the fission yeast *Schizosaccharomyces pombe*, *FEBS Lett.*, 375, 183, 1995.

46. Talmont, F., Sidobre, S., Demange, P., Milon, A., and Emorine, L. J., Expression and pharmacological characterization of the human μ-opioid receptor in the methylotrophic yeast *Pichia pastoris*, *FEBS Lett.*, 394, 268, 1996.

47. Mollaaghababa, R., Davidson, F. F., Kaiser, C., and Khorana, H. G., Structure and function in rhodopsin: expression of functional opsin in *Saccharomyces cerevisiae*, *Proc. Natl. Acad. Sci. U.S.A.*, 93, 482, 1996.

48. Abdulaev, N. G., Popp, M. P., Smith, W. C., and Ridge, K. D., Functional expression of bovine opsin in the methylotrophic yeast *Pichia pastoris*, *Protein Exp. Purif.*, 10, 61, 1997.

49. Price, L. A., Kajkowski, E. M., Hadcock, J. R., Ozenberger, B. A., and Pausch, M. H., Functional coupling of a mammalian somatostatin receptor to the yeast pheromone response pathway, *Mol. Cell. Biol.*, 15, 6188, 1995.

50. Erickson, J. R., Wu, J. J., Goddard, G., Tigyi, G., Kawanishi, K., Tomei, L. D., and Kiefer, M. C., Edg-2/Vzg-1 couples to the yeast pheromone response pathway selectively in response to lysophosphatidic acid, *J. Biol. Chem.*, 273, 1998.

51. Kajkowski, E. M., Price, L. A., Pausch, M. H., Young, K. H., and Ozenberger, B. A., Investigation of growth hormone releasing hormone receptor structure and activity using yeast expression technologies, *J. Receptor Signal Transduction Res.*, 17, 293, 1997.

52. David, N. E., Gee, M., Andersen, B., Naider, F., Thorner, J., and Stevens, R. C., Expression and purification of the *Saccharomyces cerevisiae* α-factor receptor (Ste2p), a 7-transmembrane-segment G protein-coupled receptor, *J. Biol. Chem.*, 272, 15553, 1997.

53. Broach, J. R., Pringle, J. R., and Jones, E. W., *The Molecular and Cellular Biology of the Yeast Saccharomyces*, Vol. 1, Cold Spring Harbor Laboratory Press, Cold Spring Harbor, NY, 1991.

54. Jones, E. W., Pringle, J. R., and Broach, J. R., *The Molecular and Cellular Biology of the Yeast Saccharomyces*, Vol. 2, Cold Spring Harbor Laboratory Press, Cold Spring Harbor, NY, 1992.

55. Pringle, J. R., Broach, J. R., and Jones, E. W., *The Molecular and Cellular Biology of the Yeast Saccharomyces*, Vol. 3, Cold Spring Harbor Laboratory Press, Cold Spring Harbor, NY, 1997.

56. Giga-Hama, Y. and Kumagai, H., Foreign gene expression in fission yeast *Schizosaccharomyces pombe*, Springer-Verlag, Berlin and Landes Bioscience, Georgetown, TX, 1997.

57. Maundrell, K., Thiamine-repressible expression vectors pREP and pRIP for fission yeast, *Gene,* 123, 127, 1993.

58. Higgins, D. R. and Cregg, J. M., *Methods in Molecular Biology,* Vol. 103, *Pichia* Protocols, Humana Press, Totowa, NJ, 1998.

59. Waterham, H. R., Digan, M. E., Koutz, P. J., Lair, S. V., and Cregg, J. M., Isolation of the *Pichia pastoris* glyceraldehyde-3-phosphate dehydrogenase gene and regulation and use of its promoter, *Gene,* 186, 37, 1997.

60. Cregg, J. M., Vedvick, T. S., and Raschke, W. C., Recent advances in the expression of foreign genes in *Pichia pastoris, Bio/Technology,* 11, 905, 1993.

61. Romanos, M., Advances in the use of *Pichia pastoris* for high-level gene expression, *Curr. Opin. Biotechnol.,* 6, 527, 1995.

62. Cregg, J. M., Expression in the methylotrophic yeast *Pichia pastoris,* in *Gene Expression Systems,* Fernandez, J. and Hoeffler, J. P., Eds., Academic Press, San Diego, CA, 1998.

63. Stratowa, C., Himmler, A., and Czernilofsky, A. P., Use of a luciferase reporter system for characterizing G protein-linked receptors, *Curr. Opin. Biotechnol.,* 6, 574, 1995.

64. Kirsch, D. R., Development of improved cell-based assays and screens in *Saccharomyces* through the combination of molecular and classical genetics, *Curr. Opin. Biotechnol.,* 4, 543, 1993.

65. Kurjan, J., The pheromone response pathway in *Saccharomyces cerevisiae, Annu. Rev. Genet.,* 27, 147, 1993.

66. Levin, D. E. and Errede, B., The proliferation of MAP kinase signaling pathways in yeast, *Curr. Opin. Cell. Biol.,* 7, 197, 1995.

67. Pausch, M. H., G protein-coupled receptors in *Saccharomyces cerevisiae*: high-throughput screening assays for drug discovery, *Trends Biotechnol.,* 15, 487, 1997.

68. Stadel, J. M., Wilson, S., and Bergsma, D. J., Orphan G protein-coupled receptors: a neglected opportunity for pioneer drug discovery, *Trends Pharmacol. Sci.,* 18, 430, 1997.

69. Amatruda, T. T., III, Steele, D. A., Slepak, V. Z., and Simon, M. I., $G\alpha16$, a G protein α subunit specifically expressed in hematopoietic cells, *Proc. Natl. Acad. Sci. U.S.A.,* 88, 5587, 1991.

70. Milligan, G., Marshall, F., and Rees, S., G_{16} as a universal G protein adapter: implications for agonist screening strategies, *Trends Pharmacol. Sci.,* 17, 235, 1996.

71. Medici, R., Bianchi, E., Di Segni, G., and Tocchini-Valentini, G. P., Efficient signal transduction by a chimeric yeast-mammalian G protein α subunit Gpa1-Gsα covalently fused to the yeast receptor Ste2, *EMBO J.,* 16, 7241, 1997.

72. Bertin, B., Freissmuth, M., Jockers, R., and Strosberg, A. D., Cellular signaling by an agonist-activated receptor/G$_s\alpha$ fusion protein, *Proc. Natl. Acad. Sci. U.S.A.,* 91, 8827, 1994.

73. Watson, S. and Arkinstall, S., *The G protein Linked Receptor,* Academic Press, Harcourt Brace, London, 1994.

74. Summers, R. J. and McMartin, L. R., Adrenoceptors and their second messenger systems, *J. Neurochem.,* 60, 10, 1993.

75. Strosberg, A. D., Structure, function and regulation of adrenergic receptors, *Protein Sci.,* 2, 1198, 1993.

76. Marullo, S., Delavier-Klutchko, C., Eshdat, Y., Strosberg, D., and Emorine, L., Human β_2-adrenergic receptors expressed in *Escherichia coli* membranes retain their pharmacological properties, *Proc. Natl. Acad. Sci. U.S.A.,* 85, 7551, 1988.

77. Marullo, S., Delavier-Klutchko, C., Guillet, J.-G., Charbit, A., Strosberg, A. D., and Emorine, L. J., Expression of human β_1 and β_2 adrenergic receptors in *E. coli* as a new tool for ligand screening, *Bio/Technology*, 7, 923, 1989.

78. Chapot, M. P., Eshdat, Y., Marullo, S., Guillet, J. G., Charbit, A., Strosberg, A. D., and Delavier-Klutchko, C., Localization and characterization of three different β adrenergic receptors expressed in Escherichia coli, *Eur. J. Biochem.*, 187, 137, 1990.

79. Hampe, W., Voss, R.-H., Haase, W., Boege, F., Michel, H., and Reiländer, H., Engineering of a proteolytically stable human β_2-adrenergic receptor/Maltose-binding protein fusion and its production in *Escherichia coli* and baculovirus-infected insect cells, submitted, 1999.

80. Söhlemann, P., Soppa, J., Oesterhaelt, D., and Lohse, M. J., Expression of β_2-adreno-ceptors in halobacteria, *Naunyn-Schmiedeberg's Arch. Pharmacol.*, 355, 150, 1997.

81. George, S. T., Arbabian, M. A., Ruoho, A. E., Kiely, J., and Malbon, C. C., High efficiency expression of mammalian β-adrenergic receptors in baculovirus-infected insect cells, *Biochem. Biophys. Res. Commun.*, 163, 1265, 1989.

82. Parker, E. M., Kameyama, K., Higashijima, T., and Ross, E. M., Reconstitutively active G protein-coupled receptors purified from baculovirus infected insect cells, *J. Biol. Chem.*, 266, 519, 1991.

83. Reiländer, H., Boege, F., Vasudevan, S., Maul, G., Hekmann, M., Dees, C., Hampe, W., Helmreich, E. J. M., and Michel, H. Purification and functional characterization of the human β_2-adrenergic receptor produced in baculovirus-infected insect cells, *FEBS Lett.*, 282, 441, 1991.

84. Lohse, M. J., Stable overexpression of human β_2-adrenergic receptors in mammalian cells, *Naunyn-Schmiedeberg's Arch. Pharmacol.*, 345, 444, 1991.

85. Kleymann, G., Boege, F., Hahan, M., Hampe, W., Vasudevan, S., and Reiländer, H., Human β_2-adrenergic receptor produced in stably transformed insect cells is functionally coupled via endogenous GTP-binding protein to adenylyl cyclase, *Eur. J. Biochem.*, 213, 797, 1993.

86. Kwatra, M. M., Schreurs, J., Schwinn, D. A., Innis, M. A., Caron, M. G., and Lefkowitz, R. J., Immunoaffinity purification of epitope-tagged human β_2 adrenergic receptor to homogeneity, *Protein Exp. Purif.*, 6, 717, 1995.

87. Kobilka, B. K., Amino and carboxyl terminal modifications to facilitate the production and purification of a G protein-coupled receptor, *Anal. Biochem.*, 231, 269, 1995.

88. Caron, M. G., Srinivasan, Y., Pitha, J., Kociolek, K., and Lefkowitz, R. J., Affinity purification of the β-adrenergic receptor, *J. Biol. Chem.*, 254, 2923, 1979.

89. Marjamäki, A., Pohjanoska, K., Ala-Uotila, S., Sizmann, D., Oker-Blom, C., Kurose, H., and Scheinin, M., Similar ligand binding in recombinant human α_2C2-adrenoceptors produced in mammalian, insect and yeast cells, *Eur. J. Pharmacol.*, 267, 117, 1994.

90. Scorer, C. A., Clare, J. J., McCombie, W. R., Romanos, M. A., and Sreekrishna, K., Rapid selection using G418 of high copy number transformants of *Pichia pastoris* for high-level foreign gene expression, *Bio/Technology*, 12, 181, 1994.

91. Breyer, R. M., Strosberg, A. D., and Guillet, J.-G., Mutational analysis of ligand binding activity of β_2 adrenergic receptor expressed in *Escherichia coli*, *EMBO J.*, 9, 2679, 1990.

92. Loisel, T. P., Ansanay, H., St-Onge, S., Gay, B., Boulanger, P., Strosberg, A. D., Marullo, S., and Bouvier, M., Recovery of homogenous and functional β_2 adrenergic receptor from extracellular baculovirus particles, *Nature Biotechnol.*, 15, 1300, 1997.

93. Parker, E. M. and Ross, E. M., Truncation of the extended carboxyl-terminal domain increases the expression and regulatory activity of the avian beta-adrenergic receptor, *J. Biol. Chem.*, 266, 9987, 1991.

94. Fraser, C. M., Chung, F. Z., and Venter, J. C., Continuous high density expression of human beta 2-adrenergic receptors in a mouse cell line previously lacking beta-receptors, *J. Biol. Chem.*, 263, 14843, 1987.

95. Wess, J., Molecular basis of muscarinic acetylcholine receptor function, *Trends Pharmacol. Sci.*, 14, 308, 1993.

96. Felder, C. C., Muscarinic acetylcholine receptors: signal transduction through multiple effectors, *FASEB J.*, 9, 619, 1995.

97. Bonner, T. I., Buckley, N. J., Young, A. C., and Brann, M. R., Identification of a family of muscarinic acetylcholine receptor genes, *Science*, 237, 527, 1987.

98. Hulme, E. C., Birdsall, N. J. M., and Buckley, N. J., Muscarinic receptor subtypes, *Annu. Rev. Pharmacol. Toxicol.*, 30, 633, 1990.

99. Peralta, E. G., Ashkenazi, A., Winslow, J. W., Smith, D. H., Ramachandran, J., and Capon, D. J., Distinct primary structures, ligand-binding properties and tissue specific expression of four human muscarinic acetylcholine receptors, *EMBO J.*, 6, 3923, 1987.

100. Ashkenazi, A., Peralta, E. G., Winslow, J. W., Ramachandran, J., and Capon, D. J., Functional diversity of muscarinic receptor subtypes in cellular signal transduction and growth, *Trends Pharmacol. Sci. Suppl.*, 16, 1989.

101. Fukuda, K., Kubo, T., Akiba, I., Maeda, A., Mishina, M., and Numa, S., Molecular distinction between muscarinic acetylcholine receptor subtypes, *Nature*, 327, 623, 1987.

102. Savarese, T. M., Wang, C-D., and Fraser, C. M., Site directed mutagenesis of the rat m1 muscarinic acetylcholine receptor, *J. Biol. Chem.*, 267, 11439, 1992.

103. Lee, N. H. and Fraser, C. M., Cross -talk between m1 muscarinic acetylcholine and β_2-adrenergic receptors. cAMP and the third intracellular loop of m1 muscarinic receptors confer heterologous regulation, *J. Biol. Chem.*, 268, 7949, 1993.

104. Jakubik, J., Bacakova, L., El-Fakahany, E. E., and Tucek, S., Constitutive activity of the M1-M4 subtypes of muscarinic receptors in transfected CHO cells and of muscarinic receptors in the heart cells revealed by negative antagonists, *FEBS Lett.*, 377, 275, 1995.

105. Conklin, B. R., Brann, M. R., Buckley, N. J., Ma, A. L., Bonner, T. I., and Axelrod, J., Stimulation of arachidonic acid release and inhibition of mitogenesis by cloned genes for muscarinic receptor subtypes stably expressed in A9l cells, *Proc. Natl. Acad. Sci. U.S.A.*, 85, 8698, 1988.

106. Shapiro, R. A., Scherer, N. M., Habecker, B. A., Subers, E. M., and Nathanson, N. M., Isolation, sequence, and functional expression of the mouse M1 muscarinic acetylcholine receptor gene, *J. Biol. Chem.*, 263, 18397, 1988.

107. Parker, I. and Yao, Y., Regenerative release of calcium from functionally discrete subcellular stores by inositol trisphosphate1991, *Proc. R. Soc. Lond. B. Biol. Sci.*, 256, 269, 1991.

108. Rinken, A., Subtype-specific changes in ligand binding properties after solubilization of muscarinic receptors from baculovirus-infected Sf9 insect cell membranes, *J. Pharmacol. Exp. Ther.*, 272, 8, 1995.

109. Kukkonen, J. P., Nasman, J., Ojala, P., Oker-Blom, C., and Akerman, K. E. O., Functional properties of muscarininc receptor subtypes Hm1, Hm3 and Hm5 expressed in Sf9 cells using the baculovirus expression system, *J. Pharmacol. Exp. Ther.*, 279, 593, 1996.

110. Buckley, N. J., Bonner, T. I., Buckley, C. M., and Brann, M. R., Antagonist binding properties of five cloned muscarinic receptors expressed in CHO-K1 cells, *Mol. Pharmacol.*, 35, 469, 1989.

111. Gies, J. P., Ilien, B., and Landry, Y., Muscarinic acetylcholine receptor: thermodynamic analysis of the interaction of agonists and antagonists, *Biochim. Biophys. Acta*, 889, 103, 1986.

112. Giraldo, E., Martos, F., Gomez, A., Garcia, A., Vigano, M. A., Ladinsky, H., and Sanchez de La Cuesta, F., Characterization of muscarinic receptor subtypes in human tissues, *Life Sci.*, 43, 1507, 1988.

113. Vasudevan, S., Reiländer, H., Maul, G., and Michel, H., Expression and cell membrane localization of rat M3 muscarinic acetylcholine receptor produced in Sf9 insect cells using the baculovirus system, *FEBS Lett.*, 283, 52, 1991.

114. Vasudevan, S., Premkumar, L., Stowe, S., Gage, P. W., Reiländer, H., and Chung, S.-H., Muscarinic acetylcholine receptor produced in recombinant baculovirus infected Sf9 insect cells couples with endogenous G proteins to activate ion channels, *FEBS Lett.*, 311, 7, 1992.

115. Vasudevan, S., Hulme, E. C., Bach, M., Haase, W., Pavia, J., and Reiländer, H., Characterization of the rat m3 muscarinic acetylcholine receptor produced in insect cells infected with recombinant baculovirus, *Eur. J. Biochem.*, 227, 466, 1995.

116. Rinken, A., Formation of the functional complexes of m2 muscarinic acetylcholine receptors with GTP-binding regulatory proteins in solution, *J. Biochem.*, 120, 193, 1996.

117. Shapiro, R. A. and Nathanson, N. M., Deletion analysis of the mouse muscarinic acetylcholine receptor: effects on phosphoinositide metabolism and down-regulation, *Biochemistry*, 28, 8946, 1989.

118. Bujo, H., Nakai, J., Kubo, T., Fukuda, K., Akiba, I., Maeda, A., Mishina, M., and Numa, S., Different sensitivities to agonist of muscarinic acetylcholine receptor subtypes, *FEBS Lett.*, 40, 95, 1988.

119. Felder, C. C., Kanterman, R. Y., Ma, A. L., and Axelrod, J., A transfected m1 muscarinic acetylcholine receptor stimulates adenylate cyclase via phophatidylinositol hydrolysis, *J. Biol. Chem.*, 264, 20356, 1989.

120. Richards, M. H. and van Giersbergen, P. L. M., Human muscarinic receptors expressed in A9L and CHO cells: activation by full and partial agonists, *Br. J. Pharmacol.*, 114, 1241, 1995.

121. Lee, T., Seeman, P., Rajput, A., Farley, I. J., and Hornykiewicz, O., Receptor basis for dopaminergic supersensitivity in Parkinson's disease, *Nature*, 273, 59, 1978.

122. Seeman, P., Dopamine receptors and the dopamine hypothesis of schizophrenia, *Synapse*, 1, 133, 1987.

123. Kebabian, J. W. and Calne, D. B., Multiple receptors for dopamine, *Nature*, 227, 93, 1979.

124. Civelli, O., Bunzow, J. R., and Grandy, D. K., Molecular diversity of the dopamine receptors, *Annu. Rev. Pharmacol. Toxicol.*, 32, 281, 1993.

125. Gingrich, J. A. and Caron, M. G., Recent advances in the molecular biology of dopamine receptors, *Annu. Rev. Neurosci.*, 16, 299, 1993.

126. Reynolds, G. P., Developments in the drug treatment of schizophrenia, *Trends Pharmacol. Sci.*, 13, 116, 1992.

127. Seeman, P., Dopamine receptor sequences: therapeutic levels of neuroleptics occupy D_2 receptors, Clozapine occupies D_4. *Neuropsychopharmacology*, 7, 261, 1992.

128. Grünewald, S., Haase, W., Reiländer, H., and Michel, H., Glycosylation, palmitoylation and localization of the human D_{2S} receptor in baculovirus-infected insect cells, *Biochemistry*, 35, 15149, 1996.

129. Sander, P., Grünewald, S., Bach, M., Haase, W., Reiländer, H., and Michel, H., Heterologous expression of the human D_{2S} dopamine receptor in protease-deficient *Saccharomyces cerevisiae* strains, *Eur. J. Biochem.*, 226, 697, 1994.

130. Hildebrandt, V., Ramezani-Rad, M., Swida, U., Wrede, P., Grzesiek, S., Primke, M., and Büldt, G., Genetic transfer of the pigment bacteriorhodopsin into the eukaryote *Schizosaccharomyces pombe*, *FEBS Lett.*, 243, 137, 1989.

131. Hildebrandt, V., Polakowski, F., and Büldt, G., Purple fission yeast: overexpression and processing of the pigment bacteriorhodopsin in *Schizosaccharomyces pombe*, *Photochem. Photobiol.*, 54, 1009, 1991.

132. Hildebrandt, V., Fendler, K., Heberle, J., Hoffmann, A., Bamberg, E., and Büldt, G., Bacteriorhodopsin expressed in *Schizosaccharomyces pombe* pumps protons through the plasma membrane, *Proc. Natl. Acad. Sci. U.S.A.*, 90, 3578, 1993.

133. Ng, G. Y. K., Mouillac, B., George, S. R., Caron, M., Dennis, M., Bouvier, M., and O'Dowd, B. F., Desensitization, phosphorylation and palmitoylation of the human dopamine D_1 receptor, *Eur. J. Pharmacol.*, 267, 7, 1994.

134. Grandy, D. K., Marchionni, M. A., Makam, H., Stofko, R. E., Alfano, M., Frothingham, L., Fischer, J. B., Burke-Howie, K. J., Bunzow, J., Server, A. C., and Civelli, O., Cloning of the cDNA and gene for a human D_2 dopamine receptor, *Proc. Natl. Acad. Sci. U.S.A.*, 86, 9762, 1989.

135. Boundy, V. A., Lu, L., and Molinoff, P. B., Differential coupling of rat D2 dopamine receptor isoforms expressed in *Spodoptera frugiperda* insect cells, *JPET*, 276, 784, 1996.

136. Giros, B., Sokoloff, P., Martres, M.-P., Riou, J.-F., Emorine, L. J., and Schwartz, J.-C., Alternative splicing directs the expression of two D_2 dopamine receptor isoforms, *Nature*, 342, 1989.

137. Bunzow, J. R., Van Tol, H. H. M., Grandy, D. K., Albert, P., Salon, J., Christie, M., Machida, C. A., Neve, K. A., and Civelli, O., Cloning and expression of a rat D_2 dopamine receptor, *Nature*, 336, 1988.

138. Todd, R. D., Khurana, T. S., Sajovic, P., Stone, K.R., and O'Malley, K. L., Cloning of ligand-specific cell lines via gene transfer: identification of a D_2 dopamine recptor subtype, *Proc. Natl. Acad. Sci. U.S.A.*, 86, 10134, 1989.

139. Javitch, J. A., Kaback, J., Li, X., and Karlin, A., Expression and characterization of human dopamine D_2 receptor in baculovirus-infected insect cells, *J. Receptor Res.*, 14, 99, 1994.

140. Filtz, T. M., Artymyshyn, R. P., Guan, W., and Molinoff, P. B., Paradoxical regulation of dopamine receptors in transfected 293 cell lines, *Mol. Pharmacol.*, 44, 371, 1993.

141. Julius, D., Biology of serotonin receptors, *Annu. Rev. Neurosci.*, 14, 335, 1991.

142. Peroutka, S. J., Molecular biology of serotonin (5-HT) receptors, *Synapse*, 18, 241, 1994.

143. Grisshammer, R., Little, J., and Aharony, D., Expression of rat NK-2 (neurokinin A) receptor in *E. coli*, *Receptors and Channels*, 2, 295, 1994.

144. Aharony, D., Little, J., Powell, S., Hopkins, B., Brundell, K. R., McPheat, W. L., Gordon, R. D., Hassall, G., Hockney, R., Griffin, R., and Graham, A., Pharmacological characterization of cloned human NK-2 (neurokinin A) receptor expressed in a baculovirus/Sf-21 insect cell system, *Mol. Pharmacol.*, 44, 356, 1993.

145. Eistetter, H. R., Mills, A., and Arkinstall, S. J., Signal transduction mechanisms of recombinant bovine neurokinin-2 receptor stably expressed in baby hamster kidney cells, *J. Cell Biochem.*, 52, 84, 1993.

146. Gether, U., Marray, T., Schwartz, T. W., and Johansen, T. E., Stable expression of high affinity NK$_1$ (substance P) and NK$_2$ (neurokinin A) receptors but low affinity NK$_3$ (neurokinin B) receptors in transfected CHO cells, *FEBS Lett.*, 296, 241, 1992.

147. Turcatti, G., Ceszkowski, K., and Chollet, A., Biochemical characterization and solubilization of human NK2 receptor expressed in Chinese hamster ovary cells, *J. Receptor Res.*, 13, 639, 1993.

148. Lundstrom, K., Vargas, A., and Allet, B., Functional activity of a biotinylated human neurokinin 1 receptor fusion expressed in the *Semliki Forest* virus system, *Biochem. Biophys. Res. Commun.*, 208, 260, 1995.

149. Yamamoto, M., The molecular control mechanisms of meiosis in fission yeast, *Trends Biochem. Sci.*, 21, 18, 1996.

150. Knapp, R. J., Malatynska, E., Collins, N., Fang, L., Wang, J. Y., Hruby, V. J., Roeske, W. R., and Yamamura, H. I., Molecular biology and pharmacology of cloned opioid receptors, *FASEB J.*, 9, 516, 1995.

151. Massotte, D., Baroche, L., Simonin, F., and Yu, L., Characterization of δ, κ and μ human opioid receptors overexpressed in baculovirus-infected insect cells, *J. Biol. Chem.*, 272, 19987, 1997.

152. Guarna, M. M., Lesnicki, G. J., Tam, B. M., Robinson, J., Radziminski, C. Z., Hasenwinkle, D., Boraston, A., Jervis, E., MacGillivray, R. T. A., Turner, R. F. B., and Kilburn, D. G., On-line monitoring and control of methanol concentration in shake-flask cultures of *Pichia pastoris*, *Biotechnol. Bioeng.*, 56, 279, 1997.

153. Gellissen, G., Melber, K., Janowicz, Z. A., Dahlems, U. M., Weydemann, U., Piontek, M., Strasser, A. W. M., and Hollenberg, C. P., Heterologous protein production in yeast, *Antonie van Leeuwenhoek*, 62, 79, 1992.

154. Hirsch, H. H., Rendueles, P. S., and Wolf, D. H., *Molecular and Cell Biology of Yeasts*, Van Nostrand Reinhold, New York, 1989.

155. Cregg, J. M., Barringer, K. J., Hessler, A. Y., and Madden, K. R., *Pichia pastoris* as a host system for transformations, *Mol. Cell. Biol.*, 5, 3376, 1985.

156. Scorer, C. A., Buckholz, R. G., Clare, J. J., and Romanos, M. A., The intracellular production and secretion of HIV-1 envelope protein in the methylotrophic yeast *Pichia pastoris*, *Gene*, 136, 111, 1993.

157. Thill, G. P., Davis, G. R., Stillman, C., Holz, G., Brierley, R., Engel, M., Buckholz, R., Kennedy, J., Provow, S., Vedvick, T., and Siegel, R. S., *Proceedings of the 6th International Symposium on Genetics of Microorganisms*, Vol. II, Heslot, H., Davies, J., Florent, J., Bobichon, L., Durand, G. and Penasse, L., Eds., Societe Franscaise de Microbiologie, Paris, 1990, 477.

158. Cregg, J. M. and Madden, K. R., *Biological Research on Industrial Yeasts,* Vol. II, CRC Press, Boca Raton, FL, 1987, 1.

159. White, C. E., Hunter, M. J., Meininger, D. P., White, L. R., and Komives, E. A., Large-scale expression, purification and characterization of small fragments of thrombomodulin: the roles of the sixth domain and of methionine 388, *Protein Eng.*, 8, 1177, 1995.

160. Higgins, D. R., Busser, K., Comiskey, J., Whittier, P. S., Purcell, T. J., and Hoeffler, J. P., Small vectors for expression based on dominant drug resistance with multicopy selection, *Methods Mol. Biol.*, 103, 41, 1998.

12 Electron Crystallography of Membrane Proteins

Yoshinori Fujiyoshi

CONTENTS

12.1 INTRODUCTION

In recent years, signal transduction pathways occurring in living cells have been the focus of interest for many biologists, and the intense study of signal transduction cascades involving G protein-coupled receptors, for example, has led to the elucidation of detailed pathways and to the discovery of a variety of new proteins involved in the respective pathways. To date, nearly 2000 primary structures of G protein-coupled seven-transmembrane-segment receptors have been determined,[1] but as yet no three-dimensional (3D) structure for any of these receptors is available at an atomic level. This is in contrast to the water-soluble, heterotrimeric G proteins themselves, for which atomic structures have already been determined.[2,3] Sigler and coworkers could even analyze the structure of two different functional species of the transducin α subunit, namely the GTPγS-bound form[4] and the GDP-bound form.[5] The only structural information available so far for G protein-coupled receptors comes from electron crystallographic studies of rhodopsin by Schertler and coworkers.[6] Recently,

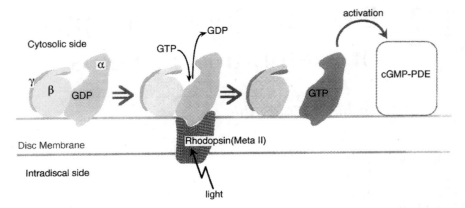

FIGURE 12.1 Schematic diagram of photosignal transduction in rod membranes based on the structures of transducin and rhodopsin, determined by X-ray and electron crystallography, respectively. Information from our direct observations of these complexes by electron cryomicroscopy has also been included.

their studies resulted in a 3D structure of rhodopsin at an effective resolution of 7.5 Å in the membrane plane and 16.5 Å normal to it. In combination with extensive sequence analysis of the 200 known sequences, this resolution was sufficient to propose a model for the arrangement of the transmembrane α-helices in G protein-coupled receptors. The signal transduction in photoreceptor cells of the retina has been thoroughly investigated and Figure 12.1 shows the schematic diagram for the transduction of a photosignal in the rod membranes. This model is based on the structures of the involved proteins determined by X-ray and electron crystallography as well as on our own observations of the complexes by electron cryomicroscopy (unpublished data). The available structural information on rhodopsin and the heterotrimeric G proteins helped considerably in the understanding of the detailed mechanism of the signal transduction pathway in photoreceptor cells. However, the current resolution of the density map for rhodopsin is not yet sufficient to build an atomic model. Therefore, we need to await high-resolution structures for rhodopsin as well as a rhodopsin–transducin complex for a more complete understanding of the signal transduction mechanism in photoreceptors.

Bacteriorhodopsin (bR) is a light-driven proton pump found in the membrane of *Halobacterium salinarium*.[7] Elucidation of its 3D structure by electron crystallography[8] revealed a seven-transmembrane helix bundle, which became the basis for almost every model of the structure of any seven-helix G protein-coupled receptor. To date, electron crystallography allowed only two atomic structures of membrane proteins to be determined, namely bR[8-10] and the light-harvesting complex II (LHC-II) from higher plants.[11] bR functions as a very efficient light-driven proton pump and absorption of a single photon by the bound retinal molecule suffices to transport a proton from the inside of the cell to the outside. Many studies have been performed on bR and have revealed the functional intermediates of its photocycle.[12] These studies have laid the basis for the proposed proton pumping mechanism of bR. However, details of the proton transport from the inside to the outside are still unknown. It is not clear how protons are efficiently guided to the opening of the bR channel, how they are carried across the channel, or how the protons are released from the other side of the protein. High-resolution electron crystallography has been recognized as one of the most promising tools to address these challenging questions, because of four fundamental advantages of this technique in regard to membrane proteins:

1. Electron crystallography uses 2D crystals of membrane proteins which are produced by reconstitution of the membrane protein into lipid bilayers at a low lipid-to-protein ratio. The lipid bilayer strongly resembles the native environment of a membrane protein and thus the native or close-to-native structure of the membrane protein can be analyzed.

2. Membrane proteins often form 2D crystals much more readily than 3D crystals and 2D crystallization also usually requires less material than 3D crystallization trials. This is a very important feature of 2D crystallization experiments because it is generally difficult to express and/or purify large amounts of membrane proteins.

3. While only the outermost molecules are easily accessible in a 3D crystal, every single molecule in a 2D crystal exposes some surface area. By using a spray technique to apply a biologically active molecule to the surface and in combination with rapid freezing, 2D crystals or tubular crystals can be used to trap and analyze structural intermediates of a protein, as shown by Unwin.[13]

4. Electron crystallography has the potential to determine the charge status of an atom because the atomic scattering factors for electrons are significantly different for a neutral and an ionized atom, especially in the low-resolution range, typically below 5 Å.

Although an almost endless number of biologically interesting membrane proteins have already been discovered and their functions described, to this day the analysis of their atomic structures remains a difficult challenge, mainly because of the difficulty to grow well-ordered 3D crystals which are needed in X-ray crystallographic structure analysis. As their name suggests, *in situ* membrane proteins are found in lipid bilayers. Accordingly, it is usually easier to crystallize this type of proteins in 2D by reconstitution of the protein into lipid bilayers than it is to produce 3D crystals of detergent-solubilized protein. Electron cryomicroscopy is coming to be an ever more important method for the structural study of membrane proteins as well as large macromolecular complexes, because it not only allows high-resolution structures to be analyzed from 2D crystals but also from helical arrays and even single particles. Because electron microscopy of 2D crystals has already produced atomic models for three proteins, namely for bR, LHC-II, and tubulin,[14] the bottleneck for structure analysis by electron crystallography is now the production of suitable 2D crystals.

Some biological molecules form helical arrays rather than planar 2D crystals. Using a tubular crystal, one type of such helical arrays, Unwin succeeded in analyzing the structure of the nicotinic acetylcholine receptor to a resolution of 9 Å.[15]

Unwin and his collaborators are currently using an electron microscope equipped with a helium-cooled specimen stage[16] to record many more high-resolution images. From these images, they hope to determine a 3D density map of the receptor at near-atomic resolution.

Crowther and coworkers studied the structure of the icosahedral core particle of the hepatitis B virus. From these single particles which were embedded in a thin vitreous film of rapidly frozen buffer solution, they determined the structure at 7.4 Å resolution and proposed a model for the chain fold of the core protein.[17] If it becomes possible in the future to determine the atomic structure of a protein from such symmetrical particles without the need of crystallization, electron cryomicroscopy will become even more powerful for structural studies, because it might be possible to form symmetrical assemblies of receptors by cross-linking them to an artificial core which has an appropriate symmetry and/or periodicity. However, in this chapter only electron crystallography of 2D crystals is discussed, because many membrane proteins can already be crystallized in 2D and because electron crystallography of 2D crystals is so far the only electron microscopic technique that has been proven to be capable of producing an atomic model for a protein.

Although of the three types of specimens used in electron cryomicroscopy 2D crystals are the most challenging to prepare, electron crystallography of 2D crystals is the most promising approach to produce an atomic model for a protein. Moreover, progress is being made in every step of electron crystallographic structure determination: experience in expression, purification, and 2D crystallization of membrane proteins is continually growing, specimen preparation techniques for high-resolution data collection are improving, electron microscopes with helium-cooled specimen stages are available,[16] and computer analysis is still developing. In this chapter recent advances in electron crystallography and its future prospects are presented, with reference mainly to bR as an example and also briefly to seven transmembrane G protein-coupled receptors.

12.2   ELECTRON CRYSTALLOGRAPHY

The main steps in the determination of a protein structure by electron crystallography are depicted in the flowchart of Figure 12.2. The computer programs that are used in the analysis of electron crystallographic data were developed mainly by Henderson and coworkers and the procedure was published in their paper.[8] A prerequisite for electron crystallography is the availability of a 2D crystal of the protein of interest, which can be produced by various techniques (see 12.4). For 2D crystallization, a large amount of protein is needed, because it is the least controllable step and, in general, many crystallization trials are required before a suitable 2D crystal can be produced (see Table 12.1). Although this step has been omitted in the flowchart shown in Figure 12.2, every electron crystallographic structure determination actually starts with the purification of the membrane protein of interest from native sources or from an expressing cell line. This presents the first inevitable obstacle for any structural study of a membrane protein. Once suitable amounts of pure protein can be produced, 2D crystallization presents the next major difficulty in the structure determination of a biological macromolecule. For high-resolution data

TABLE 12.1
Parameters for 2D Crystallization of Membrane Proteins from Detergent Solution

Protein	Protein Concentration (mg ml⁻¹)	Lipid Added (mg ml⁻¹)	Concentration of Detergent (% wt/vol)	Buffer	pH	T (°C)	Time (h)	Additive	Resolution by Electron Diffraction (Å)	Ref.
			Crystallization by Detergent Removal							
PS-I RC	1.6	0.4 DMPC	0.5 OTG	10 mM HEPES	7.0	25–37	60			Ford et al. (1990)
PS-II RC	n.r		0.03 LM	20 mM bis-Tris	6.5	n.r.	n.r.	1.5% taurine		Dekker et al. (1990)
BRC	1–2		0.1 LDAO	100 mM phosphate	5.3	23	>24			Miller & Jacob (1983)
Complex I	2–3	Soybean PC	0.05 TX-100	50 mM Tris 0.1–0.2 M NaCl	8–8.5	20	40			Leonard et al (1987)
Cytochrome reductase	1–5	1–5 Soybean PC	0.5–2 various	50 mM Tris	5.5–6	25	n.r.			Hovmöller et al. (1983); Wingfield et al. (1979)
CaATPase	2	0.3–2 dioleyl PC or egg yolk lecithin	0.03–0.2 $C_{12}E_8$	0.5 mM EDTA 20 mM MES	6.0	4	>100		4.0	Stokes & Green, (1990)
Rhodopsin	1.0	0.06–0.6 PC	0.2 C_8E_4	100 mM KCl 3 mM MgCl$_2$ 20 mM HEPES 0.1 M NaCl 10 mM MgCl$_2$ 1 mM dithiothreitol	7.0	20	120–480		5.0	Schertler et al. (1998)
AQP1	0.5	0.05–1.0 E. coli lipids	1.2 OG	20 mM Tris-HCl 0.25 mM NaCl 1 mM DTT 1 mM NaN$_3$	5.8–8.8	25–35	30	10–100 mM Mg^{2+} or Ca^{2+}	3.0	Walz et al. (1994)
Glutathione transferase	1.0	1.4 bovine liver lecithin	1.0 TX-100	10 mM potassium phosphate 0.1 mM EDTA 1 mM glutathione	n.r.	20	114 ~240		3.0	Hebert et al. (1995)

TABLE 12.1 (CONTINUED)
Parameters for 2D Crystallization of Membrane Proteins from Detergent Solution

Protein	Protein Concentration (mg ml⁻¹)	Lipid Added (mg ml⁻¹)	Concentration of Detergent (% wt/vol)	Buffer	pH	T (°C)	Time (h)	Additive	Resolution by Electron Diffraction (Å)	Ref.
				0.1 M KCl 20% glycerol						
Crystallization by the Batch Method										
bR	1–3		0.2 TX-100	80–500 mM phosphate	5.5	26	>16	DTAC	3.5	Michel et al. (1980)
LHC-II	1.5		0.11 TX-100	10 mM glycine	7.0	25–40	2–48		2.9	Wang and Kühlbrandt (1991)
			0.24 NG	40% glycerol						

Ford, R. C., Hefti, A. and Engel, A. (1990). Ordered arrays of the photosystem I reaction centre after reconstitution: projections and surface reliefs of the complex at 2 nm resolution, *EMBO J.*, 9, 3067.

Dekker, J. P., Betts, S. D., Yocum, C. F. and Boekema, E. J. (1990). Characterization by electron microscopy of isolated particles and two-dimensional crystals of the CP₄₇-D₁-D₂-cytochrome b-559 complex of photosystem II, *Biochemistry*, 29, 3220.

Miller, K. R. and Jacob, J. S. (1983). Two-dimensional crystals formed from photosynthetic reaction centers, *Cell Biol.*, 97, 1266.

Leonard, K., Haiker, H. and Weiss, H. (1987). Three-dimensional structure of NADH: ubiquinone reductase (complex I) from *Neurospora* mitochondria determined by electron microscopy of membrane crystals, *Mol. Biol.* 194, 277.

Hovmöller, S., Slaughter, M., Berriman, J., Karlsson, B., Weiss, H. and Leonard, K. (1983). Structural studies of cytochrome reductase. Improved membrane crystals of the enzyme complex and crystallization of a subcomplex, *Mol. Biol.*, 165, 401.

Wingfield, P., Arad, T., Leonard, K. and Weiss, H. (1979). Membrane crystals of ubiquinone: cytochrome c reductase from *Neurospora* mitochondria, *Nature*, 280, 696.

Stokes, D. L. and Green, N. M. (1990). Three-dimensional crystals of CaATPase from sarcoplasmic reticulum. Symmetry and molecular packing, *Biophysics*, 57, 1.

Schertler, G. F. X., Krebs, A., Villa, C. and Edwards, P. C. (1998). Characterization of an improved two-dimensional p22₁2₁ crystal from bovine rhodopsin, *Mol. Biol.*, 282, 991.

Walz, T., Smith, B. L., Zeidel, M. L., Engel, A. and Agre, P. (1994). Biologically active two-dimensional crystals of aquaporin CHIP, *Biol. Chem.*, 269, 1583.

Hebert, H., Schmidt-Krey, I. and Morgenstern, R. (1995). The projection structure of microsomal glutathione transferase, *EMBO J.*, 14, 3864.

Michel, H., Oesterhelt, D. and Henderson, R. (1980). Orthorhombic two-dimensional crystal form of purple membrane, *Proc. Natl. Acad. Sci. U.S.A.*, 77, 338.

Wang, D. N. and Kühlbrandt, W. (1991). High-resolution electron crystallography of light-harvesting chlorophyll a/b protein complex in three different media, *Mol. Biol.*, 217, 691.

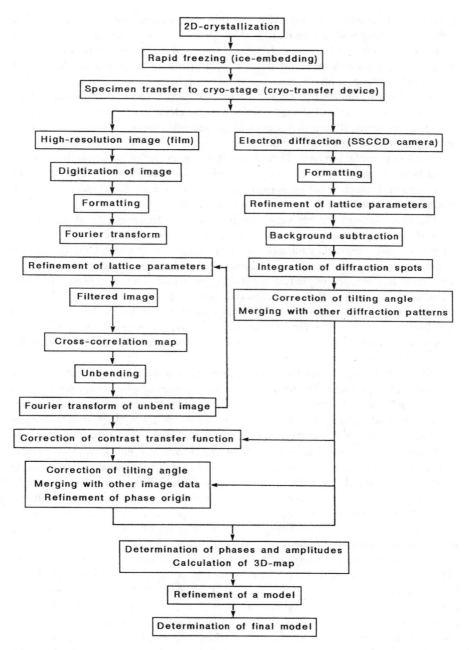

FIGURE 12.2 Flowchart showing the major steps in the procedure of structure analysis by electron crystallography. The computer programs used in the analysis of electron crystallographic data were developed mainly by Henderson and coworkers.[8]

collection, 2D crystals are embedded in a thin layer of amorphous ice and/or sugar solution, as illustrated in Figure 12.3. The specimen is then mounted in the cryo-stage of an electron microscope using a cryo-transfer device. During this transfer the specimen must be kept below the critical temperature of around 140 K at which the water undergoes a phase transition from amorphous ice to cubic ice.

Many high-resolution images and electron diffraction patterns of specimens have to be recorded at various tilt angles with minimal electron exposure. Only a single image or electron diffraction pattern can be recorded from an individual crystal, because the recording of a single image or electron diffraction pattern causes serious damage to the protein crystal. This means that the resolution of an electron micrograph is not limited by the resolution provided by the electron microscope but by the radiation damage to the biological specimen. Therefore, a full electron crystallographic data set has to be collected from many 2D crystals.

Electron micrographs are very important for the determination of the phases. Because the images are recorded on electron microscopic (EM) films which produce the best figure of merit, they must be carefully digitized with a high DQE (detectable quantum efficiency) using a high-performance densitometer.[18] Many recording systems, including charge-coupled device (CCD) cameras and imaging plates (IP) were examined for recording electron microscopic images, but so far EM films still display the best recording characteristics. EM film requires a long time to be digitized and therefore an analog system (an optical diffractometer) is essential for efficient pre-selection of good films and also to select the good areas of a film based on the Fourier transform.

Radiation damage limits the electron dose that can be used to record an image of a biological specimen. The resulting low-dose images display a very poor signal-to-noise (S/N) ratio. Therefore, it is usually impossible to see any structural features in an image that includes data to a resolution better than 3 Å. Accordingly, Fourier transformation of such a noisy image produces a Fourier pattern with an equally poor S/N ratio, but the diffraction spots can still clearly be seen. By applying a filter which only allows the information concentrated in the diffraction spots to pass while the statistically distributed noise in the Fourier transform is masked off, it is possible to calculate a noise-filtered (or averaged) image of the structure. A noise-filtered image of bR is shown in Figure 12.4 where the projections of the α-helices can clearly be seen in the bR trimer while no features of the bR structure could be seen in the original, unfiltered image.

In the computational image analysis, an important step to improve the quality of the Fourier components that can be extracted from an image is the digital unbending of the crystal. In the case of a biological specimen, 2D crystals are generally not perfect and always show a certain degree of lattice distortion. Furthermore, adsorption of the crystal to the carbon film of the grid and illumination of the crystalline array induce further distortions. Therefore, Henderson et al. developed a procedure based on correlation analysis of the crystal to correct for such lattice distortions. A successful crystal unbending can improve the achievable resolution of a given image by a factor of almost two.[8] Image Quality (IQ) values are calculated as a measure for the S/N ratio of each individual Fourier component.[19] An example is shown in Figure 12.5, which represents the image of a bR 2D crystal recorded at

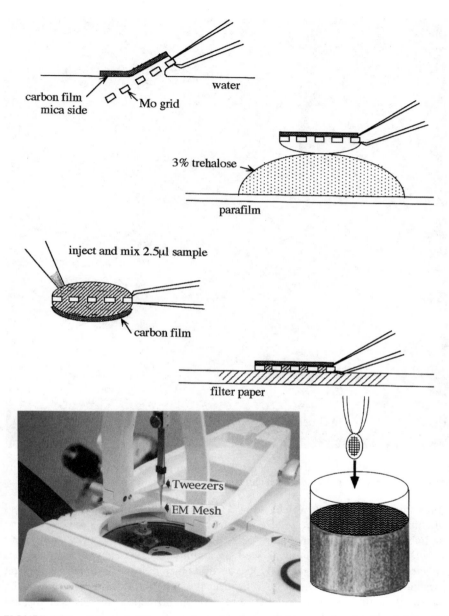

FIGURE 12.3   Schematic diagram of specimen preparation and sketch of a plunger used for rapid freezing. A carbon film is floated on an air–water interface and picked up with a molybdenum grid. bR 2D crystals, suspended in citrate buffer (pH 5.5) containing 3% (w/v) trehalose, are then applied to the carbon film on the molybdenum grid and the grid mounted on a plunger (Reichert KF 80) with tweezers. Within 5 s after blotting off the excess bR suspension with a filter paper, the specimen is plunged into liquid ethane at about −160°C. The frozen specimen is transferred into our electron cryomicroscope and observed at an acceleration voltage of 300 kV.

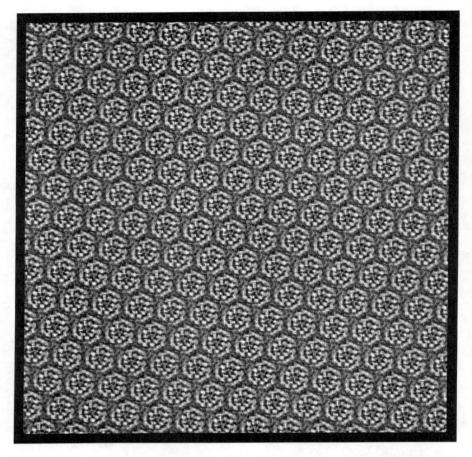

FIGURE 12.4 Noise-filtered image of bR. The white circles represent the projected density of α-helices in the bR molecules. Each unit contains three bR molecules with seven α-helices.

a tilt angle of 70°. The IQ values allow assessment of the quality of an image and only high-quality images are selected for the merging procedure. As indicated in the left line of the flowchart in Figure 12.2, images are used for Fourier analysis to extract crystallographic phase as well as amplitude data. However, amplitudes extracted from images are generally less accurate than those determined by electron diffraction experiments. Therefore, whenever possible, the amplitudes used for the final structure analysis are calculated from electron diffraction patterns, as indicated in the right line of the flowchart in Figure 12.2.

In the case of electron diffraction patterns, a slow-scan CCD camera displays better recording characteristics than EM film because of the significantly higher dynamic range of 10^5 compared to only 10^3 of EM film.[20] The resolution limit of the recorded data can be estimated from the linear region of the Wilson plot, as shown in Figure 12.6. In the case of electron diffraction patterns recorded with a CCD camera, significant information is found up to a resolution of about 2.8 Å,

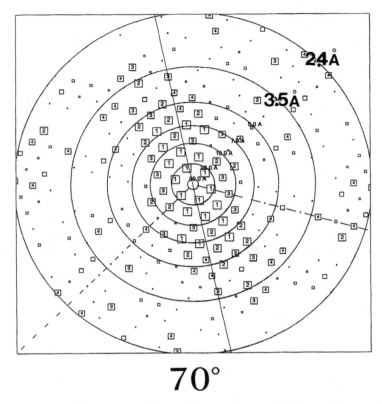

FIGURE 12.5 Fourier transform calculated from an image of bR that was recorded at a tilt angle of 70°. The solid line indicates the tilt axis and the boxed numbers represent the IQ values of the Fourier components. The IQ values, which are a measure of the signal-to-noise ratio, allow assessment of the quality of an image.

even for specimens tilted at 60°. The resolution of electron diffraction patterns recorded on EM film was usually limited to only 3.2 Å. Although the values recorded by the CCD cells correspond linearly to the electron dose, the sensitivity of each cell in a CCD array varies. However, the individual cells can be calibrated using an evenly exposed and a nonexposed frame to determine the gain of each cell. As illustrated by the electron diffraction pattern shown in Figure 12.7, using a CCD camera we were able within only a few weeks to collect a high-quality 3D amplitude data set for bR, which includes more than 200 electron diffraction patterns from specimens tilted up to 70°. Compared to the use of EM film to record electron diffraction patterns, the use of a CCD camera saves a significant amount of time because the diffraction patterns are already recorded in a digitized form. Diffraction patterns collected on EM-film have to be developed, fixed, washed, dried, and then digitized, which requires several hours for each film. More importantly, the CCD camera enables us to relate the quality of a diffraction pattern with the shape of the corresponding 2D crystal which can be seen on a low magnification image. This helps tremendously to make the collection of diffraction patterns more

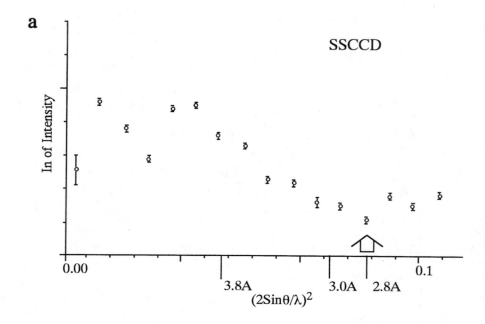

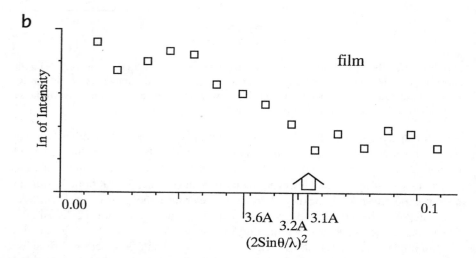

FIGURE 12.6   Collection of electron diffraction data with a slow-scan charge-coupled device (SSCCD) camera and electron microscopic (EM) film. Wilson plots of electron diffraction patterns of 60° tilted bR crystals recorded by SSCCD camera (a) and EM film (b). From the end point of the linear range of the curve, it is possible to estimate the resolution limit of the data (open arrows). The values are 2.8 Å for the CCD camera and only 3.2 Å for the EM film.

efficient because it allows selection of only those 2D crystals for recording of an electron diffraction pattern that exhibit a promising shape.

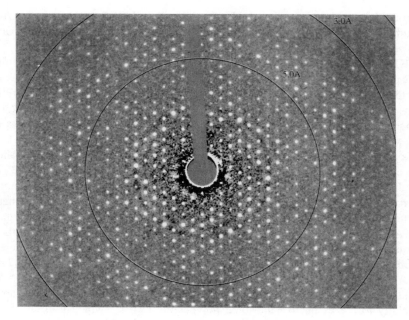

FIGURE 12.7 Electron diffraction pattern of bR recorded with a 2k × 2k SSCCD camera. The circles indicate resolutions of 5.0 and 3.0 Å. The unscattered beam heavily saturates the CCD cells in the center of an electron diffraction pattern and causes overflows to adjacent cells of the CCD array. Therefore, a beam stopper is used to remove the unscattered beam from the electron diffraction pattern to avoid these overflow effects.

Both phase and amplitude data calculated from many images and electron diffraction patterns of untilted and tilted specimens are merged to produce a consistent 3D data set. The merged data set is then used to calculate a 3D density map, into which an atomic model is built. Model building and refinement of the model, the last three boxes in the flowchart of Figure 12.2, are done in the same way as is routinely done in X-ray crystallographic structure analysis.

12.3 EXPRESSION AND PURIFICATION OF MEMBRANE PROTEINS

Recent progress in structural biology is strongly linked to methodological innovations, especially in the recombinant DNA techniques, which enable us nowadays to produce sufficient amounts of protein (including mutant proteins) for structural studies. When *E. coli* can be used as an expression system, large amounts of pure protein can readily be produced and used for crystallization trials. Unfortunately, *E. coli* is rarely useful as an expression system for membrane proteins, because the expressed protein rarely folds in the correct way. Therefore, more reliable expression systems for membrane proteins need to be developed. A good candidate for such an expression system is Sf9 insect cells which we used as an expression system for

G protein-coupled receptors, namely the human endothelin B receptor (ET_BR) including mutant receptors,[21] and the human pituitary adenylate cyclase-activating polypeptide receptor (PACAPR).[22]

ET_BR is one of two subtypes of endothelin receptors which have been identified as G protein-coupled receptors.[23] A ligand to ET_BR, endothelin-1 (ET-1), was first isolated from endothelial cells as a potent vasoconstrictor peptide.[24] Two more iso-forms (ET-2, ET-3) have been found and ET-1, -2, and -3 have now been identified in many tissues and cell types. The three ligands for ET_BR are peptides of 21 amino acids with similar primary sequences and containing two disulfide bonds. While ET_BR binds all three isopeptides equally efficiently, another subtype receptor, ET_AR, shows high affinities only for ET-1 and -2 and a two orders lower affinity for ET-3. The endothelins, which are widely distributed in many tissues and cell types, play impor-tant roles in the modulation of hemodynamics, cardiac, pulmonary, renal, and neural functions, and these work related to mitogenesis as well as neural development.[25-27]

PACAP, a ligand to PACAPR, was first identified as a novel hypothalamic hormone that increases adenylate cyclase activity in pituitary cells.[28] Molecular cloning studies revealed that PACAP, which exists in two forms, PACAP37 and PACAP38, has a highly conserved structure among rat, sheep, and humans.[29,30] The ligand has structural similarities to peptides of the secretin/glucagon peptide family, especially to the vasoactive intestinal polypeptide (VIP),[28] which is found in the central nervous system and various peripheral organs.[31] PACAP also plays important roles in a wide variety of biological systems, such as neuroprotective action against gp120-induced cell death,[32] protection of cerebellar granule neurons from apopto-sis,[33] secretion of pituitary hormones,[28] secretion of interleukin-6 from astrocytes or folliculo-stellate cells,[34] secretion of catecholamines from chromaffin cells[35] or adre-nal glands,[36] and insulin release.[37]

Detailed procedures for expression and purification of ET_BR and PACAPR can be found in publications by Doi et al.[21] and Ohtaki et al.,[22] respectively. A putative assignment for secondary structure elements for ET_BR is shown in Figure 12.8a, and Figure 12.8b summarizes constructed ET_BR mutants as well as ^{125}I-ET-1-binding activity of the receptors solubilized in 1% digitonin. While the wild-type ET_BR showed the highest binding activity of about 100 pmol/mg membrane protein, [H57-H62,G63-G65] ET_BR showed a binding activity of 60 to 70 pmol/mg membrane protein.

Significant amounts of both the ligand-bound and the ligand-free form of ET_BR were purified using two types of elution system, as shown in Figure 12.9. A mutant of the receptor [H57-H62,G63-G65] ET_BR was purified by Ni affinity column chromatography. Typical yields for each step of both ligand-free and ligand-bound forms of the [H57-H62,G63-G65] ET_BR expressed from a 3-l culture of Sf9 cells are shown in parentheses in Figure 12.9. All aiming receptors, ligand-bound ET_BR, ligand-free ET_BR, and PACAPR for large-scale preparations can be expressed in Sf9 cells grown in 3-l medium in a spinner flask equipped with a porous Teflon tube around the paddle and dissolved oxygen, as shown in Figure 12.10. Milligram amounts of these receptors were sufficiently purified by two affinity column steps, which preserved the ligand binding activity as well as the G protein coupling activity. The final yields of ligand-free and ligand-bound forms of [H57-H62,G63-G65] ET_BR

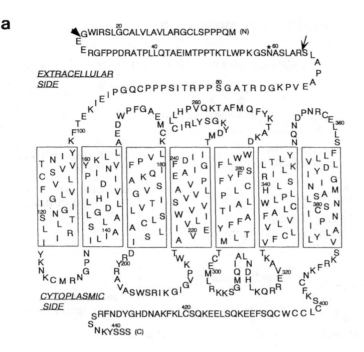

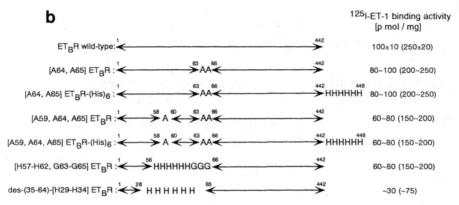

FIGURE 12.8 Secondary structure of human ET_BR (a) and constructed ET_BR mutants (b). The expression in Sf9 cells is attached in (b). The arrowhead indicates the signal peptidase cleavage site, the asterisk indicates the N-linked glycosylation site, and the arrow indicates the site of proteolysis observed in purified placental ET_BR.[21]

were about 0.3 and 0.6 mg for a 3-1 culture of Sf9 cells (Figure 12.9). PACAPR was purified more efficiently and the final yields were about 0.9 mg for the same 3-1 culture of Sf9 cells as described in a paper by Ohtaki et al. (2.4 mg could be purified from an 8-1 culture).[22] We are now using the expressed and purified receptors for 2D crystallization trials and tubular crystals have already been produced for both

Purification procedure

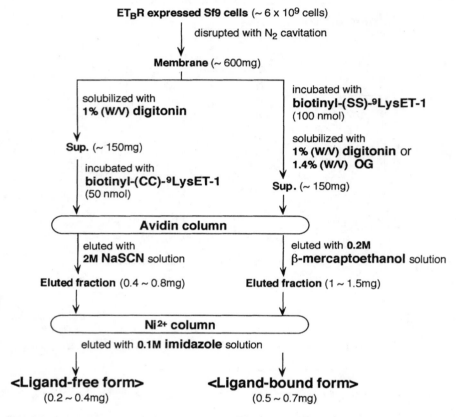

FIGURE 12.9   Purification procedure for the ligand-free form (left) and the ligand-bound form (right) of ETBR. The yield of each step is indicated in parentheses.

ligand-bound and ligand-free ET_BR, although the crystal formation is so far not very reproducible.

Protocol 12-1: Preparation of Sf9 Membranes

Buffer A: 20 mM Tris/HCl (pH 7.5), 10 mM ethylene diamine tetraacetic acid (EDTA), mixtures of protease inhibitors (1 mM phenylmethylsulfonyl fluoride [PMSF], 10 μg/ml aprotinin, 10 μg/ml leupeptin, 100 μg/ml trypsin inhibitor).

Buffer B: 20 mM Tris/HCl (pH 7.5), 5 mM EDTA, 0.2 M NaCl, protease inhibitors.

Buffer C: 20 mM Tris/HCl (pH 7.5), 5 mM EDTA, protease inhibitors.

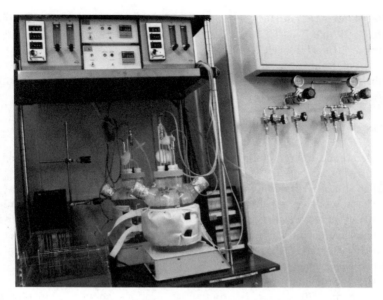

FIGURE 12.10 Sf9 cells, which are a promising candidate for a more reliable expression system for membrane proteins, are grown in a 3-l spinner flask.

All operations were carried out at 4°C.

1. Infected cells were thawed and suspended at 2 to 5×10^7 cells/ml in buffer A.
2. Cell suspensions were disrupted by nitrogen cavitation (Parr Bomb) under 1500 Pa for 30 min. When small-scale cell preparations were disrupted, the cell suspensions, at 2×10^7 cells/ml, were homogenized in a Potter-Elvehjem homogenizer with 10 to 20 strokes.
3. The cell lysate was centrifuged at $800 \times g$ for 10 min and the resulting supernatant was centrifuged at $100,000 \times g$ for 1 h.
4. The membrane pellet was washed in half the original volume of buffer B.
5. The membrane pellet was washed in half the original volume of buffer C.
6. The pellet was suspended at a final concentration of about 20 mg/ml membrane protein in buffer C, frozen in liquid nitrogen, and stored at −70°C.

Protocol 12-2: Purification of Ligand-Free ET$_B$R

Buffer D: 20 mM Tris/HCl (pH 7.5), 5 mM EDTA, 0.3 M NaCl, 1% (w/v) digitonin.

Buffer E: 20 mM Tris/HCl (pH 7.5), 5 mM EDTA, 0.3 M NaCl, 0.1% (w/v) digitonin.

Buffer F: 20 mM Tris/HCl (pH 7.5), 5 mM EDTA, 1 M NaCl, 0.1% (w/v) digitonin.

Buffer G: 20 mM Tris/HCl (pH 7.5), 0.5 M NaSCN, 0.1% (w/v) digitonin.

Buffer H: 50 mM Tris/ HCl (pH 7.5 or 9.5), 2 M NaSCN, 0.1% (w/v) digitonin.
Buffer I: 20 mM Tris/ HCl (pH 7.5), 0.1 M NaCl, 0.1% (w/v) digitonin.
Buffer J: 20 mM Tris/HCl (pH 7.5), 0.4 M NaCl, 10 mM imidazole, 0.1% (w/v) digitonin.
Buffer K: 20 mM Tris/HCl (pH 8.5), 0.4 M NaCl, 100 mM imidazole, 0.1% (w/v) digitonin.

All operations were carried out at 4°C.

1. Membranes were solubilized for 1 h at 5 mg/ml membrane protein in buffer D.
2. The solubilized membranes were centrifuged at 100,000 × g for 1 h.
3. The solubilized supernatant was incubated for 1 h with 10 mg/ml or 20 mg/ml biotinylated endothelin-1 to the ^{125}I-endothelin-1 binding activity.
4. The supernatant was incubated for 2 to 3 h with inactivated Affi-gel 10 agarose to remove the proteins that would nonspecifically bind to the matrix.
5. The supernatant was centrifuged at 1200 × g for 10 min.
6. The supernatant was mixed with the avidin-agarose and was incubated overnight with rotation. For routine purifications, about 1200 to 1400 mg membrane proteins from a 6-l culture were processed with 100 nmol biotinylated ligand, 5 ml inactivated agarose, and 10 ml avidin-agarose.
7. After centrifugation at 1200 × g for 10 min, the pelleted agarose resin was packed into a column.
8. The resin was washed with buffer E.
9. The resin was washed with buffer F.
9. The resin was washed with buffer G.
9. The ET$_B$R proteins were eluted with buffer H.
10. After immediate neutralization, the eluate was concentrated to about 1 ml in a Centriprep 30 filter (Amicon), diluted to 5–10 ml with buffer I and concentrated to about 1 ml.
11. The concentrated proteins were incubated with a nickel-chelate resin (Ni-nitrilotriacetic acid; Qiagen) for 30 min.
12. The resin was washed with buffer J.
13. The resin was eluted with buffer K.
14. The eluate was desalted and concentrated in a Centriprep 30 filter. For a typical preparation, 0.2 ml Ni-nitrilotriacetic acid resin was used for 1 mg protein eluted from the avidin resin.

Protocol 12-3: Purification of Ligand-Bound ET$_B$R

Buffer L: 20 mM Tris/HCl (pH 7.5), 5 mM EDTA, 0.3 M NaCl.
Buffer M: 20 mM Tris/HCl (pH 7.5), 5 mM EDTA, 0.3 M NaCl, 0.8% (w/v) n-octyl-β-D-glucopyranoside or 0.1% (w/v) digitonin.
Buffer N: 20 mM Tris/HCl (pH 7.5), 5 mM EDTA, 1 M NaCl, 0.8% (w/v) n-octyl-β-D-glucopyranoside or 0.1% (w/v) digitonin.

Buffer O: 20 mM Tris/HCl (pH 7.5), 0.3 M NaCl, 0.8% (w/v) n-octyl-β-D-glucopyranoside or 0.1% (w/v) digitonin.

Buffer P: 20 mM Tris/HCI (pH 7.5), 0.2 M 2-mercaptoethanol, 0.3 M NaCl, 0.8% (w/v) n-octyl-β-D-glucopyranoside or 0.1% (w/v) digitonin.

1. When prepared in digitonin, solubilization and ligand binding were performed with biotinyl-SS-endothelin-1, as described for the ligand-free ET$_B$R.

2. For preparation in n-octyl-β-D-glucopyranoside, the membranes, suspended at 5 mg/ml membrane protein in buffer L, were incubated with biotinyl-SS-endothelin-1 for 1 to 2 h at room temperature.

3. After confirming by the ligand-binding assay that all of the ligand-binding sites were occupied with the added biotinyl ligand, the mixture was cooled to 4°C, and 1.4% n-octyl-β-D-glucopyranoside was added.

4. In both n-octyl-β-D-glucopyranoside and digitonin, the mixture was centrifuged at 100,000 × g for 1 h after solubilization for 1 h.

5. The resulting supernatant was processed with inactivated Affi-gel agarose and avidin-agarose, as described above.

6. The resin was packed into a column.

7. The resin was washed with buffer M.

8. The resin was washed with buffer N.

9. The resin was washed with buffer O.

10. The ET$_B$R proteins were eluted with buffer P.

11. The eluate was concentrated and the bulk of the 2-mercaptoethanol in the eluate was removed by a Centriprep 30 filter, as described above.

12. The ligand-bound receptor was further purified with a Ni-niotrilotriacetic acid column, as described above. When required, G-50 gel filtration was performed after the nickel-resin elution.

12.4 TWO-DIMENSIONAL CRYSTALLIZATION

As already mentioned, 2D crystallization currently constitutes the bottleneck for the electron crystallographic structure analysis of membrane proteins. Although the history of 2D crystallization is still young, already various techniques have been employed for the production of 2D crystals depending on the type of protein under investigation. In this chapter only a few examples of the 2D crystallization of a membrane protein are mentioned. For further information on 2D crystallization, the reader is referred to a very good and comprehensive review by Kühlbrandt.[38]

12.4.1 FUSION OF SMALL 2D CRYSTALS

Large 2D-crystals of bR, suitable for high-resolution structure analysis, were prepared by fusion of purple membrane patches[39] which were isolated from *Halobacterium salinarium* JW 5, a retinal deficient mutant strain.[40] Functional bR molecules were reconstituted by addition of the retinal to the culture medium in the dark. This also induced the formation of nice although small bR crystals (less than 1000 Å in

	final
purple membrane	3 mg/mℓ
potassium phosphate buffer (pH 5.1)	0.1 M
OG	6 mM
DTAC	200 μM

OG (n-Octyl-β-D-glucoside)

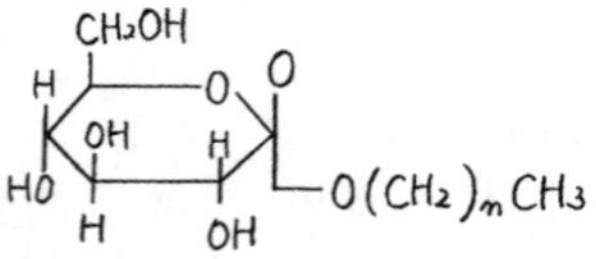

MW	292.4
CMC	25mM

DTAC (Dodecyltrimethylammonium chloride)

$$C_{12}H_{25}\text{-}N^{+}Cl^{-}$$
$$|$$
$$(CH_3)_3$$

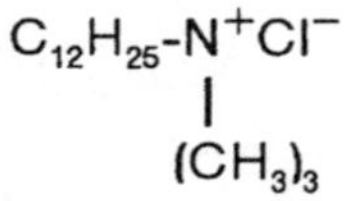

MW	263.5
CMC	16mM

FIGURE 12.11 Conditions for the fusion of bR patches to large 2D crystals and chemical structures of the two detergents used, OG and DTAC.

diameter). The small bR crystals could subsequently be fused to larger patches by incubation in a detergent mixture of the cationic dodecyltrimethylammonium chloride (DTAC) and n-octyl-β-D-glucoside (OG) (Figure 12.11). The fused crystals can grow to a size of more than 5 μm in diameter which is ideal for high-resolution electron crystallographic data collection. In the future, fusion techniques might be used more generally for other small crystals, many of which have already been analyzed by electron crystallography at low resolution.

12.4.2   2D Crystallization from Semipurified Membrane

Some 2D crystals as well as tubular crystals are formed from semipurified native membranes. In this procedure, the membrane protein of interest is not solubilized but kept within membrane patches that are purified from cells or tissues. The mere concentration of the membrane proteins in these semipurified membrane vesicles or patches led to formation of 2D crystals of Na^+, K^+-ATPase,[41] Kdp-ATPase,[42] and reaction center-light harvesting 1 complex[43] and tubular crystals of Ca^+-ATPase[44] and nicotinic acetylcholine receptor.[45] An essential condition for the formation of a crystal is that the protein is fixed in a single conformation, for example, by inhibiting the function of the cation pump of the ATPases mentioned above. This technique enables us to grow 2D crystals even of large and complex proteins which are generally fragile and easily denatured by detergent solubilization. Unfortunately, this technique can only be used for a few exceptional proteins and 2D crystals formed in this way are usually not of sufficient quality for high-resolution structure analysis. Tubular crystals that were produced by this technique, on the other hand, could be used for high-resolution structure analysis by electron cryomicroscopy.

12.4.3   2D Crystallization of Detergent-Solubilized Proteins

2D crystals suitable for high-resolution electron crystallography are normally grown from detergent-solubilized and extensively purified membrane proteins. The purified membrane protein is mixed with an appropriate detergent-solubilized lipid and the detergent removed by dialysis in order to reconstitute the protein into a lipid bilayer. This method is only suitable for membrane proteins that are reasonably stable in detergent solution, but many mild detergents are available and reconstitution is so far the most universal method to make a 2D crystal. The 2D crystals of light-harvesting complex II (LHC-II) from higher plants is only one example of 2D crystals produced by dialysis and it allowed the structure of LHC-II to be analyzed to a resolution of 3.4 Å.[11]

12.4.4   Possible Future Technique for 2D Crystallization

In some biological systems membrane proteins or functional complexes of proteins assemble spontaneously. Examples of such assemblies are the agrin-musk-rapsyn-AChR complex at the neuromuscular junction[46] and the transducisome.[47] The nature of the events taking place at and in the vicinity of the chemical synapse has been explored intensively over the last few years by a range of molecular biological, cell biological, and electrophysiological approaches. Kleinman and Reichardt explain in their review[46] the importance of the clustering of components of the neuromuscular junction as well as the agrin-dependent retrograde signaling for presynaptic differentiation of the junction in terms of the "Agrin Hypothesis". Froehner and Fallon[48] also explain the retrograde signal by rapsyn-induced clustering of AChR associated with agrin based on a paper published in *Nature*.[49] The clustering of components of the synapse is central to our understanding of the pre-synapse formations and the process of cell–cell communication in the nervous system. A sketch of the neuromuscular junction is shown in Figure 12.12. Clustering is not only found in the

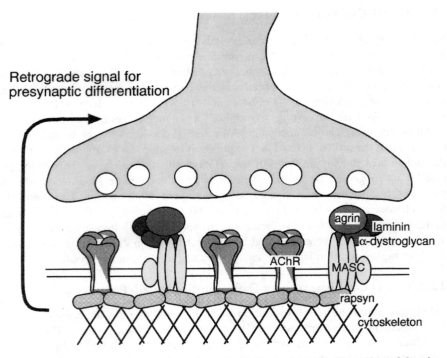

Retrograde signal for
presynaptic differentiation

agrin
laminin
α-dystroglycan
AChR
MASC
rapsyn
cytoskeleton

FIGURE 12.12 Schematic diagram of the neuromuscular junction. Clustering of the nicotinic acetylcholine receptor (AChR) and retrograde signaling for the presynaptic differentiation are marked. Most important for the clustering at the neuromuscular junction are
interactions in an agrin-MASC-musk-cytoskeleton-rapsyn-AChR complex.

agrin-musk-rapsyn-AChR complex at the neuromuscular junction but can also be
seen in some other systems, such as the transducisome, for example.[47] A new
crystallization method could be developed from understanding of these naturally
occurring clustering events. The most important point of this approach is that it is
applicable to very complicated and fragile complexes because the complexes do not
need to be solubilized for purification but are crystallized *in situ*. So far, all crystals
grown *in situ* have been of relatively poor quality but in future it might be possible
to determine the atomic structure of a protein from such poor crystals. Electron
cryomicroscopy and computational methods are still developing and the fusion
technique described in Section 12.4.1 may also help in future.

12.5 STRUCTURE OF BACTERIORHODOPSIN

12.5.1 HISTORY OF THE STRUCTURE ANALYSIS OF THE PROTON PUMP

This chapter provides an example of the use of electron crystallography in the high-
resolution structure analysis of a membrane protein. The structure of bR was determined with the highest resolution in recent electron crystallographic analysis and,
therefore, I would like to show the result of our analysis of the bR structure.

Continuing from his pioneering work with Unwin,[50] Henderson and coworkers determined an electron crystallographic density map of bR at a resolution of 3.5 Å. This 3D map enabled Henderson to build an atomic model of the protein[8] and the atomic coordinates were deposited in the Brookhaven Protein Data Bank (PDB). Later on, the atomic model was improved by crystallographic refinement of electron diffraction data corrected for diffuse scattering and also by additional phase information calculated from 30 new images of tilted specimens.[9] The atomic coordinates of the resulting new atomic model for bR have also been registered in the PDB (2BRD). Confirmation of the reliability of structure analysis by electron crystallography is, however, still thought to be usef*l and thus we set out to analyze the structure of bR at a higher resolution, independent of the work by Henderson and coworkers. While Henderson et al. could propose a model for the proton transport mechanism within the bR channel,[8] we were also interested in the mechanisms by which protons are efficiently guided to the opening of the bR channel on the cytoplasmic surface and by which the protons are released from the extracellular surface. Another goal of our work was to confirm the possibility that electron crystallography can detect the ionization state of charged amino acid residues. If the ionization conditions could actually be detected, electron crystallography would be a very promising technique to trace the proton movement along essential amino acids in the bR channel. Detection of the ionization states would tell us exactly the localization of the amino acids to which the proton is bound in the various intermediates of the bR photocycle. The discrimination of ionization conditions is also important to understand the detailed relationship between structure and function of proteins in general.

12.5.2 Structure of the Proton Pump

The experimental 3D density map calculated from 366 electron diffraction patterns and 129 electron micrographs enabled us to build an atomic model for bR including all surface loops. The resulting R-merge and phase residual were 15.5% and 26.7°, respectively. The resolution of 3.0 Å could only be achieved by using our specially developed electron microscope which is equipped with a liquid helium-cooled top-entry specimen stage.[16] The very low specimen temperature reduces the effects of radiation damage to the specimen by factors of about 2 and 10 compared to liquid nitrogen and room temperature, respectively (Figure 12.13). Using our electron cryomicroscope, we could also collect high-resolution electron crystallographic data from bR 2D crystals from specimens tilted up to an angle of 70°. In practice, it is very difficult to collect data from specimens tilted to a higher angle and thus a cone-shaped region in the reciprocal space is not sampled. In electron crystallography this is known as the missing cone problem. However, because we could collect data up to a tilt angle of 70°, the residual missing cone was reduced to less than 5% of the entire Fourier space. The reduction of the missing cone dramatically improved the quality of the 3D map, especially in the direction perpendicular to the membrane plane. This is illustrated in Figure 12.14*, which shows the region of our map around

---

* All color plates appear after p.358

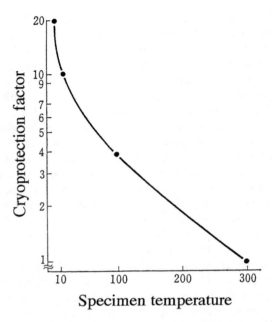

Specimen temperature

FIGURE 12.13 Plot of the cryoprotection factor as a function of the specimen temperature. The values were determined for catalase and tRNA crystals at specimen temperatures lower than 100, 20, and 8 K. The cryoprotection factor for room temperature is defined as 1. This plot demonstrates that a further reduction of the specimen temperature from liquid nitrogen to liquid helium temperature results in an improvement of the cryoprotection factor by a factor of 2 or more.

the retinal molecule (Figure 12.14a) and side chains close to the β-ionon ring of the retinal molecule (Figure 12.14b). The high quality of our 3D map enabled us not only to trace the seven transmembrane α-helices but also to interpret all surface loops in terms of the bR amino acid sequence. The loops displayed distinct structures, such as an anti-parallel β-sheet and quick turns at the end of extended α-helices (Figure 12.15).

12.5.3 LIPID MOLECULES

Our experimental 3D map revealed eight lipid molecules related to one bR molecule. In the crystallographic refinement these lipid molecules were modeled as phosphatidyl glycerophosphate monomethyl ester with dihydrophytol chains (Figure 12.16), which is a major phospholipid found in purple membranes.[51] The arrangement of the lipid molecules with respect to the bR trimer is very similar to the positions of the lipid molecules found by Grigorieff et al.[9] The bR trimer is stabilized by the interaction of the B-helix of one molecule with helices D and E of an adjacent bR molecule in the hydrophobic core of the bilayer. Interestingly, one lipid molecule between helices B and E is only present in the cytoplasmic leaflet and has no

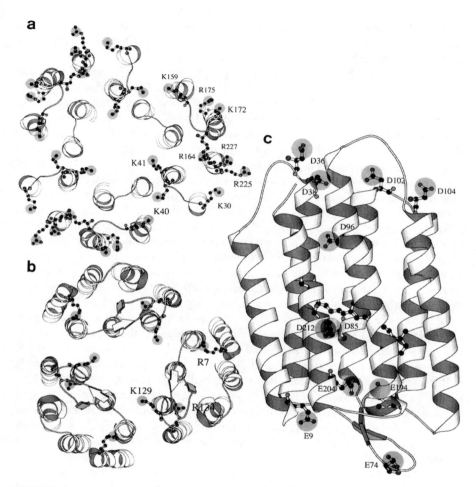

FIGURE 12.17 About 35 Å slices of the cytoplasmic (a) and the extracellular half (b) of bR together with basic amino acid residues and overall view together with acidic residues and retinal (c).

counterpart in the extracellular leaflet of the bilayer, while helices B and D have normal hydrophobic–hydrophobic contacts.

12.5.4 SURFACE STRUCTURE

The potential map analyzed from our data revealed a characteristic distribution of charged amino acids on both surfaces of the protein, as shown in Figures 12.17a and 12.17b, which highlight basic amino acid residues on the cytoplasmic and extracellular surface, respectively. Figure 12.17c depicts acidic residues in the whole molecule. The distributions of acidic and basic amino acid residues encouraged us to propose hypothetical mechanisms for efficient guidance of protons to the channel opening on the cytoplasmic surface of the protein and for efficient release of the protons from the extracellular surface.

The program GRASP was used to highlight the charge distribution on the cytoplasmic (Figures 12.18a and 12.18b) and the extracellular surfaces (Figures 12.18c and 12.18d) of the bR trimer and monomer. Only one channel entrance (indicated by an arrowhead in Figure 12.18a) can be seen in the three-fold symmetric bR trimer on the cytoplasmic surface, but this is only due to the illumination direction used to shade the structure in Figure 12.18. The entrance is located in a negatively charged plane (the red area in Figures 12.18a and b) surrounded by mountain-like protrusions which include positive charges (shown in blue in Figures 12.18a and b). Negatively charged grooves can be observed at the boundaries of the bR molecules and the lipid molecules, which are caught in these grooves, also have a negatively charged head group. Because the structure analyzed by electron crystallography represents an intact or nearly intact trimer, this characteristic charge distribution on the cytoplasmic surface of the bR trimer suggests an attractive mechanism for efficient channeling of the protons to the cytoplasmic entrance of the transmembrane channel. In our proposed mechanism, the positively charged protons are initially adsorbed on the negatively charged surfaces at the cytoplasmic side of the bR trimer and the lipid molecules. Protons adsorbed on the lipid surfaces outside of the bR trimer would migrate along the negatively charged grooves formed between the individual bR monomers into the center of the bR trimer owing to the head group of the lipid molecule inserted into helices B and E on the cytoplasmic side. Protons that have initially adsorbed to the lipid surface inside the trimer, as shown in Figure 12.19, as well as those that were guided into the center of the trimer, are trapped there until they finally enter the channel because of the positively charged mountain-like protrusions which enclose the plane with the channel entrance and faster surface migration of the protons than diversion into the bulk solution.[52]

On the extracellular surface of the bR trimer, a completely different surface structure and charge distribution is observed. The center of the trimer is dominated by positive charges and thus protons leaving the negatively charged channel (the red area in Figures 12.18c and d) tend to be released from the surface of the bR trimer.

Thus, our electron crystallographic analysis of the intact trimer structure of bR led to an attractive explanation for the mechanisms governing the efficient accumulation of protons on the cytoplasmic side and the release of protons from the extracellular side.

12.5.5 DETECTION OF THE IONIZATION STATE

While the scattering factors of a neutral and a charged oxygen atom are very similar for X-rays (Figure 12.20b), they differ greatly for electrons in the low-resolution range (Figure 12.20a). The electron scattering factor of a negatively charged oxygen atom has even a negative value at low resolution (Figure 12.20a). Therefore, the density of a negatively charged atom in the simulation of a potential map appears weak or even with a negative contrast, while a neutral oxygen atom creates a clear

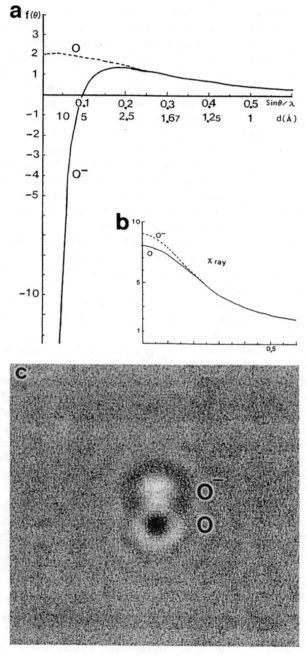

FIGURE 12.20 Scattering factors of a neutral and an ionized oxygen atom for electrons (a; broken line and continuous line, respectively) and for X-rays (b; broken line and continuous line, respectively), as well as computer simulated potential maps of a charged oxygen ion (c; upper) and a neutral oxygen atom (c; lower).

positive contrast in the potential map (Figure 12.20c). However, such effects should disappear in a potential map where the low-resolution data are omitted. This is actually helpful because it allows discrimination between a negatively charged atom and an atom that is neutral but disordered, which would both appear as weak densities in an experimental map. For example, when we recalculated a map for bR using only data from 7 to 3 Å, the initially weak density of a negatively charged atom became much stronger, while the density of a neutral but disordered atom remained weak.

In our 3D map of bR which includes the low-resolution data, we could see no density for Asp 85 (red map in Figure 12.21a); however, when we recalculated the map using only data from 7 to 3 Å, the density for this residue became visible (blue map in Figure 12.21a). On the other hand, the density for Asp 96 could be seen in maps calculated with and without the low-resolution data (Figure 12.21b; red and blue maps were calculated with and without the low-resolution data, respectively). Because it was already shown spectroscopically that Asp 85 is charged while Asp 96 is uncharged,[53] by these analyses we could confirm that electron crystallography is capable of discriminating between charged and uncharged Asp residues. This result shows that high-resolution electron crystallography, taking advantage of our refined electron cryomicroscope and an improved specimen preparation technique,[54] can actually be used to detect protons in the structure of a membrane protein.

The various intermediate states of bR during the photocycle are well characterized[55] and helped Henderson to propose a model for proton translocation across the channel in bR based on the atomic model.[8] However, to really understand the proton pumping mechanism in every detail, the high-resolution 3D structure of all the intermediates in the bR photocycle must also be determined.

If the electron crystallographic analysis of the intermediates are really achieved in practice, a movie of the proton translocation could then be created. Since electron crystallography can discriminate between the charged and uncharged states of the key amino acid residues in the transmembrane helices of bR based on the characteristic difference of the electron scattering factors, we would be able to follow the proton on its passage through the bR channel. Therefore, high-resolution electron crystallography will be able to provide us with new and exciting insights into the structure and function of bR and many other membrane proteins.

Results achieved in structure analysis by high-resolution electron crystallography, such as those described in this chapter, strongly encourage us to take on the challenge of structure analysis of receptors and/or receptor complexes involved in G protein-coupled signal transduction pathways.

ACKNOWLEDGMENTS

These studies were performed in collaboration with T. Doi, Y. Hiroaki, and I. Arimoto for ET$_B$R; T. Ohtaki, H. Onda, M. Fujino, and K. Mitsuoka for PACAPR; and K. Mitsuoka, T. Hirai, K. Murata, A. Miyazawa, A. Kidera, and Y. Kimura for bR. I would like to express sincere thanks to Drs. T. Walz and I. Schmidt-Krey for critical reading of the manuscript. Parts of this work (ET$_B$R and bR) have been supported by the Japan Society for the Promotion of Science (JSPS-RFTF96L00502). I thank

Elsevier Science B.V. for its permission to reproduce figures from papers published in the *European Journal of Biochemistry*.

REFERENCES

1. Ji, T. H., Grossmann, M., and Ji, I., G Protein-coupled receptors, *J. Biol. Chem.*, 273, 17299, 1998.
2. Lambright, D. G., Sondek, J., Bohm, A., Skiba, N. P., Hamm, H. E., and Sigler P. B., The 2.0 Å crystal structure of a heterotrimeric G protein, *Nature*, 379, 311, 1996.
3. Wall, M. A., Coleman, D. E., Lee, E., Iniguez-Lluhi, J. A., Posner, A., Gilman, A. G., and Sprang, S. R., The structure of G protein heterotrimer $G_{i\alpha1}\beta_1\gamma_2$, *Cell*, 83, 1047, 1995.
4. Noel, J. P., Hamm, H. E., and Sigler, P. B., The 2.2 Å crystal structure of transducin-α complexed with GTPγS, *Nature*, 366, 654, 1993.
5. Lambright, D. G., Noel, J. P., Hamm, H. E., and Sigler, P. B., Structural determinants for activation of the α-subunit of a heterotrimeric G protein, *Nature*, 369, 621, 1994.
6. Unger, V. M., Hargrave, P. A., Bldwin, J. M., and Schertler, F. X., Arrangement of rhodopsin transmembrane a-helices, *Nature*, 389, 203, 1997.
7. Oesterhelt, D. and Stoeckenius, W., Rhodopsin-like protein from the purple membrane of Halobacterium halobium, *Nature New Biol.*, 233, 149, 1971.
8. Henderson, R., Baldwin, J. M., Ceska, T. A., Zemlin, F., Beckmann, E., and Downing, K. H., Model for the structure of bacteriorhodopsin based on high-resolution electron cryo-microscopy, *J. Mol. Biol.*, 213, 899, 1990.
9. Grigorieff, N., Ceska, T. A., Downing, K. H., Baldwin, J. M., and R. Henderson, Electron-crystallographic refinement of the structure of bacteriorhodopsin, *J. Mol. Biol.*, 259, 393, 1996.
10. Kimura, Y., Vassylyev, D. G., Miyazawa, A., Kidera, A., Matsushima, M., Mitsuoka, K., Murata, K., Hirai, T., and Fujiyoshi, Y., Surface of bacteriorhodopsin revealed by high-resolution electron crystallograpy, *Nature*, 389, 206, 1997.
11. KŸhlbrandt, W., Wang, D. N., and Fujiyoshi, Y., Atomic model of plant light-harvesting complex by electron crystallography, Nature, 367, 614, 1994.
12. Lanyi, J. K., Proton translocation mechanism and energetics in the light-driven pump bacteriorhodopsin, *Biochim. Biophys. Acta*, 1183, 241, 1993.
13. Unwin, N., Acetylcholine receptor channel imaged in the open state, *Nature*, 373, 37, 1995.
14. Nogales, E., Wolf, S. G., and Downing, K., Structure of the $\alpha\beta$ tubulin dimer by electron crystallography, *Nature*, 391, 199, 1998.
15. Unwin, N., Nicotinic acetylcholine receptor at 9 Å resolution, *J. Mol Biol.*, 229, 1101, 1993.
16. Fujiyoshi, Y., Mizusaki, T., Morikawa, K., Yamagishi, H., Aoki, Y., Kihara, H., and Harada, Y., Development of a superfluid helium stage for high-resolution electron microscopy, *Ultramicroscopy*, 38, 241, 1991.
17. Böttcher, B., Wynne, S. A., and Crowther, R. A., Determination of the fold of the core protein of hepatitis B virus by electron cryomicroscopy, *Nature*, 386, 88, 1997.
18. Mituoka, K., Murata, K., Kimura, Y., Namba, K., and Fujiyoshi, Y., Examination of the LeafScan 45, a line-illuminating micro-densitometer, for its use in electron crystallography, *Ultramicroscopy*, 68, 109, 1997.

19. Henderson, R., Baldwin, J. M., Downing, K. H., Lepault, J., and Zemlin, F., Structure of purple membrane from Halobacterium halobium: recording, measurement and evaluation of electron micrographs at 3.5 Å resolution, *Ultramicroscopy*, 19, 147, 1986.

20. Krivanek, O. L. and Mooney, P. E., Applications of slow-scan CCD cameras in transmission electron microscopy, *Ultramicroscopy*, 49, 95, 1993.

21. Doi, T., Hiroaki, Y., Arimoto, I., Fujiyoshi, Y., Okamoto, T., Satoh, M., and Furuichi, Y., Characterization of human endothelin B receptor and mutant receptors expressed in insect cells, *Eur. J. Biochem.*, 248, 139, 1997.

22. Ohtaki, T., Ogi, K., Masuda, Y., Mitsuoka, K., Fujiyoshi, Y., Kitada, C., Sawada, H., Onda, H., and Fujino, M., Expression, purification, and reconstitution of receptor for pituitary adenylate cyclase-activating polypeptide, *J. Biol. Chem.*, 273, 15464, 1998.

23. Sakurai, T., Yanagisawa, M., and Masaki, T., Molecular characterization of endothelin receptors, *Trends Pharmacol. Sci.*, 13, 103, 1992.

24. Yanagisawa, M., Kurihara, H., Kimura, S., Tomobe, Y., Kobayashi, M., Mitsui, Y., Yazaki, Y., Goto, K., and Masaki, T., A novel potent vasoconstrictor peptide produced by vascular endothelial cells, *Nature*, 332, 411, 1988.

25. Yanagisawa, M. and Masaki, T., Molecular biology and biochemistry of the endothelins, *Trends Pharmacol. Sci.*, 10, 374, 1989.

26. Rubanyi, G. M. and Polokoff, M. A., Endothelins: molecular biology, biochemistry, pharmacology, and pathophysiology, *Pharmacol. Rev.*, 46, 325, 1994.

27. Baynash, A. G., Hosoda, K., Giaid, A., Richardson, J. A., Emoto, N., Hammer, R. E., and Yanagisawa, M., Interaction of endothelin-3 with endothelin-B receptor is essential for development of epidermal melanocytes and enteric neurons, *Cell*, 79, 1277, 1994.

28. Miyata, A., Arimura, A., Dahl, R. R., Minamino, N., Uehara, A., Jiang, L., Culler, M. D., and Coy, D. H., Isolation of a novel 38 residue-hypothalamic polypeptide which stimulates adenylate cyclase in pituitary cells, *Biochem. Biophys. Res. Commun.*, 164, 567, 1989.

29. Kimura, C., Ohkubo, S., Ogi, K., Hosoya, M., Itoh, Y., Onda, H., Miyata, A., Jiang, L., Dahl, R. R., Stibbs, H. H., Arimura, A., and Fujino, M., A novel peptide which stimulates adenylate cyclase: molecular cloning and characterization of the ovine and human cDNAs, *Biochem. Biophys. Res. Commun.*, 166, 81, 1990.

30. Ogi, K., Kimura, C., Onda, H., Arimura, A., and Fujino, M., Molecular cloning and characterization of cDNA for the precursor of rat pituitary adenylate cyclase activating polypeptide (PACAP), *Biochem. Biophys. Res. Commun.*, 173, 1271, 1990.

31. Arimura, A., Somogyvari-Vigh, A., Miyata, A., Mizuno, K., Coy, D. H., and Kitada, C., Tissue distribution of PACAP as determined by RIA: highly abundant in the rat brain and testes, *Endocrinology*, 129, 2787, 1991.

32. Arimura, A., Somogyvári-Vigh, A., Weill, C., Fiore, R. C., Tatsuno, I., Bay, V., and Brenneman, D. E., PACAP functions as a neurotrophic factor, *Ann. N.Y. Acad. Sci.*, 739, 228, 1994.

33. Cavallaro, S., Copani, A., D'Agata, V., Musco, S., Petralia, S., Ventra, C., Stivara, F., Travali, S., and Canonico, P. L., Pituitary adenylate cyclase activating polypeptide prevents apoptosis in cultured cerebellar granule neurons, *Mol. Pharmacol.*, 50, 60, 1996.

34. Tatsuno, I., Somogyvari-Vigh, A., Mizuno, K., Gottschall, P. E., Hidaka, H., and Arimura, A., Neuropeptide regulation of interleukin-6 production from the pituitary: stimulation by pituitary adenylate cyclase activating polypeptide and calcitonin gene-related peptide, *Endocrinology*, 129, 1797, 1991.

35. Watanabe, T., Masuo, Y., Matsumoto, H., Suzuki, N., Ohtaki, T., Masuda, Y., Kitada, C., Tsuda, M., and Fujino, M., Pituitary adenylate cyclase activating polypeptide provokes cultured rat chromaffin cells to secrete adrenaline, *Biochem. Biophys. Res. Commun.*, 182, 403, 1992.

36. Watanabe, T., Shimamoto, N., Takahashi, A., and Fujino, M., PACAP stimulates catecholamine release from adrenal medulla: a novel noncholinergic secretagogue, *Am. J. Physiol.*, 269, E903, 1995.

37. Yada, T., Sakurada, M., Ihida, K., Nakata, M., Murata, F., Arimura, A., and Kikuchi, M., Pituitary adenylate cyclase activating polypeptide is an extraordinarily potent intra-pancreatic regulator of insulin secretion from islet beta-cells, *J. Biol. Chem.*, 269, 1290, 1994.

38. Kühlbrandt, W., Two-dimensional crystallization of membrane proteins, *Q. Rev. Biophys.*, 25, 1, 1992.

39. Baldwin, J. and Henderson, R., Measurement and evaluation of electron diffraction patterns from two-dimensional crystals, *Ultramicroscopy*, 14, 319, 1984.

40. Oesterhelt, D. and Stoeckenius, W., Isolation of the cell membrane of Halobacterium halobium and its fractionation into red and purple membrane, *Methods Enzymol.*, 31, 667, 1974.

41. Tahara, Y., Ohnishi, S., Fujiyoshi, Y., Kimura, Y., and Hayashi, Y., A pH induced two-dimensional crystal of membrane-bound Na^+, K^+-ATPase of dog kidney, *FEBS Lett.*, 320, 17, 1993.

42. Iwane, A. H., Ikeda, I., Kimura, Y., Fujiyoshi, Y., Altendorf, K., and Epstein, W., Two-dimensional crystal of the Kdp-ATPase of Escherichia coli, *FEBS Lett.*, 396, 172, 1996.

43. Yamasaki, I. I., Odahara, T., Mitsuoka, K., Fujiyoshi, Y., and Murata, K., Projection map of the reaction center-light harvesting 1 complex from Rhodopseudomonas viridis at 10 Å resolution, *FEBS Lett.*, 425, 505, 1998.

44. Dux, L. and Martonosi, A., Two-dimensional arrays of proteins in sarcoplasmic reticulum and purified Ca^{2+}-ATPase vesicles treatd with vanadate, *J. Biol. Chem.*, 258, 2599, 1983.

45. Brisson, A. and Unwin, P. N. T., Tubular crystals of acetylcholine receptor, *J. Cell Biol.*, 99, 1202, 1984.

46. Kleiman, R. J. and Reichardt, L. F., Testing the agrin hypothesis, *Cell*, 85, 461, 1996.

47. Tsunoda, S., Sierralta, J., Sun, Y., Bodner, R., Suzuki, E., Becker, A., Socolich, M., and Zuker, C. S., A multivalent PDZ-domain protein assembles signalling complexes in a G protein-coupled cascade, *Nature*, 388, 243, 1997.

48. Froehner, S. C. and Fallon, J. R., Synapses take the rep, *Nature*, 377, 195, 1995.

49. Gautam, M., Noakes, P. G., Mudd, J., Nichol, M., Chu, G. C., Sanes, J. R., and Merlie, J. P., Failure of postsynaptic specialization to develop at neuromuscular junctions of rapsyn-deficient mice, *Nature*, 377, 232, 1995.

50. Henderson, R. and Unwin, P. N. T., Three-dimensional model of purple membrane obtained by electron microscopy, *Nature*, 257, 28, 1975.

51. Kates, M., Moldoveanu, N., and Stewart, L. C., On the revised structure of the major phospholipid of Halobacterium salinarium, *Biochim. Biophys. Acta*, 1169, 46, 1993.

52. Heberle, J., Riesle, J., Thiedemann, G., Oesterhelt, D., and Dencher, N. A., Proton migration along the membrane surface and retarded surface to bulk transfer, *Nature*, 370, 379, 1994.

53. Braiman, M. S., Mogi, T., Marti, T., Stern, L. J., Khorana, H. G., and Rothschild, K. J., Vibrational spectroscopy of bacteriorhodopsin mutants: light-driven proton transport involves protonation changes of aspartic acid residues 85, 96, and 212, *Biochemistry*, 27, 8516, 1988.

54. Fujiyoshi, Y., The structural study of membrane proteins by electron crystallography, *Adv. Biophys.*, 35, 25, 1998.
55. Lozier, R. H., Bogomolni, R. A., and Stoeckenius, W., Bacteriorhodopsin: a light-driven proton pump in Halobacterium halobium, *Biophys. J.*, 15, 955, 1975.

13 X-Ray Crystallography of Membrane Proteins from Detergent Solution: The Structures of a Ligand-Gated Bacterial Membrane Receptor in Two Conformations; a Gallery at Atomic Resolution

Kaspar P. Locher, Bernard Rees, Dino Moras, and Jung P. Rosenbusch

CONTENTS

13.I INTRODUCTION

The events of signal transduction in ligand-gated channel proteins at an atomic level are of great interest for the understanding of the mechanism of signaling, both in

0-8493-3384-9/00/$0.00+$.50

terms of transmission and coordination. Due to the difficulty of growing three-dimensional crystals of eukaryotic membrane proteins, we have selected a bacterial receptor, the FhuA protein, as a model system that reveals similarities to porins with regard to stability and structure. The protein spans the outer membrane of the Gram-negative bacterium *Escherichia coli* and is essential for iron transport. Although quite different in many respects, the FhuA protein appears as an excellent conceptual model for receptor proteins also in eukaryotic systems.

The first step of iron import into Gram-negative organisms is the transport of ferric ions, in complex with siderophores, across the bacterial outer membrane.[1] For all siderophores, and for vitamin B12, this translocation step is mediated by a substrate-specific, ligand-gated channel protein.[2] The state of the channel, open or closed, is controlled by a protein complex anchored in the plasma membrane and serving as an energy transduction module for the siderophore transport proteins in the outer membrane.[3–7] This module comprises at least three gene products, of which the TonB protein appears the most prominent component. The system has therefore been named TonB complex.[8–10] These transport proteins in the outer membrane are collectively dubbed TonB-dependent receptors. They contain a conserved sequence motif near their N-termini, the TonB box, which is essential for their interaction with the TonB-complex.[11] The translocation mechanism across the outer membrane involves two processes: binding of the iron-siderophore complex to the extracellular face of the receptor with concomitant signal transmission to the periplasmic face, and TonB complex-mediated, reversible channel opening across the outer membrane receptor.[12]

For our studies, we have chosen the ferric hydroxamate uptake protein (encoded by the *fhuA* gene) which catalyzes ferrichrome transport into *Escherichia coli* cells.[13] As with most outer membrane proteins, the FhuA protein serves also as receptor for bacterial viruses and toxins, among which bacteriophage T5 is unique in not depending on the TonB complex for infection. The FhuA protein (714 amino acid residues, 77.8 kDa) is monomeric in detergent solution, but ligand binding studies have revealed a convex Scatchard plot and a Hill coefficient of 2.4.[14] This positive cooperativity suggests the presence of oligomeric species in detergent solution (with a minimum of three subunits on average), and chemical cross-linking studies have indeed revealed the existence of dimeric, trimeric, and higher species. This finding is interesting, as it may illustrate that the subunit association in detergent solutions need not reflect the oligomeric state of a protein in the lipid bilayer.

The architecture of the FhuA protein has been predicted to be that of a β-barrel, analogous to the porins.[15–17] On the basis of topological studies, 32 membrane-spanning β-strands were postulated, connected by long extracellular loops and short periplasmic turns.[18] The seemingly largest loop was proposed to fold backwards into the barrel lumen and assume the function of the gating device,[19] again in analogy to the porins, in which a constriction loop folds into the channel.[15,17] Evidence supporting this model came from a mutant protein, deleted in this large loop, and forming an open channel with a single channel conductance similar to that observed upon addition of phage T5 to isolated wild-type FhuA protein.[19–21] As will be seen below, the X-ray analysis reveals a rather divergent result, with the lumen of the β-barrel obstructed by a substantial N-terminal domain that snugly

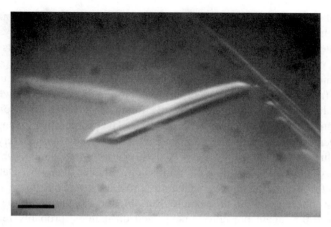

FIGURE 13.1   Crystal of FhuA protein grown in sitting drops. The bar corresponds to 0.2 mm.

fits into the barrel interior, thus controlling its accessibility. It seems only fair to note at this time that predicting the structure correctly was not possible without a structural homolog.

13.2   CRYSTALLIZATION AND STRUCTURE DETERMINATION BY X-RAY CRYSTALLOGRAPHY

Full-length, wild-type FhuA protein was expressed in an *E. coli* strain from the plasmid pHK763, which bears the *fhua* gene under the control of a T7 promoter.[22] Extraction of isolated outer membranes was performed using *n*-octyl-β-D-glucopyranoside as detergent, and the protein was purified using octyl-polyoxyethylene (octyl-POE) as detergent in standard chromatographic techniques at room temperature, as described previously.[14] FhuA protein was concentrated either using an Amicon cell and PM30 membranes, or by Centrikon 50 concentrators. Prior to crystallization, the detergent was exchanged by dialysis against a buffer of 15 mM NaP$_i$, 2 mM NaN$_3$, 0.1 M NaCl, and 0.6% (w/v) *n*-octyl-2-hydroxyethyl-sulfoxide (CHESO) at a final pH of 6.2. The protein was subsequently diluted to 12.3 mg/ml and mixed with reservoir solution (33% PEG 2000, 0.45 M NaCl, 0.15 M NaP$_i$, pH 6.2, 0.5% CHESO) in a volume ratio of 1:2, with a total drop volume of 12 to 24 µl. Crystals up to $0.2 \times 0.2 \times 2$ mm grew in sitting drops within 1 week (Figure 13.1). The FhuA protein-ferrichrome complex was crystallized in the presence of excess siderophore (final concentration of 1 mM) which was added prior to crystallization.

As described previously, the crystals obtained belonged to the monoclinic space group C2 (one monomer in the asymmetric unit), and revealed surprising stability in the X-ray beam, allowing room temperature data sets to be collected at the synchrotron.[23] Despite intensive efforts, isomorphous heavy atom derivatives could not be obtained. We therefore decided to tackle the phase problem by incorporating quantitatively selenomethionine into the bacterial protein *in vivo*, followed by multiwavelength anomalous diffraction (MAD).[24] Due to the low symmetry of our crystals, collecting data at low temperature (100 K) was essential. Freezing FhuA crystals, however, caused a large and anisotropic increase of the crystal mosaicity, precluding a precise determination of the intensities of the diffracted X-rays. Attempts to optimize the cooling procedure included different freezing techniques (direct exposure to N_2 stream, immersion in ethane and liquid N_2, annealing cycles), and the use of cryoprotectants (methylpentanediol, glycerol, low-molecular-weight polyethylene glycol, sucrose, trehalose, glucose). Alone or in combination, these protocols failed to improve significantly the freezing of the crystals. The problem was eventually overcome by perseverance: nearly 100 selenomethionine FhuA protein crystals were frozen by direct immersion into liquid N_2 (without cryoprotectants), which yielded a total of two apo-FhuA crystals with a mosaicity smaller than 1.5°. Data collected from the better crystal at three wavelengths provided excellent phase information allowing the solution of a 3.0 Å structure. Refining the structure with the MAD data collected at 100 K again presented unexpected difficulties, probably reflecting the damage that freezing had caused to the crystals. However, the data sets collected earlier at room temperature now allowed refining the structures of the apo-protein and of the complex with ferrichrome to resolutions of 2.75 and 2.6 Å, respectively.[23]

In retrospect, the reason for the difficulties experienced in attempting to freeze FhuA crystals are likely to be due to the packing of the protein. Although crystal contacts between FhuA monomers involve mainly their membrane-embedded hydrophobic faces (Figure 13.2), no stacking of membrane-like sheets of protein is present. The crystals are built from columns consisting of FhuA protein monomers, with their hydrophobic zones in lateral contact, and oriented alternatively up and downward, analogous to the packing of the trigonal OmpF-crystals.[17] These columns are kept in position by interactions in narrowly confined areas between surface-exposed loops and periplasmic turns. Upon freezing, columns may shift and rotate with respect to each other. Such an interpretation tallies with the observation that the frozen crystals had a slightly altered unit cell.

13.3 OVERALL ARCHITECTURE

The FhuA protein consists of two major domains. The long C-terminal segment (residues 160 to 714) forms a hollow barrel consisting of 22 β-strands forming a channel with a bean-shaped cross section (35 × 25 Å). The domain comprising residues 18 to 159 is a transmembrane plug that is positioned in and obstructs the channel lumen. Its secondary structure shows a central, four-stranded β-sheet that is slightly tilted relative to the membrane plane, and two additional β-strands as well as five short α-helices with variable orientations (Figure 13.3*). The N-terminal segment (residues 1 to 18) contains the TonB box and is of functional relevance. In the crystal structure, however, it is disordered.

The plug domain snugly fits the barrel lumen, yet cavities are present between the barrel and the plug (see Figures 13.4A, 13.5A, 13.6B), of which none connects the extracellular with the periplasmic space. On the extracellular side, a large protuberance of the barrel is formed by four loops, which bulge into the aqueous

---

* All color plates appear after p.358

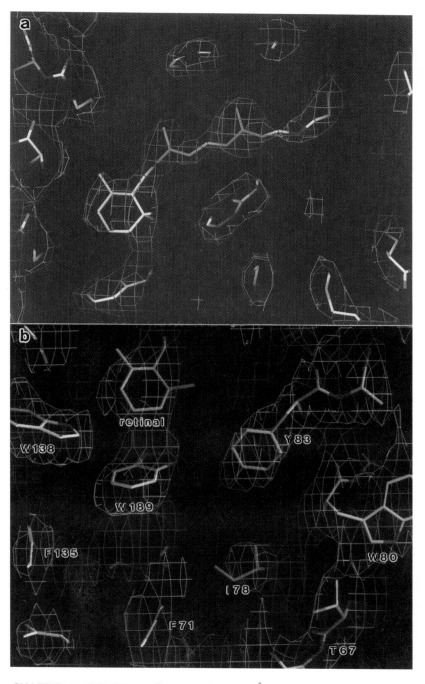

CHAPTER 12 FIGURE 14. Extracts of our 3.0 Å density map and the atomic model of bR fitted into the map. (a) Density for the retinal which is located almost in the center of the bR molecule. (b) The side-chain densities close to the β-ionon ring of the retinal molecule.

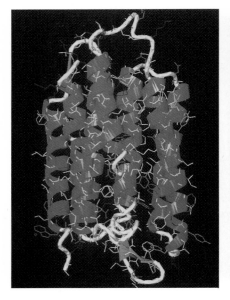

CHAPTER 12 FIGURE 15. Overall view of a bR monomer depicting the seven membrane-spanning α-helices (red ribbons), the retinal in the center of the molecule, and the loop regions including an anti-parallel β-sheet structure (green arrows) between helices B and C on the extracellular side.

CHAPTER 12 FIGURE 16. Density map of a lipid molecule, which was modeled as a phosphatidyl glycerophosphate monomethyl ester with dihydrophytol chains.

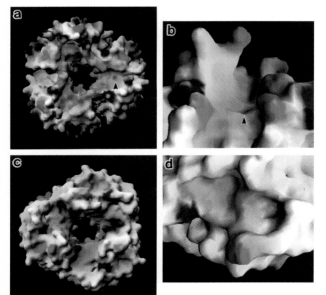

CHAPTER 12 FIGURE 18. Surface-rendered views of the bR trimer (a) and enlarged monomer viewed from a different direction (b) at the cytoplasmic side, and also the views of the bR trimer (c) and enlarged monomer from similar direction with the trimer (d) at extracellular side. Blue and red areas indicate positive and negative surface charges, respectively. Due to the unidirectional illumination used for the calculation of the surface representation with the program GRASP, only one opening of the three proton channels at cytoplasmic side in trimer can clearly be seen (marked by arrowhead).

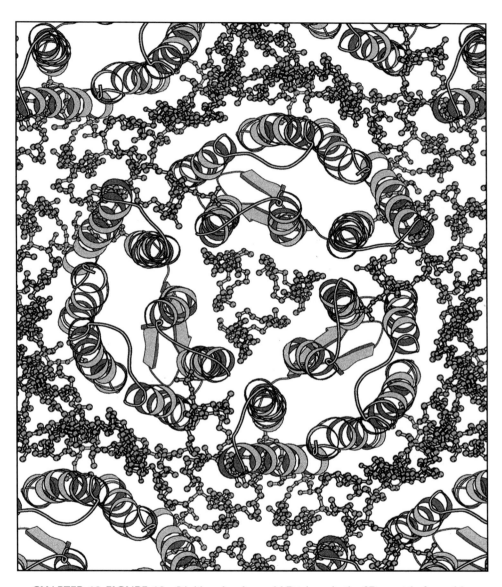

CHAPTER 12 FIGURE 19. Lipid molecules and bR trimer in the 2D crystals formed by bR. The bR molecules are shown in ribbon representation (green) while the lipid molecules are represented by ball-and-stick models (red).

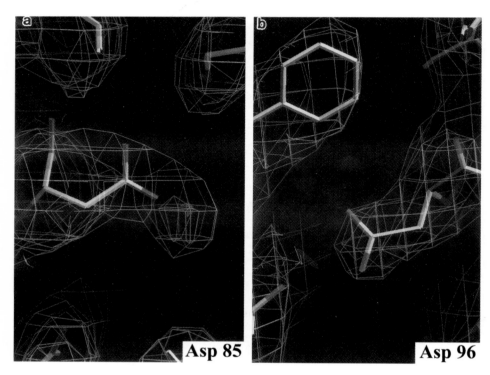

Asp 85

Asp 96

CHAPTER 12 FIGURE 21. Potential maps of bR around Asp 85 (a) and Asp 96 (b). The map shown in red contour lines was calculated using data from 54 Å to 3.0 Å while the map shown in blue contour lines was calculated using only data from 7.0 Å to 3.0 Å, omitting the low-resolution data which are strongly affected by the charge of the side chains.

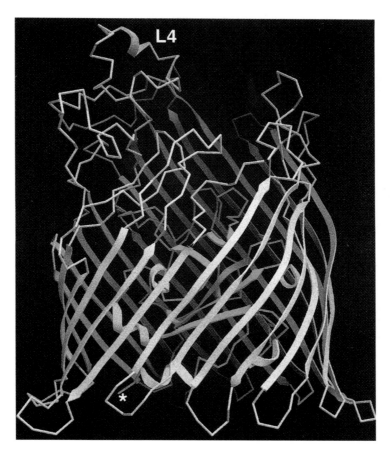

CHAPTER 13 FIGURE 3. Architecture of the FhuA protein in its unliganded state. The barrel and loops are shown in white, the plug in color (yellow: α-helices, green: β-strands, red: non-periodic segments). The view is side-on and shows the lipid-exposed barrel, which extends by loops into the extracellular space (top), and the short periplasmic turns (bottom). Loop L4, previously called "gating loop," is indicated. The asterisk indicates the first residue (S20) that is ordered in the ligand-free state. This and the following figures were produced using the program dino.

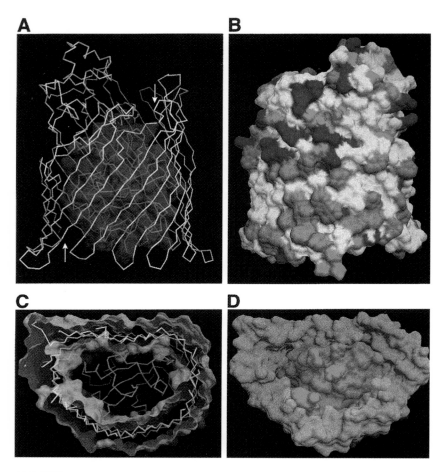

CHAPTER 13 FIGURE 4. Side views (panels A and B) and view from the bottom (panels C and D) of the FhuA protein. A, C_α-trace of the ferrichrome-bound state of the FhuA protein. The view is similar to that in Fig. 3. Barrel and plug are colored yellow and red, respectively. The shading corresponds to the molecular surface calculated for the plug. Ferrichrome in its binding site is shown in green. The arrows show entrances or exits of cavities between the plug and the barrel. The cavity at the periplasmic face of the FhuA protein is ferrichrome-induced (compare with Fig. 3). B, The total protein surface is shown and colored according to the nature of the amino acid residues (green, aromatic; red, negatively charged; blue, positively charged; white, hydrophobic). The protein has been rotated by 180° around the vertical axis with respect to panel A. Notice the two aromatic belts at the water-lipid interface and the aromatic patch in the membrane-embedded region. C, The surface of the protein has been calculated for the barrel only. The backbone of the protein has been colored red (plug) and yellow (barrel). D, Barrel and plug surfaces of the liganded state of the FhuA protein have been calculated independently, showing the firm fit of the plug in the channel.

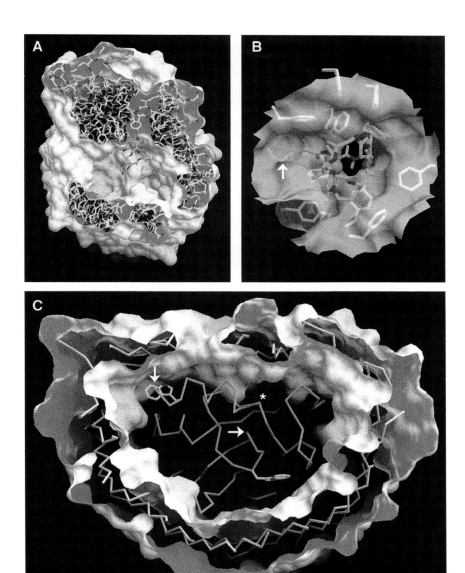

CHAPTER 13 FIGURE 5. Ligand binding and the results of transmembrane signaling. A, Ferrichrome (green, with iron as a red sphere) is shown in the binding site. The view is from the extracellular side of the protein into the channel. The protuberance, formed by extracellular loops, is visible at the top left of the picture. B, Close-up of the binding site: The protein surface is shown in white, the residues involved in forming the binding pocket are depicted in ball-and-stick representation (white arrow points towards R81, in pink). C, Conformational transition at the opposite channel entrance, facing the periplasmic space. The surface corresponds to that of the barrel, calculated in the absence of the plug, and the backbone of the plug in the unliganded (green) and liganded state (red) are shown. The asterisk indicates the Cα atom of residue 30. The arrows point towards the first ordered residues in the two conformational states. W22 undergoes the largest movement, the Cα atoms being 17 Å apart in the two states.

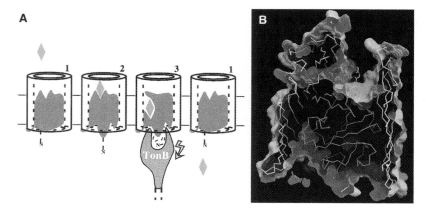

CHAPTER 13 FIGURE 6. Working hypothesis of the mechanism of ferrichrome translocation. FhuA protein is schematized by a hollow tin, with the plug drawn as its green content.
Ferrichrome is indicated by a red diamond. The view is from the side and corresponding to
the view in panel B as well as in Fig. 3. State 1 corresponds to the unliganded protein, state
2 represents the ferrichrome-bound FhuA protein. These two structures are based on the two
experimentally determined states. They provide the structures on which the hypotheses of
transmembrane signaling and ligand translocation are based. The translocation of ferrichrome (state 3) is drawn on the "left side" of the plug because (i) conspicuously few plug-
barrel interactions occur in this region, (ii) the region contains mobile/soft protein elements
that are responsible for transmembrane signaling, (iii) the region contains indications of cavities which may be involved directly, or indirectly (by compressibility) in translocation, (iv)
upon ferrichrome binding, the X-ray structure reveals a cavity which is open to the periplasmic channel entrance (Fig. 4A, 5C, 6B). B, Side view of the protein with the calculated surface of the entire protein. The backbone of plug (red) and barrel (yellow), as well as the ferrichrome molecule (green) are shown.

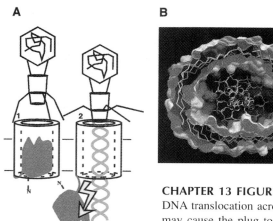

CHAPTER 13 FIGURE 7. Working hypothesis of phage
DNA translocation across the FhuA channel. A, Phage T5
may cause the plug to be displaced irreversibly from the
channel lumen. Panel B shows the barrel of the FhuA pro-
tein with its calculated surface, and an oligonucleotide (B-form DNA) modeled into the
channel lumen to illustrate the possibility of DNA being translocated through the FhuA
channel, with the plug displaced. The view is from the periplasmic side, and the channel is
cut at about one third of its total height, i.e., within the membrane-embedded part.

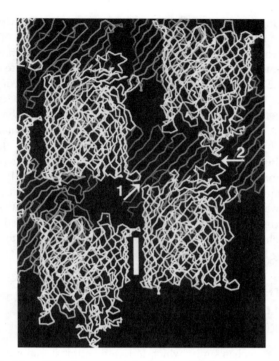

FIGURE 13.2   Packing of FhuA monomers in the crystal lattice. Two types of contacts, major and minor, are visible. The arrows indicate the minor contacts (loop–loop and loop–turn interactions), with a salt bridge between K271 and D369 (arrow 1), and van der Waals contacts between residues 330 and 334 that belong to the surface-exposed loop L4 in the two symmetry-related molecules (arrow 2). The major contact zone is between the hydrophobic zones of the barrels (white bar) which are located within membrane boundaries. Several ordered detergent molecules are present in this area (not shown). This figure was produced using the program O.[32]

space and appear to restrict the access of the channel entrance from the outside. On the periplasmic side of the membrane, short turns bulge but little into the aqueous phase. The area in contact with the membrane lipids consists of short, aliphatic side-chain residues and of aromatic patches (Figure 13.4B). The presence of the latter may mediate protein–protein interactions in the membrane *in vivo* and in detergent solutions *in vitro*, and could cause cooperative ligand-binding behavior (see above). The boundaries of the membrane-embedded surface of the FhuA protein is marked by two conspicuous belts of aromatic residues at the water–lipid interface, as it has been observed also in the porins. Charged residues occur almost exclusively on the water-exposed surfaces of the protein (Figure 13.4B).

13.4   SUBSTRATE BINDING AND TRANSMEMBRANE SIGNALING

Ferrichrome binds to a cavity that is located on top of the plug just above the level of the outer surface of the lipid bilayer (Figure 13.5A). The binding pocket is padded by aromatic residues and an arginyl residue, R81 (Figure 13.5B), which is conserved in ferrichrome-translocating TonB-dependent receptor proteins. The siderophore consists of a "head" containing the ferric ion coordinated by three hydroxamate groups.[25] This moiety is buried deeply in the binding pocket, whereas the "tail" part of the ferrichrome molecule remains accessible to solvent. Six hydrogen bonds and extensive van der Waals contacts with the aromatic residues hold ferrichrome in its binding site. Ligation causes minor conformational changes in the binding pocket on the FhuA protein surface. The local change of residue R81, which shifts towards the siderophore by ~1.5 Å, is propagated asymetrically across the plug domain (represented in Figures 13.3 and 13.4A on the left side), and markedly amplified towards the periplasmic channel mouth. As a result, the first α-helix (residues 22 to 30) unfolds completely, with the corresponding peptide segment swinging to the opposite barrel wall. The Cα atom of the tryptophanyl residue W22 is now 17 Å apart from its original site (Figure 13.5C). The drastically altered plug surface at the periplasmic channel side is likely to allow the TonB protein to sense the effect of ligand binding to the extracellular face of the FhuA protein.

13.5 HYPOTHESES CONCERNING THE TRANSLOCATION OF FERRICHROME AND BACTERIOPHAGE T5 DNA

At a first glance, a mechanism by which ferrichrome translocation is facilitated by a cyclic expulsion and reinsertion of the plug appears attractive. It would represent a structural correlate of a ball-and-chain mechanism, as it was suggested for the inactivation of sodium channels.[26] In view of the extensive and firm interactions of the plug with the interior barrel wall, with over 60 hydrogen bonds and 9 salt bridges, such a notion appears implausible, however, on energetic grounds. Moreover, a complete plug displacement seems incommensurate with the dimensions of ferrichrome and may open a thoroughfare also for toxic substances. Our current working model of ferrichrome transport (Figure 13.6A) therefore calls for the plug to remain in the channel, with conformational changes forming a locally constrained pathway for the translocation of ferrichrome. Although such a transition, which is likely to occur as a consequence of TonB binding to the altered periplasmic surface of the FhuA protein, appears plausible in view of the cavities present between plug and barrel (Figure 13.4A and left in panel B of Figure 13.6), a detailed pathway cannot be predicted.

Provided phage DNA transfer occurs through the FhuA protein, a channel large enough to accommodate double helical B-form DNA[27] could only be provided if the plug is displaced completely from the barrel lumen (Figure 13.7B). The forces involved in phage DNA ejection from the capsid may well be sufficient to overcome those involved in plug–barrel interactions. An ejection of the plug by the phage is consistent with the conversion of FhuA protein to a permanently open channel upon reaction with phage T5 *in vitro*.[21]

13.6   EXAMINING HYPOTHESES PAST AND PRESENT

The scaffold of the FhuA protein is a β-barrel that forms a wide channel, a motif typical for outer membrane proteins. Unusual and unexpected in the FhuA protein is the presence of a plug, consisting of α/β structures and located in the channel lumen. This observation thwarts the notion that only single domains, either α-helical bundles or β-barrels, may be present in membrane proteins[28] and reveals that constrictions in outer membrane channels can be due to motifs other than extracellular loops. The structure of the FhuA protein, solved to atomic resolution by X-ray crystallography, demonstrates that earlier models, based on sequence and mutant analysis as well as on spectroscopic and topological studies, are incorrect. The loop 4 (L4), proposed to form the constriction on the basis of an assumed structural homology with the constriction loop in OmpF-porin, actually protrudes 35 Å into the extracellular space, consistent with its function in phage recognition. As it is sandwiched between the two adjacent loops, deletion of L4 may rather cause the collapse of the columnar structure above the extracellular channel mouth, thus possibly preventing the proper insertion of the plug into the channel lumen, but without having detrimental effects on the barrel. This would rationalize the channel conductance measurements, performed with a L4 deletion mutant, which revealed a large, permanently open channel.[19] As the conductance of the mutant protein is comparable to that observed upon addition of phage T5 to wild-type FhuA protein, we hypothesize that in both cases the barrel is preserved, whereas the plug is either expelled or prevented from assuming its proper location.

What has been mentioned about the prediction of the gating device is also true for the ferrichrome binding site. Although deletion mutants, found to affect ferrichrome binding, have often included residues that we have now shown to form part of the binding pocket, in no instance were the specific predictions correct.[22,29,30] Without a good *in vitro* assay system and detailed knowledge of ligand interactions, it is indeed not surprising that neither the arginyl residue R81 nor the dozen or so aromatic residues padding the binding pocket remained unidentified.

How may the working model for ferrichrome translocation, illustrated in Figure 13.6, be tested? The main prerequisite for the investigation of these hypotheses is a new, sensitive *in vitro* assay system for FhuA protein. The assay we have developed earlier suffers from the fact that, by substituting phage T5 for the TonB protein in triggering channel opening, reversibility is lost.[31] It will be essential, but not trivial, to reconstitute the physiologically relevant energy transduction cascade *in vitro*. And although *in vivo* experiments allow certain conclusions about the mechanism to be drawn, the most direct approach appears to be structure determinations, at high resolution, of the components involved. Complex formation and crystallization of the FhuA protein with the TonB complex (or a moiety thereof, such as a truncated TonB protein) appear as logical next steps in unveiling the molecular events affecting ferrichrome permeability. In the case of phage T5 DNA infection, straightforward methods to verify the hypothesis of plug ejection appear to exist: the plug may be tethered to the barrel wall by means of disulfide bridges engineered on the basis of the atomic coordinates, or by genetic insertion of specific proteolytic target sites in the hinge region between the plug and barrel. Investigating the effects of addition of phage T5, or of protease susceptibility in the presence of the virus could then provide the desired information.

In conclusion, the determination of the FhuA structure led us to revise our concepts of the architecture as well as of the mechanism of the ligand-gated FhuA channel. The results suggest that predictions based on structural homology require very cautious interpretations. The presence of two types of secondary structures within membrane boundaries may not be restricted to this protein, rendering not only predictions, but also the interpretation of spectroscopic results of membrane proteins even more difficult.[28] At this still early stage of membrane protein structure determination, the elucidation by high-resolution structure analyses thus becomes still more urgent. The structure of the ligand-gated FhuA protein may have heuristic value for the concepts of structure and mechanistic properties of eukaryotic membrane receptors with similar functions.

ACKNOWLEDGMENT

The support of the EU-BIOMED (BMH4 CT96-0990) to D.M. and J.P.R. is gratefully acknowledged.

REFERENCES

1. Neilands, J. B., Siderophores: structure and function of microbial iron transport compounds, *J. Biol. Chem.*, 270, 26723, 1995.
2. Guerinot, M. L., Microbial iron transport, *Annu. Rev. Microbiol.*, 48, 743, 1994.
3. Braun, V., Energy-coupled transport and signal transduction through the Gram-negative outer membrane via TonB-ExbB-ExbD-dependent receptor proteins, *FEMS Microbiol. Rev.*, 16, 295, 1995.
4. Kadner, R. J., Vitamin B_{12} transport in *Escherichia coli:* energy coupling between membranes, Mol. Microbiol., 4, 2027, 1990.
5. Klebba, P. E., Rutz, J. M., Liu, J., and Murphy, C. K., Mechanisms of TonB-catalyzed iron transport through the enteric bacterial cell envelope, *J. Bioenerg. Biomembr.*, 25, 603, 1993.
6. Postle, K., TonB protein and energy transduction between membranes, *J. Bioenerg. Biomembr.*, 25, 591, 1993.
7. Skare, J. T. and Postle, K., Evidence for a TonB dependent energy transduction complex in *E. coli, Mol. Microbiol.*, 5, 2883, 1991.
8. Postle, K. and Good, R. F., DNA sequence of the *Escherichia coli* tonB gene, *Proc. Natl. Acad. Sci. U.S.A.*, 80, 5235, 1983.
9. Eick-Helmerich, K. and Braun, V., Import of biopolymers into *Escherichia coli*: nucleotide sequences of the *exbB* and *exbD* genes are homologous to those of the *tolQ* and *tolR* genes, respectively, *J. Bacteriol.*, 171, 5117, 1989.
10. Fischer, E., Günter, K., and Braun, V., Involvement of ExbB and TonB in transport across the outer membrane of *Escherichia coli*: phenotypic complementation of *exb* mutants by overexpressed *tonB* and physical stabilization of TonB by ExbB, *J. Bacteriol.*, 171, 5127, 1989.

11. Schramm, E., Mende, J., Braun, V., and Kamp, R. M., Nucleotide sequence of the colicin B activity gene *cba*: consensus pentapeptide among TonB-dependent colicins and receptors, *J. Bacteriol.,* 169, 3350, 1987.

12. Moeck, G. S., Coulton, J. W., and Postle, K., Cell envelope signaling in *Escherichia coli*. Ligand binding to the ferrichrome-iron receptor FhuA promotes interaction with the energy-transducing protein TonB, *J. Biol. Chem.,* 272, 28391, 1997.

13. Coulton, J. W., Mason, P., Cameron, D. R., Carmel, G., Jean, R., and Rode, H. N., Protein fusions of β-galactosidase to the ferrichrome-iron receptor of *Escherichia coli* K-12, *J. Bacteriol.,* 165, 181, 1986.

14. Locher, K. P. and Rosenbusch, J. P., Oligomeric states and siderophore binding of the ligand-gated FhuA protein that forms channels across *Escherichia coli* outer membranes, *Eur. J. Biochem.,* 247, 770, 1997.

15. Weiss, M. S., Abele, U., Weckesser, J., Welte, W., Schiltz, E., and Schulz, G. E., Molecular architecture and electrostatic properties of a bacterial porin, *Science,* 254, 1627, 1991.

16. Schirmer, T., Keller, T. A., Wang, Y. F., and Rosenbusch, J. P., Structural basis for sugar translocation through maltoporin channnels at 3.1 Å resolution, *Science,* 267, 512, 1995.

17. Cowan, S. W., Schirmer, T., Rummel, G., Steiert, M., Ghosh, R., Pauptit, R. A., Jansonius, J. N., and Rosenbusch, J. P., Crystal structures explain functional properties of two *E. coli* porins, *Nature,* 358, 727, 1992.

18. Koebnik, R. and Braun, V., Insertion derivatives containing segments of up to 16 amino acids identify surface- and periplasm-exposed regions of the FhuA outer membrane receptor of *Escherichia coli* K-12, *J. Bacteriol.,* 175, 826, 1993.

19. Killmann, H., Benz, R., and Braun, V., Conversion of the FhuA-transport protein into a diffusion channel through the outer membrane of *Escherichia coli, EMBO J.,* 12, 3007, 1993.

20. Rutz, J. M., Liu, J., Lyons, J. A., Goranson, J., Armstrong, S. K., McIntosh, M. A., Feix, J. B., and Klebba, P. E., Formation of a gated channel by a ligand-specific transport protein in the bacterial outer membrane, *Science,* 258, 471, 1992.

21. Bonhivers, M., Ghazi, A., Boulanger, P., and Letellier, L., FhuA, a transporter of the *Escherichia coli* outer membrane, is converted into a channel upon binding of bac-teriophage T5, *EMBO J.,* 15, 1850, 1996.

22. Killmann, H. and Braun, V., An aspartate deletion mutant defines a binding site of the multifunctional FhuA outer membrane receptor of *Escherichia coli* K-12, *J. Bacteriol.,* 174, 3479, 1992.

23. Locher, K. P., Rees, B., Koebnik, R., Mitschler, A., Moulinier, L., Rosenbusch, J. P., and Moras, D., Transmembrane signaling across the ligand-gated FhuA receptor: crystal structures of free and ferrichrome-bound states reveal allosteric changes, *Cell,* 95, 771, 1998.

24. Hendrickson, W., Determination of macromolecular structures from anomalous dif-fraction of synchrotron radiation, *Science,* 254, 51, 1991.

25. van der Helm, D., Baker, J. R., Eng-Wilmot, D. L., Hossain, M. B., and Loghry, R. A., Crystal structure of ferrichrome and a comparison with the structure of fer-richrome A, *J. Am. Chem. Soc.,* 102, 4224, 1980.

26. Armstrong, C. M. and Benzanilla, F., Inactivation of the sodium channel. II. Gating current experiments, *J. Gen. Physiol.,* 70, 567, 1977.

27. Drew, H. R., Samson, S., and Dickerson, R. E., Structure of a B-DNA dodecamer at 16K, *Proc. Natl. Acad. Sci. U.S.A.,* 79, 4040, 1982.

28. Cowan, S. W. and Rosenbusch, J. P., Folding pattern diversity of integral membrane proteins, *Science,* 264, 914, 1994.

29. Bos, C., Lorenzen, D., and Braun, V., Specific in vivo labeling of cell surface-exposed protein loops: reactive cysteines in the predicted gating loop mark a ferrichrome binding site and a ligand-induced conformational change of the *Escherichia coli* FhuA protein, *J. Bacteriol.,* 180, 605, 1998.

30. Killmann, H., Herrmann, C., Wolff, H., and Braun, V., Identification of a new site for ferrichrome transport by comparison of the FhuA proteins of *Escherichia coli, Salmonella paratyphi* B, *Salmonella typhimurium,* and *Panoea agglomerans, J. Bacteriol.,* 180, 3845, 1998.

31. Letellier, L., Locher, K. P., Plançon, L., and Rosenbusch, J. P., Modeling ligand-gated receptor activity: FhuA-mediated ferrichrome efflux from lipid vesicles triggered by phage T5, *J. Biol. Chem.,* 272, 1448, 1997.

32. Jones, T. A., Zou, J. Y., Cowan, S. W., and Kjeldgaard, M., Improved methods for building protein models in electron density maps and the location of errors in these models, *Acta Crystallogr. A,* 47, 110, 1991.

33. http://www.bioz.unibas.ch/~xray/dino.

14

X-Ray Crystallography of Membrane Proteins: Concepts and Applications of Lipidic Mesophases to Three-Dimensional Membrane Protein Crystallization

Michele Loewen, Mark Chiu, Christine Widmer, Ehud M. Landau, Jurg P. Rosenbusch, and Peter Nollert

CONTENTS

0-8493-3384-9/00/$0.00+$.50
© 2000 by CRC Press LLC

14.1 INTRODUCTION

With a few exceptions, membrane proteins have failed to yield crystals of adequate sizes and sufficient order that allow their three-dimensional structures to be determined to high resolution by crystallographic analysis. Thus, while well over 1,000,000 nucleotide sequences have been entered into the EMBL databank, and many thousands of high-resolution structures of soluble proteins have been determined (Protein-Databank), the number of structures of membrane proteins currently available at atomic resolution is almost three orders of magnitude lower (~10). This presents a serious handicap for the understanding, at the molecular level, of the vital processes occurring in cell and organelle membranes. What are the reasons for this scarcity?

The most obvious difference between soluble and membrane proteins concerns the nature of their surfaces. Globular proteins exhibit monotonous envelopes. In contrast, membrane proteins which are integrated in a lipid bilayer (dielectric constant $\varepsilon = 2$ to 8), exhibit a hydrophobic belt, while their polar surfaces are exposed on either side of the membrane to aqueous solutions ($\varepsilon = 80$). The dielectric constants at the hydrophobic–hydrophilic interface is about $\varepsilon \sim 10$. These discontinuities secure the proper vectorial orientation of the membrane proteins in the membrane that is necessary for their biological function, while rendering them insoluble in aqueous solutions. In conventional crystallization, membrane proteins are transformed into an isotropic state by means of solubilization using detergents in order to achieve the homogeneity that is necessary for crystallization.[1,2] The detergents most apt to allow crystallization usually have short alkyl chains.[3] Due to their high critical micelle concentration, these are highly dynamic,[4] which might expose the membrane zone of the protein to aqueous solution and hence cause uncontrolled perturbation and precipitation. It may therefore be suggested that only robust proteins can crystallize readily, while apparently labile ones will resist crystallization, since removal from the native membrane and exposure to detergent solution is difficult to control. This notion is supported by the observation that the few protein families that have been crystallized and the structures of which have been solved by X-ray analysis are very stable.[5-7] A paradigm of a very labile membrane protein is lactose permease: it is stable in the plasma membrane, extremely labile in detergent solutions but, upon reconstitution into proteoliposomes, regains its biological activity.[8,9] A tentative conclusion is that if duration of the purification step in the detergent solubilized form can be minimized, thus maintaining the native state of the protein, then one would increase the success of crystallization of labile proteins.

Lipids are the native environment for the hydrophobic regions of membrane proteins and it is therefore not surprising that they are closely associated with them.[10] This notion of annular lipids has attracted much attention although not without controversy. Based on association-dissociation rate constants of lipids from a membrane protein, it has been suggested[11,12] that a subpopulation of lipid molecules is associated with a protein. Recent studies have shown that cholesterol is tightly associated with the nicotinic acetylcholine receptor,[13] and that specific lipids are bound to protein with a high degree of order in crystals of cytochrome c oxidase[14,15] and bacteriorhodopsin.[16] In outer membrane proteins from Gram-negative bacteria, immobilized amphiphile molecules have been found at subunit interfaces in crystals.[17–20] The advantage of such associations for the purpose of crystallization may be that protein–lipid complexes are more stable and native-like. Moreover, they have been suggested to play a major role in biological function such as intracellular protein sorting.[10] These observations point to the significance of the lipids within membrane boundaries.

The inference thus seems clear: if formation of three-dimensional membrane protein crystals is aimed at, the conditions prevailing in two-dimensional membranes should be extended to three-dimensional matrices. This notion may seem curious, though polymorphism of lipids[21,22] has been an intensely studied field of physical chemists.[23–27] Indeed, nonplanar lipidic polymorphs have been implicated in biological processes such as membrane fusion.[27–31] We have recently set out to exploit a specific class of these materials, the lipidic cubic phases, for crystallization of membrane proteins. The working hypothesis that we tested was that water-soluble proteins, such as lysozyme,[32] would crystallize irrespective of the lipid structure and packing arrangement, and that membrane proteins, such as bacteriorhodopsin, could diffuse laterally to a site of nucleation and crystal growth only in a continuous structure.[33] In this chapter we present an introduction to the complex geometry of cubic phases, followed by an experimental survey of the compatibilities of lipidic mesophases and their relevance to this novel crystallization approach.

14.2 UNDERSTANDING BICONTINUOUS LIPIDIC CUBIC PHASES

The fluid mosaic model of biological membranes[34] has proven a valid concept and has had an enormous impact on our understanding of membranes. In this model membranes are conceived of as planar bilayer structures. Such planar structures can be found among the phases existing in lipid polymorphism and are denoted as $L\alpha$ (Figure 14.1). Among other morphologies of bilayer architectures the bicontinuous lipidic cubic phases are of special interest, as they combine the properties of lamellar phases with a highly ordered three-dimensional structure and dynamics.[25]

14.2.1 WHAT ARE BICONTINUOUS CUBIC PHASES?

Lipidic mesophases with a cubic symmetry were discovered 40 years ago[21] and have recently become the focus of intensive investigations.[27,36] Lipidic mesophases are materials consisting of water and lipid that form the fourth state of matter, liquid crystals. Of these, cubic phases are the most structured and rigid materials that are known to form. The material properties of this soft condensed matter is characterized by very high viscosity, optical transparency, and by nonbirefringency. They exhibit a glass-like appearance but have a texture that is similar to that of cold butter.

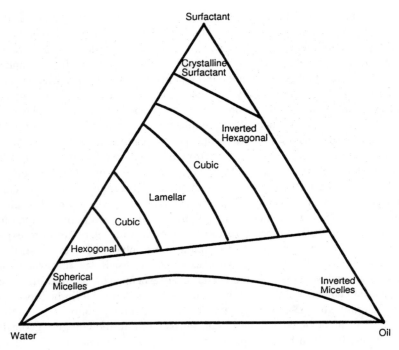

FIGURE 14.1   Schematic representation of a ternary phase diagram consisting of water, oil, and surfactant. Note that conventional crystallization of membrane proteins was carried out only in the lower left region where detergent micelles are found. Other membrane mimetic structures in this scheme are the hexagonal, cubic, lamellar, inverted hexagonal, and inverted micellar phases.

All cubic phases belong to one of two classes: the first is based on discrete micelles, the other on a bicontinuous membrane structure.[35] The latter constitute a continuous lipid membrane whose hydrophilic headgroup region is curved towards the water compartments. Three different inverted bicontinuous cubic phases with space groups of Ia3d, Pn3m, and Im3m have been identified. The chemical composition of such phases consists of an extremely large amount of lipid, usually between 50 and 80% (w/w), and 50 to 20% (w/w) water. The interface between these two components exhibits a remarkably large membrane surface: in the case of monoolein at 44% hydration it measures more than 300 m^2 per ml. This membrane is surrounded on both sides by water. Such structures are clearly different from curved multilamellar cubic phases that have been reconstructed from electron micrographs of biological samples.[37,38] Cubic phases have been suggested to be relevant in several biological events such as membrane fusion and digestion of fats. The biological significance of cubic phases is reviewed in References 27–29. Lipidic cubic phases have been investigated by several mutually complementary physical methods: small angle X-ray diffraction, nuclear magnetic resonance, freeze fracture electron microscopy, and fluorescence recovery after photobleaching.

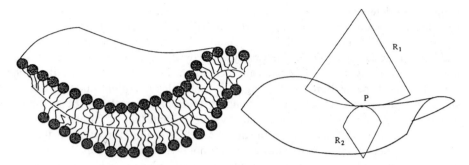

FIGURE 14.2 A model of a saddle-type curved lipid bilayer (left) and its corresponding isolated membrane midplane (right) showing the respective principal radii (R_1, R_2) drawn at a point (**P**) on this surface. (From Templer, R. H., Seddon, J. M., and Warrender, N. A., *Biophys. Chem.*, 49, 1-12, 1994. With permission.)

14.2.2 THE COMPLEX GEOMETRY OF CUBIC PHASES

In order to understand the geometry of bicontinuous cubic phases, consider first a simple curved membrane (Figure 14.2). In three-dimensional space, a surface can be curved about two independent axes. Each point on a curved surface is characterized by two principal curvatures C_1 and C_2, which are the reciprocals of the corresponding radii (R_1 and R_2) of circles that can be drawn to match the local curvature at that particular point. It is evident that both principle curvatures are necessary and sufficient to define the local geometry of a point on a surface. Two mathematical equations are defined: the mean curvature H [H = $1/2(C_1+C_2)$] and the Gaussian curvature K [K = $1/(C_1*C_2)$], of which the latter is used throughout this chapter. A specific surface, namely a saddle, is generated upon bending a plane about two axes such that the two bending directions are opposite. Because the two principal curvatures point in opposite directions, they have opposite signs and thus the Gaussian curvature of a saddle surface is always negative. When such a saddle surface is coated with a lipid monolayer on both sides, a curved bilayer membrane is obtained, as shown in Figure 14.2. Pictorially speaking, a horseback saddle constitutes such a negatively curved surface. One curvature is defined by the rider's legs, sitting on the horse back, the other (perpendicular) curvature is defined by the horse's spine in the head to tail direction. Notice that the saddle is curved towards the horse (spine) and towards the rider.

In contrast to planar bilayers, the membrane of a bicontinuous lipidic cubic phase is curved and consists of a three-dimensional periodic array of saddles, constituting a continuous space-filling membrane. The underlying configuration can be described in terms of differential geometry which deals with minimal surfaces. As an example, a curved minimal surface is formed by a thin soap film when a wire frame that deviates from planarity is withdrawn out of a soap solution. Indeed, the accepted model is that the bilayer midplane of bicontinuous lipidic cubic phases constitutes *infinite periodic minimal surfaces* (IPMS). Such surfaces exhibit negative or zero Gaussian curvature and mean curvature of zero at any point. This holds because the two principal curvatures at any point are equal in absolute value but opposite in sign.

Contrary to detergent micelles, the surface of which curves away from the water, each leaflet of membranes that are based on IPMS curve towards the water.

Why do lipid molecules like monoolein self-assemble into supramolecular structures based on IPMS? Geometrically, such lipids possess a small hydrophilic headgroup and a dynamic and kinked alkyl chain, the cross-sectional area of which exceeds that of the headgroup. Packing such molecules in a planar bilayer causes stress that can partly be relieved by bending the membrane. It has been shown that the surface area of the IPMS is always larger than that of parallel surfaces which are located at a distance from the IPMS.[39] It is therefore evident that an IPMS that is decorated on both sides with a monolayer of wedge-shaped molecules provides a favorable geometry that permits lipid molecules to pack with less stress than on a planar surface. Other geometries that accommodate similarly wedge-shaped molecules are the reverse hexagonal or reverse micellar organization, which have a larger curvature.

Schematic representations of the three different bicontinuous lipidic cubic phases are presented in Figure 14.3 to convey an intuitive understanding of their structure. All such cubic phases ideally consist of a single continuous lipid bilayer that divides space into two congruent hydrophilic compartments. The unit cells have cubic symmetry, very similar to the structure of a playhouse consisting of identical cubic Lego building blocks. Of the bicontinuous cubic phases, the Im3m, which is based on the P-surface (primitive) is the easiest to understand. In a unit cell the membrane is curved about three orthogonal axes, giving rise to six tubular outlets, which point towards the faces of a cube. The Pn3m cubic, which is based on the D-surface (diamond) contains only four of such tubular structures, which point towards the vertices of a tetrahedron. Finally, the Ia3d cubic is based on the G-surface (gyroid) and cannot be described in terms of tubes. The membrane forms around two separate hydrophilic spiral channels. In monoolein–water systems the cubic unit cell sizes depend on hydration and temperature and range from approximately 97 Å to 185 Å for the Ia3d and approximately 69 Å to 113 Å for the Pn3m cubic. The thickness of the membrane's hydrophobic core at 25°C was found to be about 34 Å and the water channel diameter can be readily calculated to be approximately 46 Å for the Pn3m phase at maximal hydration.[23] The existence of the Im3m phase in the pure water–monoolein system is presently controversial (for detailed reviews of cubic phase structure and properties see References 40 to 42). Understanding the peculiar structure of bicontinuous lipidic cubic phases requires insight into the energetics of the phase behavior of cubic phase forming lipids. Several attempts to model the lyotropic and thermotropic phase behavior of lipids have been made.[43,44] They show that the existence of such systems is mainly governed by a competition between a uniform mean interfacial curvature and the packing constraints imposed by the hydrocarbon moieties.[41] A detailed review of the thermodynamic and theoretical aspects of cubic mesophases can be found in Reference 41.

14.2.3 CONSEQUENCES OF THREE-DIMENSIONAL MEMBRANES

The complex packing of a curved bilayer surrounded by two aqueous channel systems has intriguing consequences. The materials thus formed are absolutely isotropic,

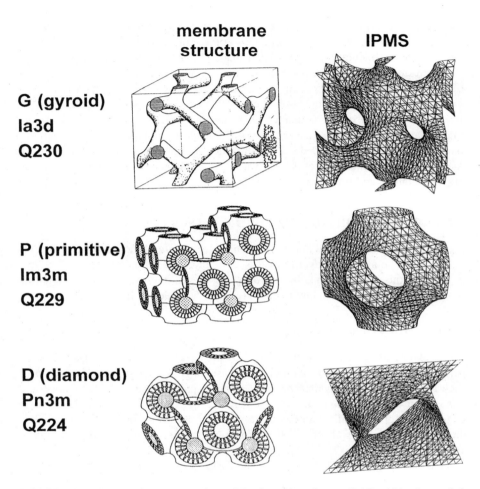

membrane structure

IPMS

G (gyroid)
Ia3d
Q230

P (primitive)
Im3m
Q229

D (diamond)
Pn3m
Q224

FIGURE 14.3   Schematic representations of the three bicontinuous lipidic cubic phases (left panel) and the unit cells of the corresponding infinite periodic minimal surfaces (IPMS). The representations of the Im3m and the Pn3m show cross sections of the tubular membrane parts. For the Ia3d, the corresponding water channels are shown. (From Chung, H. and Caffrey, M., *Nature*, 368, 224-226, 1994. With permission.)

transparent, and nonbirefringent by virtue of the cubic symmetry. Furthermore, due to the connectivity of each, the lipidic and aqueous systems, long-range diffusion is possible in both. A hydrophilic solute molecule that is membrane impermeable is confined to one aqueous channel system and may thus diffuse into half of the available aqueous space. Lipids are likewise confined to one of the respective membrane leaflets. The diffusion coefficients of lipids in these materials are of the same order of magnitude as that of lipids in planar bilayers[24,25] and in biological membranes. Detergent molecules partition between the bilayer, where they will be incorporated as part of the membrane, and the aqueous system. A coexistence of a bicontinuous cubic phase with detergent micelles structures has never been observed in lipid–water–detergent ternary phase diagrams.[45] It may therefore be assumed that detergent molecules reside in the aqueous compartments only as monomers. Small angle X-ray diffraction studies revealed that millimeter-sized cubic phases usually do not form single crystals, but consist of domains that give rise to powder X-ray diffraction patterns. In a specific case, single crystals of cubic phases have been grown.[46] A further consequence of these soft materials is that they will adopt the macroscopic shape of the container in which they are formed. They can also be manipulated at will with tools, e.g. can be filled inside a capillary or squeezed between parallel plates. Finally, embedded objects, such as crystals are immobilized inside these materials, which is useful for a variety of applications, such as spectroscopy, crystallography, and transport.

14.3 COMPATIBILITIES OF LIPIDIC MESOPHASES

As outlined earlier, the lipid matrix is amenable to structural modulation comprising water channel diameter, bilayer thickness, curvature, and diffusion in either the lipid or the aqueous compartment. In the following we outline our rationale for matrix modifications. The accompanying data are listed as tables in the appendices.

14.3.1 THE HYDROPHOBIC REGION

Cubic phases are known to form in well-defined regions of the composition–temperature phase diagrams. From published phase diagrams and structural analyses of such systems it is evident that the bilayers are capable of adapting their curvature as a response to changing hydration, temperature, or pressure, i.e., the lipids in these materials adopt an equilibrium curvature when mixed with water. Would they adapt their curvature as a result of added hydrophobic species such as lipids, detergents, and membrane proteins?

This problem can be tackled using increasing complexities:

1. Establish an empirical list of compounds compatible with the cubic phases by systematically introducing additives into cubic phases.
2. Establish detailed phase diagrams.
3. Establish the crystallizabilities of membrane proteins of various sizes in their membrane integral and peripheral parts.

We have initiated our investigations concerning the compatibility of cubic phases with the first approach, which serves as a basis for the establishment of phase diagrams and for the crystallization of membrane proteins. Increasing the lattice parameter may permit the incorporation of large membrane proteins into the matrix. It is also conceivable that phases with different diffusion rates can help to control the crystallization process by regulating the protein collision rate, which may influence the quality of the crystal. For pragmatic reasons we only consider cubic phases that form readily upon mixing. Monoglycerides, the phase diagrams of which have been established in a systematic manner,* easily form lipidic cubic phases, in contrast to dioleoylphos-phatidylethanolamine. Without special precautions[47] the latter system requires cycling through its $L\alpha \leftrightarrow H_{II}$ transition for several hundred or thousand times until it converts fully from $L\alpha$ to the cubic. Complete phase diagrams of monolgycerides have been published for monomyristolein,[48,49] monopentadecenoin,[49] monoolein,[23] monoelaidin,[50,51] monostearoylglycerol,[52] and monovaccenin.[53] Phase diagrams of other related lipids are under investigation by a number of groups.

The simplest criteria for the occurrence of cubic phases are their optical appearance (transparency), their viscosity (rigidity), and their appearance under polarizing microscope (nonbirefringency). Analogs of monoolein ($C_{18:1, c9}$), varying in the placement and number of the double bonds, including monopetroselinin ($C_{18:1, c6}$), monolinolenin ($C_{18:3, c9 c11 c15}$), and the cyclopropyl derivative 1-0-9,10 cis-methylene-octadecanoyl-rac-glycerol ($C_{18:0, c9}$) were found to give rigid, turbid phases at room temperature. Monovaccenin ($C_{18:1, c11}$) gave a rigid, transparent phase and may present a valid alternative to monoolein with larger water channels.[53] The shorter-chain monomyristolein ($C_{14:1, c9}$) does not produce a cubic phase at room temperature or at 37°C, in agreement with the prediction from its phase diagram. Neither mono-heptadecenoin, monononadecenoin, or monoerucin produced cubic phases at room temperature. Monoeicosenoin ($C_{20:1, c11}$) did produce a rigid and transparent material. The longer chain length of this lipid may be of use in the incorporation and crystallization of proteins known to coexist naturally with lipids of longer chain length.

In an attempt to produce alternative cubic phases with different lattice parameters and different diffusion rates, a variety of lipidic additives were introduced into the basic monoolein and monopalmitolein cubic systems (Appendix A, Table 14.1). Detailed phase diagrams are currently not available for most of these lipidic mixtures. Some of the mixtures have been investigated, such as the quaternary system consisting of 19.7% monoolein, 12.9% dioleylphosphatidylcholine, 1.36% dioleylphosphatidylethanolamine and water, which yields an Im3m cubic phase with a lattice parameter of 269 Å.[54] This is significantly larger than the lattice parameter of monoolein at room temperature in excess water (101.8 Å for the Pn3m cubic). Nuclear magnetic resonance (NMR) study has indicated that cholesterol decreases the diffusion rate of lipids in cubic phases.[55] Unfortunately the specific effects of many other additives remain to be characterized. At this time it can be concluded that monoolein can be mixed to produce transparent, rigid, and nonbirefringent cubic phases with up to 10% cholesterol, 10% dimyristoylphosphatidylcholine, low concentrations of dioleylphosphatidylethanolamine (<5%), and is compatible with up to 30% dioleylphosphatidylcholine. Monoolein is not compatible with palmitoyl-lysophosphatidylcholine or dipalmitoylphosphatidylethanolamine even at low concentrations at room temperature. Additives to the monopalmitolein cubic phase resulted largely in rigid but turbid materials. Such a turbidity is indicative of the presence of a phase different from the cubic one, and presumably constitutes a lamellar or hexagonal phase. These turbid

---

* A searchable web-based database of lipids (LIPIDAT) is available at www.lipidat.chemistry.ohio-state.edu. A database of phase diagrams (LIPIDAG) is currently being compiled (M. Caffrey, personal communication).

phases may still be useful as matrices for protein crystallization, albeit visualization of crystals in these phases becomes difficult.

Besides lipids and buffers, other factors such as detergents used to solubilize membrane proteins are important. The effect of a wide variety of different detergents commonly used in membrane protein biochemistry and crystallization was assessed for monoolein, monopalmitolein, and palmitoyl-lysophosphatidylcholine cubic phases (see Appendix A, Table 14.3). In general all detergents were stable in the cubic phases up to 0.4% of the total volume. In most cases, 1% detergent caused dissolution of the phase with a liquid birefringent phase appearing. These results imply that for crystallization in lipidic cubic phases only low concentrations of detergent may be used (Appendix B, Table 14.4). However, during concentration of solubilized proteins by ultrafiltration, simultaneous concentration of detergents cannot be avoided. The effects of a number of other detergents are summarized in Reference 32. Overall, in terms of modifications of the cubic phase to increase lattice parameters and change diffusion rates, thus providing a different lipidic environment, we have tested and defined a limited number of alternative conditions. Mixtures of monoolein with cholesterol or dioleoylphosphatidylcholine or the use of monoolein, monopalmitolein, monovaccenin, or monoelaidin homogeneous phases may improve incorporation of proteins into the cubic phase and the crystal quality.

14.3.2 THE HYDROPHILIC REGION

We have incorporated a limited number of water-soluble additives into cubic phases[32] that are commonly used for crystallization of proteins. With the exception of high-molecular-weight polyethylene glycols and certain organic solvents, cubic phases reveal a surprising tolerance towards the incorporation of water-soluble additives: we assume that these partition into the aqueous compartment of the cubic phases. In this report the water in cubic phases may be considered as bulk water. This is relevant as precipitants have to be added to the cubic phase in order to initiate crystallization. In the case of bacteriorhodopsin, the equivalent of up to 0.4 mg solid Sorensen's sodium/potassium phosphate was added directly to 1 mg of the lipidic matrix consisting of 60% monoolein and 40% water. Other salts that we have tested include ammonium phosphate, ammonium sulfate, sodium/potassium sulfate, zinc sulfate, ammonium chloride, sodium/potassium chloride, and lithium chloride. The cubic phase tolerates more than three times the saturation concentration of the corresponding salts in the aqueous phase. At such high concentrations large salt crystals accumulated at the bottom of the tubes upon centrifugation. Precipitants that have been used for crystallization purposes include polyethylene glycols, methyl-pentanediol, hexanediol, and heptanetriol. The methyl-pentanediol and hexanediol were stable up to 4% while the heptanetriol was stable up to 7.5% (see Appendix A and Reference 32). The cubic phase is slowly dissolved at high concentrations of hexanediol, heptanetriol, and methyl-pentanediol. This may offer a convenient way to release crystals from the surrounding matrix after they have formed inside the cubic phase (Figure 14.3). Alternatively, they can be released enzymatically by treatment with a lipase.[56] We conclude that low concentrations of standard protein precipitants, including polyethylene glycol, methyl-pentanediol, hexanediol, and heptanetriol are compatible and useful for crystallization in the lipidic cubic phase. We also suggest the use of a variety of salts as precipitants that are tolerated up to very high concentrations in the cubic system.

14.4 CRYSTALLIZATION IN LIPIDIC MESOPHASES

In the first part of this chapter we have introduced the structural aspects of lipidic cubic phases and their compatibility with respect to a variety of additives. In the following part we will describe and rationalize the events that lead to crystallization in lipidic mesophases. As discussed, cubic phases are the most structured and rigid bilayer forming lipidic materials. Initially we hypothesized that these properties would play a crucial role in the crystallizaton of membrane proteins, provided that we can establish a structural match between the unit cell of the host cubic phase and that of the protein. On the other hand, at the outset of the project we had anticipated that these structured materials would have to be "tailored" for each and every protein, en route to establishing lipid-based cubic phases as efficient and versatile crystallization matrices. Surprisingly, cubic phases proved to be exceptionally tolerant towards a large variety of added components, ranging from lipids to detergents, proteins, and inorganic salts. The molecular basis underlying this remarkable property is not yet understood, but it is clearly of great importance, both from the fundamental point of view as well as from the materials science point of view.

14.4.1 RATIONALE

We had devised the use of three-dimensionally curved membranes as crystallization matrices based on several advantageous properties of such materials. One of them is that their structures are more related to membranes, the native habitat of membrane proteins, than those of detergent micelles. The micelles are thought to provide an environment that is less related to the native membrane due to the relatively low lateral pressure that their hydrocarbons impose on the hydrophobic core of embedded membrane proteins. This may lead frequently to denaturation and aggregation of solubilized membrane proteins in a short period of time (hours), well before crystallization occurs. Conversely, in membranes or membrane mimetic systems such proteins are stable and can be stored in a functional form until they crystallize. Although the concept of lateral pressure has been described qualitatively, it has not yet been possible to quantitatively assess this membrane property.[57,58] As membranes are equilibrium structures that form spontaneously by self-assembly, the total lateral pressure is zero. However, the distribution of stress at different depths of the bilayer may vary considerably.[59,60] Unfortunately, as long as there is no firm quantitative evaluation of this property we have to rely on rather diffuse notions. Another advantageous property of such connected curved bilayers is the ability of membrane proteins to diffuse laterally within the entire space of the material. This may allow nucleation and crystal growth to occur significantly more efficiently than in micelles.

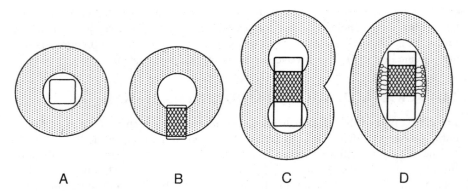

A   B   C   D

FIGURE 14.4 Schematic representations of possible locations of soluble and membrane proteins in a bicontinuous cubic phase: (A) lysozyme in the aqueous channel; (B) bacterior-hodopsin integrated in the membrane; (C) a large integral membrane protein whose hydrophilic domains penetrate two adjacent aqueous channels; (D) the same membrane protein, solubilized as a detergent/protein co-micelle, residing in a deformed cubic phase unit. Note that representations are simplified and show only a certain part of the channel system.

14.4.2 LOCATION OF PROTEINS IN CUBIC PHASES

Where do proteins reside in the lipidic cubic phase? Based on the geometrical considerations above there are two possible locations: (1) the aqueous channel, or (2) the lipidic membrane (see Figure 14.4). Hydrophilic proteins are known to be incorporated into the aqueous channel.[32] In the case of membrane proteins both possibilities are conceivable *a priori*: a detergent–protein co-micelle may be located in the aqueous channel, or an integral membrane protein that is embedded in the bilayer (Figure 14.2). We favor the latter alternative for energetic reasons and discuss our reasoning in the following paragraphs.

Hydrophobic effect accounts for partitioning into the bilayer. Both detergent and membrane protein are amphiphilic molecules. Upon mixing with a large excess of lipids that tend to form curved bilayers they both interact with and partition into the bilayer in order to reduce exposure of hydrophobic domains towards the water. The free energy associated with incorporation of a hydrophobic α-helix into a planar bilayer is considerable.[61]

Incorporation into cubic phase is a reconstitution. The incorporation of bac-teriorhodopsin into a monoolein cubic phase corresponds to a regular reconstitution of a membrane protein into a lipid membrane: the detergent-solubilized membrane protein is added with buffer to a certain amount of dry lipid in the Lc-phase, and the detergent is subsequently removed. The main differences between conventional reconstitution into lamellar membranes and reconstitution into lipidic cubic phase are the water content and the method of detergent removal. Reconstitution into lipidic cubic phases is carried out with an excess of lipid with respect to water: according to the phase diagram of monoolein[23] cubic phases are formed with water contents ranging from 20 to more than 50% (w/w) at room temperature. In conventional reconstitution experiments the water contents are usually two to three orders of magnitude higher and detergent is removed by dialysis, chromatography, or detergent-affine materials like biobeads. In the case of lipidic cubic phases the detergent is swiftly incorporated into the bilayer structure and, in specific cases is needed for the formation of a bicontinuous lipidic cubic phase. Therefore, analogously to the reconstitution into planar bilayers, the membrane protein becomes part of the nascent curved bilayer system. Indeed, crystallization experiments yielded bacteriorhodopsin crystals even in cases where the detergent concentration was lowered below the critical micellar concentration (Cmc).

Stability in membrane is higher than in micelles. A preparation containing β-octylglucoside-solubilized bacteriorhodopsin at room temperature starts to lose its initial purple color in a few weeks. After incorporation into a monoolein cubic matrix, the purple color remains unchanged for years. As the native color of this protein is a good indicator for its activity, this marked difference in stability hints that the protein is incorporated in a close to native environment.

Steric problems and diffusion in water channels. The diameters of water channels in monoolein cubic phases at room temperature are smaller than 46 Å. However, even large membrane proteins like reaction centers with large hydrophilic moieties have been successfully incorporated in lipidic cubic phases.[62] Judging from the phenomenology of bacteriorhodopsin crystallization in monoolein it is evident that diffusion of protein occurs over distances of many hundreds of micrometers in a few days. It appears difficult to envisage how bacteriorhodopsin trimers, or even monomers (approximately $45 \times 35 \times 25$ Å,[63]) could diffuse as protein–detergent mixed micelles in such small water channels (Figure 14.4). On the contrary, one could expect that such large proteins are trapped in the small water channels due to steric hindrance. Figure 14.4 illustrates that, indeed, membrane-integrated proteins may occupy an approximately threefold larger volume without exerting steric hindrance on the cubic matrix, by utilizing the space provided by two water channels and the hydrophobic core of the curved membrane.

It is conceivable that embedding of a membrane protein in the curved membrane of a bicontinuous lipidic phase should not happen without any structural consequences on the curved bilayer. Due to the size of membrane proteins and the size of cubic unit cells the formation of local, protein-induced distortions of the membrane may be expected. Such local deviations from the periodic arrangement of the curved membrane presumably affords only small bending energies. This is reasonable, as the transition between two different cubic phases, which differ from each other by the overall mean Gaussian curvature, also involve only little energy.[41] However, we are still clueless about the magnitude of this effect. In the crystallization setups the bacteriorhodopsin concentration is about 5 mg per gram of lipid, corresponding to an occupancy in the range of 0.1 protein molecules per unit cell. Most unit cells are thus not directly affected by the presence of the protein. Interestingly, the formation of some cubic bicontinuous systems have been found to be induced or promoted by hydrophobic peptides, e.g., the fusion peptide of influenza virus.[30]

14.4.3 PROTOCOL FOR BACTERIORHODOPSIN CRYSTALLIZATION IN MONOOLEIN CUBIC PHASES

The protocol for the crystallization of bacteriorhodopsin from monoolein cubic phases was recently published.[33,64] It is a straightforward procedure which consists of mixing three components—protein, lipid, and salt—to form new phases. A detailed step-by-step prodecure is given below for one single crystallization setup.

Protocol 14-1: Crystallization of Bacteriorhodopsin

1. Purple membrane from *Halobacterium salinarum* is solubilized with β-octylglucoside; monomerized bacteriorhodopsin is purified by gel filtration (Biogel 0,5 A) as described[33] and is concentrated to a final concentration of 10 mg/ml 25 mM sodium phosphate, pH 5.6.
2. 6 mg dry monoolein is weighed into a small glass vial (inner diameter 2 to 3 mm); 4 μl of the bacteriorhodopsin solution is added. These are vigorously mixed by 10 to 20 centrifugation runs at about $10,000 \times g$, while turning the tube about its long axis between runs. A purple transparent, nonbirefringent, and gel-like material is obtained and collected at the bottom of the vial.
3. Sörensen phosphate (2.7 to 4.2 mg) is added as a dry powder to the preparation, either together with the bacteriorhodopsin solution or after the formation of the cubic phase. In the latter case a further centrifugation run is necessary in order to mix the solid phosphate powder with the cubic phase material.
4. The sealed vial is stored in the dark at 20°C for several weeks and crystal growth may be observed by microscopic inspection.

14.4.4 MECHANISM OF CRYSTALLIZATION OF WATER-SOLUBLE SPECIES

We have crystallized salts, small molecules, and soluble proteins from cubic phases. The hypothesis was that if additives/precipitants are dissolved in the aqueous compartments, and if the cubic phases can accommodate the large concentrations of soluble species needed for crystallization, then the crystallization should be very similar to that from bulk water. An open and very important question was the feasibility of formation of large crystals in these highly structured cubic phase lattices. The results were straightforward: large amounts of water-soluble substances could be incorporated in these materials, and crystallization was efficient. Large crystals were grown inside the cubic phases without changing the macroscopic appearance and consistency. An analysis of the lysozyme crystals grown from two bicontinuous bilayer-based (monoolein, monopalmitolein) and one discrete micellar (palmitoyl-lysophosphatidylcholine) cubic phases revealed that they are identical to lysozyme crystals grown from bulk water.[32] These findings imply that the lipids employed to form the cubic phases act merely as "scaffolds" which "hold" the water channels and are responsible for the "gelation" of the materials. Molecularly, the lipids do not, or only weakly, interact with the dissolved species, be they small ions or large proteins, and all the interactions needed for crystallization occur only in the water compartments. The fact that large crystals can be accommodated strongly suggests that the bilayers surrounding the aqueous channels are very flexible. We thus now turn our attention to the membrane part of these materials.

The two most abundant molecular species in those crystallizations are water and monoolein, the lyotropic and thermotropic phase behavior of which has been investigated. For monoolein four distinct, thermodynamically stable phases are formed upon hydrating the lipid at room temperature: two planar lamellar phases (Lc and Lα) and two cubic phases (Ia3d and Pn3m).[23] This sets the framework in which crystallization occurs. "Additives" in this crystallization setup, such as precipitating salt, detergent, lipid, and even the protein itself, are only minor constituents. They are thus not expected to lead to the formation of completely different phases that are not already observed in the monoolein phase diagram. This assumption is indeed valid, as demonstrated in the case of a ternary mixture of ^2H_2O, palmitoyloleoyl-phosphatidylcholine and the detergent $C_{12}E_2$. Up to concentrations of about 8% (w/w) detergent, the lipid phase behavior remains essentially unchanged.[45] However, small deviations from the pure water–lipid phase diagram may be expected with the minor constituents, such as moving phase boundaries with respect to hydration level and temperature. In our crystallization setups the gel material is transparent, highly viscous, and nonbirefringent before, during, and after crystallization. These are indicative properties of lipidic cubic phases. These rather simple phase determinations were confirmed by preliminary small angle X-ray diffraction studies.

From macroscopic observation it is obvious that crystallization in the lipidic cubic phase is a well-organized phase separation process: in the case of bacteriorhodopsin the initially homogeneously purple colored material is disproportioned into a purple protein crystal and a non-colored cubic phase. The molecular situation is much more complicated because the crystallization mixture is a complex multicomponent system containing a minimum of eight different molecular species: water, monoolein, bacteriorhodopsin, octylglucoside, purple membrane lipids, as well as sodium, potassium, and phosphate ions. With the current methodological and technical limitations, it is not practical to perform a systematic phase characterization in the eight-dimensional crystallization space. Therefore, when pursuing crystallization in lipidic mesophases, the choices have to be based on phenomenology, experience, and intuition and trials to find rules that allow generalization of the effects of minor constituents on the phase behavior of the chosen lipid. Surely, a thorough theoretical understanding of the complex thermodynamics of such systems will enormously facilitate the choices and guide experimentalists who, at the present time, still have to pursue the "trial and error" approach when designing rational screens and defining key crystallization parameters.

We envisage a crystallization mechanism of membrane proteins in lipidic mesophases to be conceptually distinct from unobstructed crystallization in a homogeneous phase or from the so-called gel-acupuncture method.[65] In contrast to the latter approach of restricting diffusion in a hydrophilic environment by an obstructing polymer, the membrane proteins are thought to diffuse freely in the curved bilayer of the cubic phases. Fluorescence recovery after photobleaching experiments with bicontinuous cubic phases indeed support this notion. When doped with fluorescently labeled lipids it has unequivocally been shown that the lipids in such cubic phases are as mobile as lipids in planar membranes.[24] We speculate that crystal growth is intimately based on the interaction of curved membranes with membrane proteins. It would be attractive to extend this notion to nucleation, but experimental evidence is still lacking.

The high-resolution crystal structure of bacteriorhodopsin reveals that the packing of monomers in the a-b plane is virtually identical to that of bacteriorhodopsin in the two-dimensional array of the native purple patches: monomers are associated into trimers, which form a p3 two-dimensional lattice (a = b = 62 Å).[63,66] Such a plane is related to the adjacent one by a sixfold screw axis along the c direction. Our bacteriorhodopsin crystals belong to type I membrane protein crystals (nomenclature according to H. Michel, TIBS, Feb. 1983, 56–59) as they consist of regularly oriented membranous layers of well-ordered two-dimensional crystals. Previously attempts have been made to grow such a crystal by stacking whole purple membrane sheets,[66] but these were only a few layers and exhibited disorder along the c-axis. As the rotation and annealing of comparatively rigid and large purple membrane sheets is an exceedingly slow process, such ill-aligned crystals may be trapped kinetically. Recently it has been shown that the use of additional lipid and detergent aids in the formation of well-ordered bacteriorhodopsin crystals.[67] In contrast, crystallization of bacteriorhodopsin in lipidic mesophases is initiated from an isotropic distribution of monomers.[69] A concerted formation of the crystal in all three dimensions seems to be the most attractive mechanism, rather than an initial formation of two-dimensional sheets, which would align subsequently. Finally, the tendency of bacteriorhodopsin to form regular two-dimensional arrays might help in the formation of a well-ordered three-dimensional crystal.

14.5 CONCLUDING REMARKS

In this chapter we present our notion that three-dimensionally curved bilayers constitute complex molecular environments that may be utilized to solve one of the most difficult problems in the structural biology of membrane proteins—their resistance to crystallization. We base our approach on structural and energetic considerations of the individual players, starting by discussing the geometry of lipidic systems, going on to the peculiarities of membrane proteins, followed by a synthesis of the two. We discuss the compatibilities of such lipidic systems upon incorporation of a variety of additives into the two separate and complementary loci—the hydrophilic channels and the amphiphilic curved bilayers. The resulting tables, as fragmentary as they are, are necessary and very useful, as they enable the widespread use of cubic phases. The last section is based on the first three and shows the practical advantages of lipidic cubic phases for the solubilization, stabilization, and crystallization of proteins.

The geometry of curved bilayer systems, and specifically its curvature, is of primary importance in defining its physical and chemical properties. We have established that additives can indeed be tolerated by these systems. This versatility is brought about by the extreme flexibility of the single bilayer that is curved in three dimensions and is stabilized by the aqueous channels that pervade space. Nontheless, the crystallization of membrane proteins is highly dependent on the "signature" of individual lipids: the different crystal habits of bacteriorhodopsin crystals obtained in monoolein and monopalmitolein cubic phases, and the lack of any successful crystallizations in other lipidic systems (Table 14.5, Appendix C) clearly demonstrate the effect, unpredictable though it may be, of the specific lipidic environment on crystallization.[33] In contrast, crystallization of soluble species, be they ionic systems, small organic molecules or proteins, proved to be independent of the type of lipid employed or their packing arrangements. This clearly demonstrates that water in these complex materials behaves very much like bulk water, and the membranes do not pose a significant barrier against crystallization.

In order to better understand the geometrical requirements and the energetic implications of such systems, a combination of physicochemical characterizations need to be undertaken. These include phase characterization by means of NMR and small angle X-ray scattering, freeze-fracture electron microscopy, as well as diffusional characterization using NMR and fluorescence. The insights from such experiments will reveal the molecular events that underly these crystallizations and enable us to design more efficient matrices for biochemical and crystallographic studies of more complex and less stable membrane proteins, the paradigm for which is lactose permease. In principle, the combination of soluble protein crystallization (lysozyme) with membrane protein crystallization (bacteriorhodopsin) is conceptually the first step in the crystallization of large complexes of biological relevance. Finally, the perfect transparency and nonbirefringence of these materials will enable the investigations of membrane proteins by a combination of time-resolved spectroscopy and crystallography, which promises to yield exciting insights on structure–function relationships of membrane proteins.

ACKNOWLEDGMENTS

The expert technical assistance of G. Rummel and A. Hardmeyer is gratefully acknowledged. We are greatly indebted to Drs. H. Belrhali, E. Pebay-Peyroula, G. Büldt, R.Templer, J. Seddon, H. Qiu, M. Caffrey, G. Lindblom, S. Hyde, and V. Luzzati for stimulating discussions. These studies were supported by grants of the Swiss National Science Foundation Priority Program in Biotechnology (5002-37911 and 5002-46092) and the EU-BIOMED (BIO-CT96-0119/BBW95.0137.2) program to E.M.L. and J.P.R.

REFERENCES

1. Helenius, A. and Simons, K., Solubilization of membranes by detergents, *Biochim. Biophys. Acta, 415*, 29-79, 1975.
2. Tanford, C. and Reynolds, J. A., Characterization of membrane proteins in detergent solutions, *Biochim. Biophys. Acta, 457*, 133-170, 1976.
3. Garavito, R. M. and Rosenbusch, J. P., Isolation and crystallization of bacterial porin, *Methods Enzymol.,* 125, 309-328, 1986.

4. Zulauf, M. and Rosenbusch, J. P., Micelle clusters of octyl-oligooxyethylenes, *J. Phys. Chem.*, 87, 856-862, 1983.

5. Schindler, M. and Rosenbusch, J. P., Structural transitions of porin, a transmembrane protein, *FEBS Lett.*, 173, 85-89, 1984.

6. Kahn, T. W., Sturtevant, J. M., and Engelman, D. M., Thermodynamic measurements of the contributions of helix-connecting loops and of retinal to the stability of bacteriorhodopsin., *Biochemistry*, 31, 8829-8839, 1992.

7. Phale, P. S., Philippsen, A., Kiefhaber, T., Koebnik, R., Phale, V. P., Schirmer, T., and Rosenbusch, J. P., Stability of trimeric OmpF porin: the contributions of the latching loop L2, *Biochemistry*, 37, 15663-15670, 1998.

8. Newman, M. J., Foster, D. L., Wilson, T. H., and Kaback, H. R., Purification and reconstitution of functional lactose carrier from Escherichia coli, *J. Biol. Chem.*, 256, 11804-11808, 1981.

9. Page, M. G., Rosenbusch, J. P., and Yamato, I., The effects of pH on proton sugar symport activity of the lactose permease purified from Escherichia coli, *J. Biol. Chem.*, 263, 15897-15905, 1988.

10. Simons, K. and Ikonen, E., Functional rafts in cell membranes, *Nature*, 387, 569-572, 1997.

11. Warren, G. P., Toon, P. A., Birdsall, N. J. M., Lee, A. G., and Metcalfe, J. C., Reversible lipid titration of the activity of ATP-ase complexes, *Biochemistry*, 13, 5501-5507, 1971.

12. Seelig, A. and Seelig, J., The dynamic structure of fatty acyl chains in a phospholipid bilayer measured by deuterium magnetic resonance, *Biochemistry*, 13, 4839-4845, 1974.

13. Jones, O. T. and McNamee, M. G., Annular and nonannular binding sites for cholesterol associated with the nicotinic acetylcholine receptor, *Biochemistry*, 27, 2364-2374, 1988.

14. Iwata, S., Ostermeier, C., Ludwig, B., and Michel, H., Structure at 2.8 A resolution of cytochrome c oxidase from Paracoccus denitrificans, *Nature*, 376, 660-668, 1995.

15. Yoshikawa, S., Beef heart cytochrome c oxidase, *Curr. Opin. Struct. Biol.*, 7, 574-579, 1997.

16. Essen, L.-O., Siegert, R., Lehmann, W. D., and Oesterhelt, D., Lipid patches in membrane protein oligomers: Crystal structure of the bacteriorhodopsin-lipid complex, *Proc. Natl. Acad. Sci.*, 95, 11673-11678, 1998.

17. Cowan, S. W., Garavito, R. M., Jansonius, J. N., Jenkins, J., Karlsson, R., Koenig, N., Pai, E. F., Pauptit, R. A., Rizkallah, P. J., Rosenbusch, J. P., Rummel, G., and Schirmer, T., The structure of OmpF porin in a tetragonal crystal form, *Structure*, 3, 1041-1050, 1995.

18. Penel, S., Pevay-Peyroula, E., Rosenbusch, J. P., Rummel, G., Schirmer, T., and Timmins, P. A., Detergent binding in trigonal crystals of OmpF porin from Escherichia coli, *Biochimie*, 80, 543-551, 1998.

19. Locher, K. P., Rees, B., Koebnik, R., Mitschler, A., Moulinier, L., Rosenbusch, J. P., and Moras, D., Transmembrane signaling across the ligand-gated FhuA receptor: crystal structures of free and ferrichrome-bound states reveal, allosteric changes, *Cell*, 95, 771-778, 1998.

20. Dutzler, R., Rummel, G., Alberti, S., Hernandez-Alles, S., Phale, P. S., Rosenbusch, J. P., Benedi, V. J., and Schirmer, T., Crystal structure and functional characterization of OmpK36, the osmoporin of Klebsiella pneumoniae, *Structure*, 7, 425-434, 1999.

21. Luzzati, V., Mustacchi, M., and Skoulios, A., The Structure of the Liquid-Crystal Phases of Some Soap & Water Systems, *Discuss. Faraday Soc.*, 25, 43-48, 1958.

22. Luzzati, V., Gulik-Krzywicki, T., and Tardieu, A., Structure of the cubic phases of lipid-water systems, *Nature*, 218, 1031-1034, 1968.
23. Briggs, J., Ching, H., and Caffrey, M., The temperature-composition phase diagram and mesophase structure characterization of the monoolein/water system, *J. Phys. II France*, 6, 723-751, 1996.
24. Cribier, S., Gulik, A., Fellmann, P., Vargas, R., Deveaux, P. F., and Luzzati, V., Cubic phases of lipid-containing systems. A translational diffusion study by fluorescence recovery after photobleaching, *J. Mol. Biol.*, 229, 517-525, 1993.
25. Lindblom, G., Nuclear magnetic resonance on lipids and surfactants, *Curr. Op. Colloid Interface Struct.*, 1, 287-295, 1996.
26. Delacroix, H., Gulik-Krzywicki, T., and Seddon, J. M., Freeze fracture electron microscopy of lyotropic lipid systems: quantitative analysis of the inverse micellar cubic phase of space group Fd3m (Q227), *J. Mol. Biol.*, 258, 88-103, 1996.
27. Luzzati, V., Biological significance of lipid polymorphism: the cubic phases, *Curr. Op. Struct. Biol.*, 7, 661-668, 1997.
28. White, J. M., Membrane fusion, *Science*, 258, 917-924, 1992.
29. de Kruijff, B., Lipid polymorphism and biomembrane function, *Curr. Opin. Chem. Biol.*, 1, 564-569, 1997.
30. Colotto, A. and Epand, R. M., Structural study of the relationship between the rate of membrane fusion and the ability of the fusion peptide of influenza virus to perturb bilayers, *Biochemistry*, 36, 7644-7651, 1997.
31. Siegel, D. P., Burns, J. L., Chestnut, M. H., and Talmon, Y., Intermediates in membrane fusion and bilayer/nonbilayer phase transitions imaged by time-resolved cryo-transmission electron microscopy, *Biophysical J.*, 56, 161-169, 1989.
32. Landau, E. M., Rummel, G., Cowan-Jacob, S. W., and Rosenbusch, J. P., Crystallization of a polar protein and small molecules from the aqueous compartment of lipidic cubic phases, *J. Phys. Chem.*, B 101, 1935-1937, 1997.
33. Landau, E. M. and Rosenbusch, J. P., Lipidic cubic phases: a novel concept for the crystallization of membrane proteins, *Proc. Natl. Acad. Sci.*, 93, 14532-14535, 1996.
34. Singer, S. J. and Nicolson, G. L., The fluid mosaic model of the structure of cell membranes, *Science*, 175, 720-731, 1972.
35. Lindblom, G. and Rilfors, L., Cubic phases and isotropic structures formed by membrane lipids - possible biological relevance, *Biochim. Biophys. Acta*, 988, 221-256, 1989.
36. Luzzati, V., Delacroix, H., Gulik, A., Gulik-Krzywicki, T., Mariani, P., and Vargas, R., The Cubic Phases of Lipids, *Current Topics in Membranes*, 44, 3-25, 1997.
37. Landh, T., From entangled membranes to eclectic morphologies: cubic membranes as subcellular space organizers, *FEBS Lett.*, 369, 13-17, 1995.
38. Deng, Y. R. and Mieczkowski, M., Three-dimensional periodic cubic membrane structure in the mitochondria of amoebae Chaos carolinensis, *Protoplasma*, 203, 16-25, 1998.
39. Andersson, S., Hyde, S. T., Larsson, K., and Lidin, S., Minimal surfaces and structures: from inorganic and metal crystals to cell membranes and biopolmers, *Chem. Rev.*, 88, 221-242, 1988.
40. Larsson, K., Cubic lipid-water phases - structures and biomembrane aspects, *J. Phys. Chem.*, 93, 7304-7314, 1989.
41. Templer, R. H., Thermodynamic and theoretical aspects of cubic mesophases in nature and biological amphiphiles, *Curr. Opin. Colloid & Interface Sci.*, 3, 255-263, 1998.
42. Caffrey, M. and Cheng, A., Kinetics of lipid phase changes, *Curr. Opin. Struct. Biol.*, 5, 548-555, 1995.

43. Chung, H. and Caffrey, M., The curvature elastic-energy function of the lipid-water cubic mesophase, *Nature*, 368, 224-226, 1994.

44. Templer, R. H., Seddon, J. M., Duesing, P. M., Winter, R., and Erbes, J., Modeling the phase behavior of the inverse hexagonal and inverse bicontinuous cubic phases in 2:1 fatty acid/phosphatidylcholine mixtures, *J. Phys. Chem.*, B 102, 7262-7271, 1998.

45. Funari, S. S. and Klose, G., Phase behaviour of the ternary system POPC/C(12)E(2)/(H2O)-H-2., *Chemistry and Physics of Lipids*, 75, 145-154, 1995.

46. Clerc, M. and Dubois-Violette, D., X-ray scattering by bicontinuous cubic phases, *J. Phys. II France*, 4, 275-286, 1994.

47. Tenchov, B., Koynova, R., and Rapp, G., Accelerated formation of cubic phases in phosphatidylethanolamine dispersions, *Biophysical Journal*, 75, 853-866, 1998.

48. Briggs, J. and Caffrey, M., The temperature-composition phase diagram of mono-myristolein in water - equilibrium and metastability aspects., *Biophys. J.*, 66, 573-587, 1994a.

49. Briggs, J. and Caffrey, M., The temperature composition phase diagram and mesophase structure characterization of monopentadecenoin in water, *Biophys. J.*, 67, 1594-1602, 1994b.

50. Chung, H. and Caffrey, M., Polymorphism, mesomorphism, and metastability of monoelaidin in excess water, *Biophys. J.*, 69, 1951-1963, 1995.

51. Czeslik, C., Winter, R., Rapp, G., and Bartels, K., Temperature- and pressure-dependent phase behavior of monoacylglycerides monoolein and monoelaidin, *Biophys. J.*, 68, 1423-1429, 1995.

52. Ishiguro, T., Aomori, H., Kuwata, K., Kaneko, T., and Ogino, K., Study on thermal and structural behavior of monoacylglycerol-water systems. I. The phase behavior of monostearoylglycerol-water system, *Yukagaku*, 44, 997-1003, 1995.

53. Qiu, H. and Caffrey, M., Lyotropic and thermotropic phase behavior of hydrated monoacylglycerols: Structure characterization of monovaccenin, *J. Phys. Chem.*, B 102, 4819-4829, 1998.

54. Templer, R. H., Turner, D. C., Harper, P., and Seddon, J. M., Corrections to some models of the curvature elastic energy of inverse bicontinuous cubic phases, *J. Phys.*, 5, 1053-1065, 1995.

55. Eriksson, P. and Lindblom, G., Lipid and water diffusion in bicontinuous cubic phases measured by NMR, *Biophys. J.*, 64, 129-136, 1993.

56. Nollert, P. and Landau, E. M., Enzymic release of crystals from lipidic cubic phases, *Biochm. Soc. Trans.*, 26, 709-713, 1998.

57. Seelig, A., Local anesthetics and pressure: a comparison of dibucaine binding to lipid monolayers and bilayers, *Biochim. Biophys. Acta*, 899, 196-204, 1987.

58. Templer, R. H., Seddon, J. M., Warrender, N. A., Syrykh, A., Huang, Z., Winter, R., and Erbes, J., Inverse bicontinuous cubic phases in 2:1 fatty acid/phosphatidylcholine mixtures. The effects of chain length, hydration, and temperature, *J. Phys. Chem.*, B 102, 7251-7261, 1998.

59. Templer, R. H., Seddon, J. M., and Warrender, N. A., Measuring the elastic parameters for inverse bicontinuous cubic phases, *Biophys. Chem.*, 49, 1-12, 1994.

60. Cantor, R. S., The lateral pressure profile in membranes: a physical mechanism of general anesthesia, *Biochemistry*, 36, 2339-2344, 1997.

61. Janes, *Proc. Natl. Acad. Sci.*, 80, 3691-3695, 1983.

62. Hochkoeppler, A., Landau E.M., Venturoli, G., Zannoni, D., Feick, R., and Luisi, P.L., Photochemistry of a photosynthetic reaction center immobilized in lipidic cubic phases, *Biotechnol. Bioeng.*, 46, 93-98, 1995.

63. Pebay-Peyroula, E., Rummel, G., Rosenbusch, J. P., and Landau, E., X-ray structure of bacteriorhodopsin at 2.5 angstroms from microcrystals grown in lipidic cubic phases, *Science*, 277, 1676-1681, 1997.

64. Rummel, G., Hardmeyer, A., Widmer, C., Chiu, M. L., Nollert, P., Locher, K. P., Pedruzzi, I., Landau, E. M., and Rosenbusch, J. P., Lipidic cubic phases: New matrices for the three-dimensional crystallization of membrane proteins, *J. Struct. Biol.*, 121, 82-91, 1998.

65. Garcia-Ruiz, J. M. and Moreno, A., Investigations on protein crystal growty by the gel acupuncture method, *Acta Cryst. D.*, 50, 484-490, 1994.

66. Henderson, R. and Shotton, D., Crystallization of purple membrane in three dimensions., *J. Mol. Biol.*, 139, 99-109, 1980.

67. Takeda, K., Sato, H., Hino, T., Kono, M., Fukuda, K., Sakurai, I., Okada, T., and Kouyama, T., A novel three-dimensional crystal of bacteriorhodopsin obtained by successive fusion of the vesicular assemblies, *J. Mol. Biol.*, 283, 463-474, 1998.

68. Nollert, P., Caffrey, M., in preparation.

69. Lustig, A., personal communication.

APPENDIX A: EFFECT OF MEMBRANE SOLUBLE COMPOUNDS ON CUBIC PHASE STABILITY

TABLE 14.1
Homogeneous Lipidic Phases

Lipid		% Lipid	Cubic Phase[a]
Monomyristolein	(C14:1, c9)	40–50	−
Monopalmitolein	(C16:1, c9)	60	+
Monoheptadecenoin	(C17:1, c10)	55–65	−
Monopetroselinin	(C18:1, c6)	60	−
Monoolein	(C18:1, c9)	60	+
Monovaccenin	(C18:1, c11)	60	+
Monolinolenin	(C18:3, c9,c12,c15)	60	−
Cyclopropyl-monoolein derivative[b]	(C18:0, cyclo9)	60	+
Monononadecenoin	(C19:1, c10)	55–65	−
Monoeicosenoin	(C20:1, c11)	50–65	+
Monoerucin	(C22:1, c13)	50–70	−

[a] , at least one of the criteria: viscosity (rigid), transparency (clear), and birefringence (no) was not met.

[b] 1-0-9, 10-*cis*-methylene-octadecenoyl-*rac*-glycerol.

TABLE 14.2
Effect of Additives on MO and MP Cubic Phases

Additive	% Additive	% MO	% MP	Cubic Phase[a]
Cholesterol	5–10	50–60		+
	10–25		35–50	−
Monomyristolein	5–7	55–60		+
	7.5		57.5	−
DOPE	5–10	55–60		+
	15	60		−
	5–15		50	−
DOPC	5–30	30–60		+
	5–15		50	−
DMPC	10	50–60		+
	20–30	30–50		−
PLPC	10–30	30–60		−

[a] , at least one of the criteria: viscosity (rigid), transparency (clear), and birefringence (no) was not met.

TABLE 14.3
Effects of Detergents on Cubic Phases

Lipid	Detergent	% Detergent	Cubic Phase[a]
60% MO	DTACl	0.4	+
	Zwittergent	0.4	+
	DM	0.4	+
	DM	1.5–10	−
	CTAB	0.4	+
	β-OG	1.5–10	−
60 and 70% MO	MEGA9	0.4–1.0	+
	MEGA9	5–8	−
60% MP	DM	1.5–10	−
	β-OG	1.5–10	−
	MEGA9	1–9	−
43% PLPC	DM	1.5–10	−
	β-OG	1.5	+
	β-OG	10	−
	MEGA9	1	+
	MEGA9	1.5–5.0	−

[a] , at least one of the criteria: viscosity (rigid), transparency (clear), and birefringence (no) was not met.

APPENDIX B: EFFECT OF WATER-SOLUBLE COMPOUNDS ON CUBIC PHASE STABILITY

TABLE 14.4
Effects of Precipitants Used for Protein Crystallization on Cubic Phases

Additive	% Additive	Cubic Phase[a]
MPD	1–4	+
HT	1–7.5	+
HD	1–4	+
PEG200	4–14	+
PEG400	2–12	+
PEG600	2–12	+
PEG1000	2–6	+
PEG1000	10–14	−
PEG1500	2–12	+
PEG1500	14	−

[a] , at least one of the criteria: viscosity (rigid), transparency (clear), and birefringence (no) was not met.

APPENDIX C:

TABLE 14.5
Growth of Bacteriorhodopsin Crystals in
Various Lipid Cubic Systems

Lipid	Additive (%)	Stability of BR[a]	Stability of Phase	Crystals
MO		+	+	Hexagonal
MO	Cholesterol (5–30)	+	+	Hexagonal
MP		+	+	Rhombic
PLPC		+	+	No crystals
ME		+	– (turbid)	nd
MOcyclo		+	– (turbid)	nd
MV		+	+	
Mpet		+	+	
ML		– (bleached)	+	
MO	DOPC (30)	+	+	
MO	DMPC (10)	+	+	

[a] Stability of BR determined by maintenance of purple color.

Index

A